CHEMICAL PROCESS TECHNOLOGY

CHEMICAL PROCESS TECHNOLOGY

Jacob A. Moulijn

Michiel Makkee

Annelies van Diepen

JOHN WILEY & SONS, Ltd

Chichester • New York • Weinheim • Brisbane • Singapore • Toronto

Other Wiley Editorial Offices

John Wiley & Sons Inc., 111 River Street, Hoboken, NJ 07030, USA

Jossey-Bass, 989 Market Street, San Francisco, CA 94103-1741, USA

Wiley-VCH Verlag GmbH, Boschstr. 12, D-69469 Weinheim, Germany

John Wiley & Sons Australia Ltd, 33 Park Road, Milton, Queensland 4064, Australia

John Wiley & Sons (Asia) Pte Ltd, 2 Clementi Loop #02-01, Jin Xing Distripark, Singapore 129809

John Wiley & Sons Canada Ltd, 22 Worcester Road, Etobicoke, Ontario, Canada M9W 1L1

British Library Cataloguing in Publication Data

A catalogue record for this book is available from the British Library

ISBN 10: 0-471-63062-4 (P/B)
ISBN 13: 978-0-471-63062-3 (P/B)
ISBN 10: 0-471-63009-8 (H/B)
ISBN 13: 978-0-471-63009-8 (H/B)

Typeset in 10/12pt Times by Deerpark Publishing Services, Shannon, Ireland.
Printed and bound in Great Britain by Antony Rowe Ltd, Chippenham, Wiltshire.
This book is printed on acid-free paper responsibly manufactured from sustainable forestry
in which at least two trees are planted for each one used for paper production.

Contents

Preface

The main purpose of this book is to bring alive the concepts forming the basis of the chemical process industry and to give a solid background for innovative process development. This is done by treatment of actual practical processes, which all present one or more challenges that chemical engineers have to deal with during the development of these particular processes and which are often still challenges. It is not the intention to treat the chemical process industry in an encyclopedic way. Concepts are emphasized rather than facts. Hopefully, this approach will stimulate students in chemical engineering and also those who play a large role in the field such as chemists, biologists, and physicists. In particular, it is intended to provide students with an innovative background. The next generation should invent and develop novel unit operations and processes!

From the wealth of processes a selection had to be made. We have attempted to do this in a logical way. Knowledge of some processes is essential for the understanding of the culture of the chemical engineering discipline. Examples are the major processes in the oil refinery, the production of base chemicals from synthesis gas, and catalytic purification of exhaust gases from cars. Chemical engineers have been tremendously successful in the bulk chemicals industry. However, in some other sectors this was not the case in the past, but today they are becoming more and more important. Major examples are fine chemistry and biotechnology. Therefore, these areas are treated in separate chapters. More recently, the emphasis has shifted to sustainable technology and, related to that, process intensification. These subjects are also touched upon.

In all chapters the processes treated are represented by simplified flow schemes. For clarity these generally do not include process control systems, and valves and pumps are also omitted in most cases.

It is expected that students after having read the book will be able to think in 'conceptual process designs'.

This book can be used in different ways. It has been written as a consistent textbook, but in order to give flexibility it has not been attempted to avoid repetition in all cases. In particular, it has been written such that dependent on the local interest and the personal taste of the lecturer a selection can be made, as most chapters have been written in such a way that they can be read separately. For instance, in the chapter on biotechnology, reactors are treated in some detail, although in previous chapters on bulk chemicals production and fine chemistry similar discussions on reactors are present.

It is not trivial how much detail should be incorporated in the text of a course book like the present one. In principle, the selected processes are not treated in much detail, except when this is useful for explaining concepts. For instance, we decided to treat FCC in some detail because it is such a nice case of process development where over time catalyst improvements enabled improvements in chemical engineering and vice versa. We also decided to treat one process, viz., ammonia synthesis in some detail with respect to reactors, separation, and energy integration. If desired this can be the start of a discussion on process integration and design. The production of polyethene was chosen in order to give an example of the tremendously important polymerization industry and this specific subject was chosen because of the unusual process conditions and the remarkable development in novel processes. Fine chemistry is treated in much more detail than analogous chapters in order to expose chemistry students to reactor modeling coupled to the practice they will be interested in.

In order to stimulate students in their conceptual thinking a lot of questions are asked throughout the text.

At the Delft University of Technology, the text is the basis for a course of two credit points in the third year. The contents are Chapters 1–3 (3.1–3.5), 4–6, 8, 10, 11, and 13. Chapter 3 is the starting point for an optional course 'petroleum conversion'. All students do at least one design project. This book forms one of the bases for that. It is hoped that the text will help in giving chemical engineers sufficient feeling for chemistry and chemists for chemical engineering. Needless to say we would highly appreciate any comments from users.

Jacob A. Moulijn
Michiel Makkee
Annelies van Diepen
Delft, The Netherlands

Chapter 1

Introduction

Chemical process technology has had a long, branched road of development. Processes such as distillation, dyeing, and the manufacture of soap, wine, and glass have long been practiced in small-scale units. The development of these processes was based on chance discoveries and empiricism rather than thorough guidelines, theory, and chemical engineering principles. Therefore, it is not surprising that improvements were very slow. This situation persisted until the 17th and 18th centuries. Only then were mystical interpretations replaced by scientific theories.

It was not until the 1910s and 1920s, when continuous processes became more common, that disciplines such as thermodynamics, material and energy balances, heat transfer, and fluid dynamics, as well as chemical kinetics and catalysis became (and still are) the foundations on which process technology rests. Allied with these are the unit operations including distillation, extraction, etc. In chemical process technology various disciplines are integrated. These can be divided according to their scale:

- Scale independent

  - Chemistry, Biology, Physics, Mathematics
  - Thermodynamics
  - Physical Transport Phenomena

- Microlevel

  - Kinetics
  - Catalysis on a molecular level
  - Interface Chemistry
  - Microbiology
  - Particle Technology

- Mesolevel

  - Reactor Technology
  - Unit Operations
  - Scale-up

- Macrolevel

  - Process Technology and Process Development
  - Process Integration and Design
  - Process Control and Operation

Of course, this scheme is not complete. Other disciplines such as applied materials science, information science, process control, and cost engineering also play a role.

In the development stage of a process or product all necessary disciplines are integrated. The role and position of the various disciplines perhaps can be better understood from Figure 1.1, in which they are arranged according to their level of integration. Process development, in principle, roughly proceeds through the same sequence, and the *x*-axis also to a large extent represents the time progress in the development of a process. The initial phase depends on thermodynamics and other scale independent principles. As time passes, other disciplines become important, e.g. kinetics and catalysis on a microlevel, reactor technology, unit operations, and scale-up on the mesolevel, and process technology, process control, etc. on the macro level.

Of course, there should be intense interaction between the various disciplines. To be able to quickly implement new insights and results, these disciplines are prefer-ably applied more or less in parallel rather than in series as can also be seen from Figure 1.1. Figure 1.2 represents the relationship between the different levels of development in another way. The plant is the macrolevel. When one would like to focus on the chemical conversion, the reactor would be the level of interest. When the interest goes down to the molecules converted, the microlevel is reached.

An enlightening way of placing the discipline Chemical Engineering in a broader framework has been put forward by Villermaux [1]. He made a plot of length and time scales as is done in Figure 1.3. From this figure it can be appreciated that chemical engineering is a broad integrating discipline. On the one hand, molecules, having dimensions in the nanometer range and a vibration time on nanosecond scale, are considered. On the other hand, chemical plants may have a size of half a kilometer,

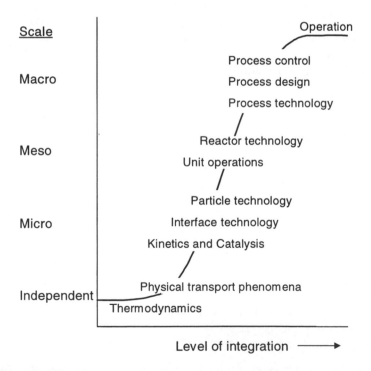

Figure 1.1 Disciplines in process development organized according to level of integration.

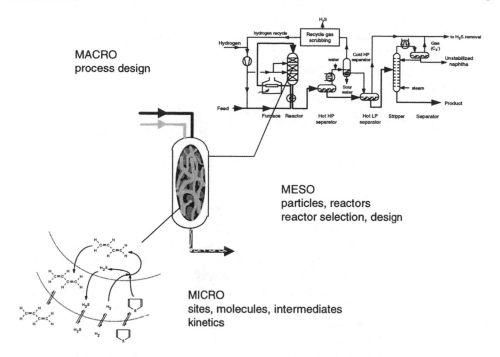

Figure 1.2 Relationship between different levels of development.

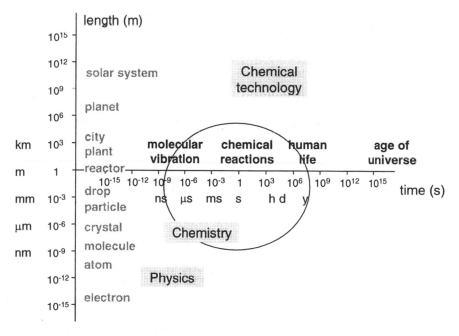

Figure 1.3 Space and time scales. After [1].

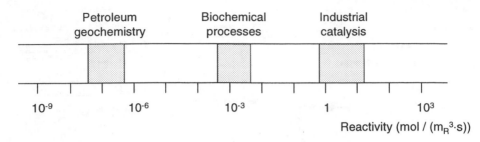

Figure 1.4   Windows on reality for useful chemical reactivity (subscript R denotes 'reactor'). After [2].

while the life expectancy of a new plant is 10–20 years. Every division has the danger of oversimplification. For instance, the atmosphere could be envisaged as a chemical reactor and chemical engineers can contribute to predictions for temperature changes etc. by modeling studies analogously to those referring to 'normal' chemical reactors. This text focuses on the Chemical Process Industry.

Rates of chemical reactions vary over several orders of magnitude. Processes in oil reservoirs might take place on a time scale of a million years, processes in nature are often slow (but not always), and reactions in Chemical Process Industry usually proceed at a rate that reactor sizes are reasonable, say smaller than 100 m³. Figure 1.4 indicates the very different productivity of three important types of processes.

It might seem surprising that despite the very large number of commercially attractive catalytic reactions, the commonly encountered reactivity is within a rather narrow range; the practically relevant rate is rarely less than 1 and seldom more than 10 mol m_R^3 s^{-1} for processes in oil refining and the bulk chemical industry. The lower limit is set by economic expectations: the reaction should take place in a reasonable amount of (space) time and in a reasonably sized reactor. What is reasonable, is determined by physical (space) and economical constraints. At first sight one might think that rates exceeding the upper limit are something to be happy about. The rates of heat and mass transport become limiting, however, when the intrinsic rate far exceeds the upper limit. In Process Intensification it is the aim to drastically decrease sizes of chemical plants. It is no surprise that the first step often is the development of better catalysts, i.e. catalysts exhibiting higher activity (reactor volume is reduced) and higher selectivity (separation section reduced in size). However, mass and heat transfer might become rate determining and equipment allowing higher heat- and mass-transfer rates is needed. For instance, a lot of attention is given to the development of compact heat exchangers that allow high heat-transfer rates on a volume basis. Novel reactors are also promising in this respect. A good example is the multiphase monolithic reactor, which allows unusually high rates and selectivities [3].

In the laboratory, transport limitations may lead to under- or overestimation of the local conditions in the catalyst particle, and hence to a wrong estimation of the actual or observed reaction rate. When neglected, the practical consequence is an over-designed, or worse, under-designed reactor. Transport limitations also can interfere with the selectivity and, as a consequence, upstream and downstream processing units, such as the separation train may be poorly designed.

QUESTIONS:

> *What would have been the consequence of much lower and of much higher reactivity of petroleum geochemistry for humanity?*
> *Which factors determine the lower and upper limit of the window for biochemical processes?*
> *Given a production rate of between 10^4 and 10^6 t/a of large volume chemicals (bulk chemicals), estimate required reactor volumes. Do the same for the production of petroleum products (10^6 to 10^8 t/a).*
> *$A \rightarrow B \rightarrow C$ kinetics in which B is the desired product is often encountered. Explain why the particle size of the catalyst influences the observed selectivity to B.*

Every industrial chemical process is designed to produce economically a desired product or range of products from a variety of starting materials (i.e. feed, feedstocks, or raw materials). Figure 1.5 shows a typical structure of such a process.

The feed usually has to be pretreated. It may undergo a number of physical treatment steps, e.g. coal has to be pulverized, liquid feedstocks may have to be vaporized, water is removed from benzene by distillation prior to its conversion to ethylbenzene, etc. Often, impurities in the feed have to be removed by chemical reaction, e.g. desulfurization of the naphtha feed to a catalytic reformer, making raw synthesis gas suitable for use in the ammonia converter, etc. Following the actual chemical conversion, the reaction products need to be separated and purified. Distillation is still the most common separation method, but extraction, crystallization, membrane separation, etc. can also be used.

In this book, emphasis is placed on the reaction section, since the reactor is the heart of each process, but feed pretreatment, and product separation will also be given attention. In the discussion of each process, the following questions will be answered:

- Which reactions are involved?
- What are the thermodynamics of the reactions, and what operating temperature and pressure should be applied?
- What is the kinetics, and what are the optimal conditions in that sense?
- Is a catalyst used, and if so, is it heterogeneous or homogeneous? Is the catalyst stable? If not, what is the deactivation time scale? What are the consequences for process design? Is regeneration required?
- Apart from the catalyst, what are the phases involved? Are mass- and heat-transfer limitations important?
- Is a gas or liquid recycle necessary?
- Is feed purification necessary?
- How are the products separated?
- What are the environmental issues?

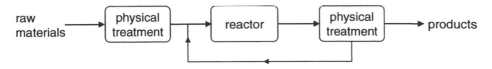

Figure 1.5 Typical chemical process structure.

The answers to these questions determine the type of reactor, and the process flow sheet. Of course, the list is not complete, and specific questions may be raised for individual processes, e.g. how to solve possible corrosion problems in the production of acetaldehyde. Other matters are also addressed, either for a specific process or in general terms:

- What are safety issues?
- Can different functions be integrated in one piece of equipment?
- What is the economics (comparison between processes)?
- Can sustainable technology be used?

References

1 Villermaux J, personal communication.
2 Weisz PB (1982) 'The science of the possible' *CHEMTECH* July 424–425.
3 Kapteijn F, Heiszwolf J, Nijhuis TA and Moulijn JA (1999) 'Monoliths in multiphase processes – aspects and prospects' *CATTECH* 3(1) 24–40.

General literature

Douglas JM (1988) *Conceptual Design of Chemical Processes*, McGraw-Hill, New York.
Gerhartz W et al. (eds.) (1988–1992) *Ullman's Encyclopedia of Industrial Chemistry* 5th ed., VCH, Weinheim.
Kroschwitz JI and Howe-Grant M (eds.) (1991–1996) *Kirk Othmer Encyclopedia of Chemical Technology*, 4th ed., Wiley, New York.
Levenspiel O (1999) *Chemical Reaction Engineering*, 3rd ed., Wiley, New York.
Seider WD, Seader JD and Lewin DR (1999) *Process Design Principles. Synthesis, Analysis, and Evaluation*, Wiley, New York.
Sinnot RK (1996) *Coulson and Richardson's Chemical Engineering*, vol. 6, 2nd ed., Butterworth-Heinemann, Oxford.
Stankiewicz AI and Moulijn JA (2000) 'Process intensification: transforming chemical engineering' *Chem. Eng. Progr.* January 22–33.
Westerterp KR, van Swaaij WPM and Beenackers AACM (1984) *Chemical Reactor Design and Operation*, 2nd ed., Wiley, New York.

Chapter 2

The Chemical Industry

2.1 INTRODUCTION

In the chemical industry, raw materials are converted into products for other industries and consumers. The basic raw materials for the chemical industry can be divided in organic and inorganic materials. Inorganic raw materials include air, water, and minerals. Minerals are mined and converted to metals (e.g. iron ore is reduced to iron in blast furnaces), building materials, etc. This field, although very important, will not be treated in detail in this book.

Fossil fuels and biomass belong to the class of organic raw materials. Originally, organic chemicals were produced from coal and plant and animal materials. In the 1800s and 1900s tremendous activities were taking place in Europe with respect to the synthesis of organic compounds from coal-derived feedstocks. Examples are aniline dyes and polymers such as 'Bakelite' (a co-polymer of phenol and formaldehyde). The use of petroleum as a raw material started in the 1920s. It was recognized in the United States in particular that petroleum-derived hydrocarbons were a superior feedstock for the chemical industry. In the next decades many novel products were discovered, e.g. nylon, PVC, polyethene, and novel processes were commercialized, for instance methanol synthesis and catalytic cracking.

2.2 STRUCTURE OF THE CHEMICAL INDUSTRY

As indicated in the introduction, the chemical industry covers a very broad range. However, we will mainly focus on the chemicals derived from hydrocarbons, i.e. the organic chemicals.

The vast majority of chemicals, about 85%, is produced from a very limited number of simple chemicals called *base chemicals*, which in turn are produced from only about ten *raw materials*, the most important hydrocarbon ones being oil, natural gas, and coal. Conversion of base chemicals can produce about 300 different *intermediates*, which are still relatively simple molecules. Both the base chemicals and the intermediates can be classified as *bulk chemicals*. A wide variety of *consumer products* can be obtained by further reaction steps. Figure 2.1 shows this tree-shaped structure of the chemical industry.

Figure 2.2 presents a survey of the petrochemical industry. Coal, oil (petroleum) fractions and natural gas are the primary raw materials for the production of most bulk chemicals. The first stage in the petrochemical industry is conversion of these raw materials into base chemicals:

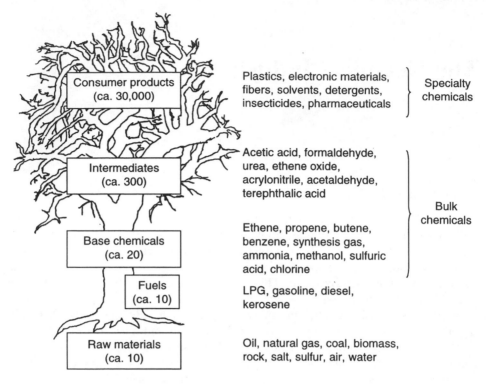

Figure 2.1 Structure of the chemical industry. After [1].

- synthesis gas (mixture of mainly hydrogen and carbon monoxide), ammonia, methanol;
- lower alkenes: ethene, propene, butadiene;
- aromatics: benzene, toluene, xylenes ('BTX').

This division also represents the most important processes for production of these chemicals. Synthesis gas, which is the feedstock for ammonia and methanol, is predominantly produced by steam reforming of natural gas (Chapter 5). The lower alkenes and aromatics are mainly produced by steam cracking of ethane or naphtha (Chapter 4), while aromatics are also produced in the catalytic reforming process (Chapter 3).

In the second stage, a new variety of chemical operations is conducted, often with the aim of introducing various hetero-atoms (oxygen, chlorine, sulfur, etc.) into the molecule. This leads to the formation of the chemical intermediates, such as acetic acid, formaldehyde, acetaldehyde, and monomers like acrylonitrile, terephthalic acid, etc. A final series of operations (often consisting of a number of steps) is needed to obtain consumer products. These products include:

- plastics: e.g. polyvinylchloride (PVC), polyacrylonitrile;
- synthetic fibers: e.g. polyesters like polyethene terephthalate (PET), nylon-6;
- elastomers: e.g. polybutadiene;
- insecticides;
- fertilizers: e.g. ammonium nitrate;

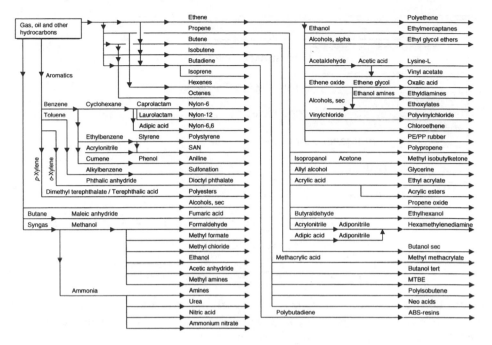

Figure 2.2   Survey of the petrochemical industry [2].

- vitamins;
- pharmaceuticals;
- detergents.

Of course, this three-stage classification shows exceptions. For instance, ethene, a base chemical formed from petroleum hydrocarbons, is a monomer for direct formation of plastics like polyethene, so the second stage does not occur. Another example is acetic acid. Depending on its application, it can be classified as an intermediate or as a consumer product.

The chemical industry can be conveniently divided into seven sectors, for which Figure 2.3 shows the 1989 sales figures.

QUESTION:

The sales figures in Figure 2.3 are not of very recent date. Which sector(s) do you expect have shown the largest growth since 1989?

It is obvious that petrochemicals, both base chemicals and intermediates, hold the largest market. This is not surprising since they are raw materials for all the other sectors except for most of the inorganic chemicals. However, even some inorganic chemicals are produced from petrochemicals: ammonia, which is a large volume inorganic compound, is produced mainly from oil or natural gas. An interesting case is sulfur. Sulfur compounds are present in fossil fuels in small quantities (see Section 3.3.3). Hence, the chemicals produced from these raw materials also contain

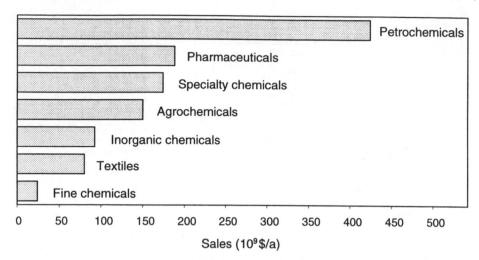

Figure 2.3   World chemical market (1989) [3].

sulfur. For many products (e.g. gasoline, diesel) maximum sulfur limits exist, mainly for environmental reasons. As a consequence, many processes have been developed to meet these standards by removing the sulfur-containing compounds. Over 50% of the sulfur demand is now produced as a by-product of oil and gas refining. In fact, with increasingly stricter demands on sulfur content, in the future the amount of sulfur produced in refineries might exceed the demand!

2.3 RAW MATERIALS AND ENERGY

2.3.1 Fossil Fuel Consumption and Reserves

From the previous section, it has become clear that raw materials and energy are closely related. Indeed, the main raw materials for the chemical industry are the fossil fuels, i.e. oil, natural gas, and coal. These are also the most important sources of energy, as is clear from Figure 2.4. Although in the 1950s and 1960s the energy consumption increased exponentially, the last decades have been a period of relative calmness. Despite continuing economic growth, energy consumption has only risen slightly, mainly due to more efficient use of energy. The consumption of oil and coal has stayed nearly constant, while the natural gas consumption tends to increase slightly. The use of other energy sources also increases, but the overall picture has not changed: fossil fuels still are the main energy carriers.

The major source of energy is oil, accounting for about 40% of the total energy consumption, while coal and natural gas account for 26 and 21%, respectively. The reserves of fossil fuels show a quite different picture, as illustrated in Figure 2.5.

Shale oil is by far the most abundant, but it is very difficult to process. It cannot be pumped from reservoirs like crude oil as it is contained in rocks. It can only be recovered by heating the rock material. It is not surprising that this is a very energy consuming process. Of the other fossil fuels the coal reserves are by far the largest. Although far less abundant, natural gas reserves are also large: they exceed the oil reserves. This is not generally realized: as a result of intensified exploration, in recent

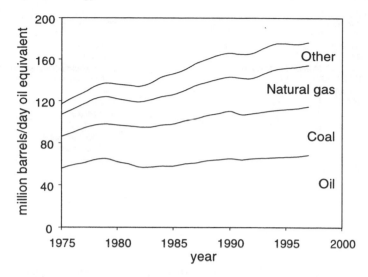

Figure 2.4   Total world energy consumption (1 barrel = 0.159 m^3 ≈ 0.136 metric ton) [5,6].

decades the estimates for natural gas have doubled every 10 years! About 10% of the natural gas reserves are located in 11 giant gas fields, one of which is in The Nether-lands.

   Probably, in the near future the energy consumption patterns will show major changes. Conversion of coal into gas (coal gasification) and production of liquid fuels from both coal gas and natural gas is technically possible and has been demon-strated on a large scale. Undoubtedly, these processes will gain importance. Natural gas also will become more important because it is convenient and well-distributed over the regions where the consumers live. Renewables will play a more important role in the future; renewables are based on biomass, but also wind and solar energy can

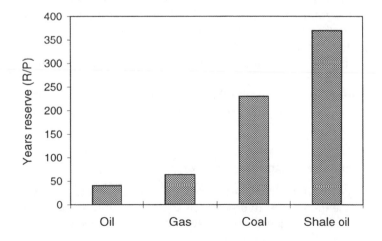

Figure 2.5   Fossil fuel reserves; R/P = Total proven reserves at end of year/production in same year (base year: 1997) [6].

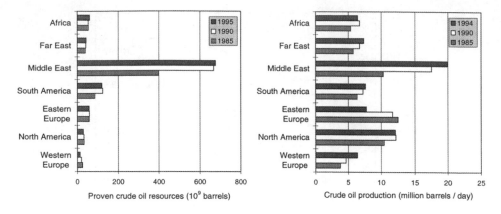

Figure 2.6   Proven crude oil resources (left) and crude oil production (right) [5,6].

be considered as part of this class. In the further future solar energy might well become a major source for electricity production.

Since oil still remains the main source of energy, its reserves, production, and utilization are examined in more detail. Figure 2.6 shows the proven resources (left) and production (right) of crude oil.

Crude oil resources are distributed over the world very unevenly. Clearly, the distribution is not very democratic: the major supplies are present in the Middle East, whereas the resources in Europe, the United States, and the Far East are very limited. The production of crude oil shows a different picture. Although the resources are limited, the production in the Western world is relatively high. The fraction of offshore reserves varies strongly in the different regions. In the former USSR, Eastern Europe, and the Middle East the onshore reserves far outweigh the offshore reserves, whereas in Europe almost 90% of the oil lies offshore.

It is not surprising that the usage of crude oil is concentrated in the developed countries. The world trade is very large. OPEC countries export over two thirds of the crude oil. The refinery capacity (expressed as crude distillation capacity) shows a similar distribution as the crude oil demand (see Figure 2.7).

A small fraction of the crude oil demand (about 8%) is used as a raw material for the petrochemical industry, while the remainder is used for fuel production, as shown in Figure 2.8.

Figure 2.8 explains why, despite the relative scarcity of oil, the driving force for finding alternative raw materials for the chemical industry is smaller than is often believed.

2.3.2 Energy and the Chemical Industry

A lot of energy is used in the chemical industry. The amount of energy used is of the same order as the quantity of hydrocarbons used as feedstock. Fuel is used in direct heaters and furnaces for heating process streams, and for the generation of steam and electricity, the most important utilities.

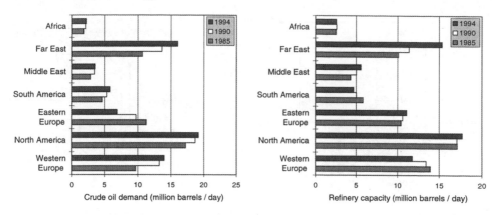

Figure 2.7 Crude oil demand (left) and refinery capacity (right) [5,6].

2.3.2.1 Fuels for direct heaters and furnaces

The fuel used in process furnaces often is the same as the raw material used. For instance, in steam reforming of natural gas (see Chapter 5), natural gas is used both as feedstock and as fuel in the reformer furnace. Fuel oil, a product of crude oil distillation, which is less valuable than crude oil itself, is often used in refineries, e.g. to preheat the feed to the crude oil fractionator. Due to its high sulfur content, measures are necessary to prevent air pollution.

2.3.2.2 Steam

The steam system is the most important utility system in most chemical plants. Steam has various applications, e.g. for heating process streams, as a reaction medium, and as a distillation aid. Steam is generated and used saturated, wet, or superheated. Saturated steam contains no moisture or superheat, wet steam contains moisture, and superheated steam contains no moisture and is above its saturation temperature. Steam generation takes place in water tube boilers (see Figure 2.9) using the most economic fuel available.

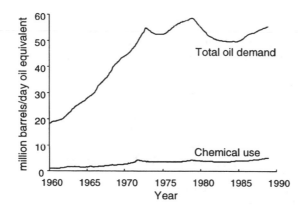

Figure 2.8 Petrochemical share of total world oil demand [3].

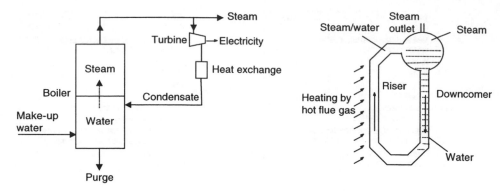

Figure 2.9 Water management in boiler system including a turbine for power generation (left) and a water tube boiler based on natural convection (right).

The amount of steam drawn off the system has to be replenished by make-up water. A water purge is necessary to prevent accumulation of salts. Circulation of the water/ steam mixture usually takes place by natural convection. Occasionally, a pump is used.

The flue gases discharged during the combustion of the fuel serve to heat the boiler feed water to produce steam. In steam reforming of natural gas, also for energy purposes usually natural gas is used, and steam is generated in a so-called waste heat boiler by heat exchange with both the furnace off gases and with the synthesis gas produced.

Steam is generally used at three different pressure levels. Table 2.1 indicates the pressure levels and their corresponding saturation temperatures.

The exact levels depend on the particular plant. In the flow sheets presented in these lecture notes, the steam pressures are generally not shown. Where an indication is given (HP, MP or LP), the values in Table 2.1 may be assumed to be valid.

The thermodynamic properties of saturated and superheated steam have been compiled in the so-called steam tables, which can be found in numerous texts, for example in [3]. Figure 2.10 shows the saturation pressure as a function of temperature.

2.3.2.3 Electricity

Electricity can either be generated on site in steam turbines, or be purchased from the local supply company. On large sites, reduction of the energy costs is possible if the

Table 2.1 Typical steam pressure levels

	Operating conditions		Saturation temperature (K)
	Pressure (bar)	Temperature (K)	
HP (high pressure) steam	40	683	523
MP (medium pressure) steam	10	493	453
LP (low pressure) steam	3	463	407

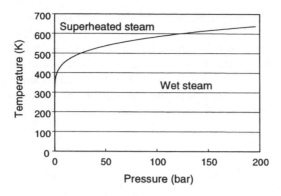

Figure 2.10 Steam saturation pressure versus temperature.

required electrical power is generated on site in steam turbines and the exhaust steam from the turbines used for process heating. It is often economical to drive large compressors, which demand much power, with steam turbines. The steam produced can be used for local process heating. A recent development is the construction of so-called co-generation plants, in which heat and electricity are generated simultaneously, usually as joint ventures between industry and public organizations. Examples are central utility boilers that provide steam for electricity generation and supply to a local heating system, and combined cycle power plants that combine coal gasification with electricity generation in gas turbines and steam turbines (see Chapter 5).

2.3.3 Composition of Fossil Fuels

All three fossil fuels (coal, petroleum, and natural gas) have in common that they mainly consist of H and C, while also a small amount of hetero-atoms like N, O, S, and metals are present. However, the ratio of these elements is very different, which

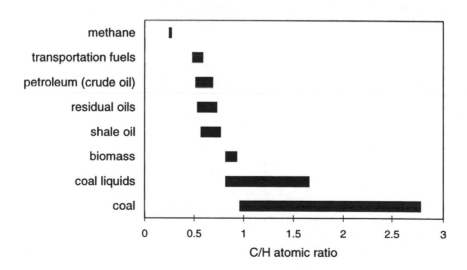

Figure 2.11 C/H atomic ratios of hydrocarbon sources and some products.

manifests itself in the very different molecular composition (size, type, etc.) and physical properties. A characteristic feature of hydrocarbons is the C/H ratio. Figure 2.11 shows the C/H ratio for the major fossil fuels and some other hydrocarbons.

Clearly, the relative amount of carbon in coal is much larger than in crude oil. Methane (CH_4) obviously has the lowest C/H ratio of all hydrocarbons. The ratio for natural gas is very similar to that of methane, because methane is the major hydrocarbon present.

QUESTION:

> *CO_2 is the most important greenhouse gas. Do the various hydrocarbons differ in their contribution to the greenhouse effect?*

2.3.3.1 Natural gas

Natural gas is a mixture of hydrocarbons with methane as the main constituent. It can be found in porous reservoirs, either associated with crude oil ('associated gas') or in reservoirs in which no oil is present ('non-associated gas'). Natural gas is of great importance not only as a source of energy but also as a raw material for the petrochemical industry. Besides hydrocarbons, natural gas usually contains small (or sometimes large) amounts of non-hydrocarbon gases such as carbon dioxide, nitrogen, and hydrogen sulfide. Tables 2.2 and 2.3 show the composition of a wide variety of natural gases.

Natural gas classifies as 'dry' or 'wet' natural gas. Dry natural gas contains only small amounts of condensable hydrocarbons (at ambient temperature). Other gases contain substantial amounts of ethane, propane, butane, and C_5^+ hydrocarbons, which liquefy on compression at ambient temperature (natural gas liquids). These are all wet natural gases. Associated gas is invariably wet, whereas non-associated gas is usually dry. The terms 'sweet' and 'sour' denote the absence or presence of H_2S and CO_2.

Non-associated gas can only be produced as and when a suitable local or export market is available.

Table 2.2 Composition of selected non-associated natural gases (volume%) [7,8]

Area	Algeria	France	Holland	New Zealand	North Sea	N. Mexico	Texas	Texas	Canada
Field	Hasi-R'Mel	Lacq	Gron.	Kapuni	West Sole	Rio Arriba	Terrell	Cliffside	Olds
CH_4	83.5	69.3	81.3	46.2	94.4	96.9	45.7	65.8	52.4
C_2H_6	7.0	3.1	2.9	5.2	3.1	1.3	0.2	3.8	0.4
C_3H_8	2.0	1.1	0.4	2.0	0.5	0.2	–	1.7	0.1
C_4H_{10}	0.8	0.6	0.1	0.6	0.2	0.1	–	0.8	0.2
C_5^+	0.4	0.7	0.1	0.1	0.2	–	–	0.5	0.4
N_2	6.1	0.4	14.3	1.0	1.1	0.7	0.2	26.4	2.5
CO_2	0.2	9.6	0.9	44.9	0.5	0.8	53.9	–	8.2
H_2S	–	15.2	Trace	–	–	–	–	–	35.8

Table 2.3   Composition of selected associated natural gases (volume%) [7]

Area	Abu Dhabi	Iran	North Sea	North Sea	N. Mexico
Field	Zakum	Agha Jari	Forties	Brent	San Juan County
CH_4	76.0	66.0	44.5	82.0	77.3
C_2H_6	11.4	14.0	13.3	9.4	11.2
C_3H_8	5.4	10.5	20.8	4.7	5.8
C_4H_{10}	2.2	5.0	11.1	1.6	2.3
C_5^+	1.3	2.0	8.4	0.7	1.2
N_2	1.1	1.0	1.3	0.9	1.4
CO_2	2.3	1.5	0.6	0.7	0.8
H_2S	0.3	–	–	–	–

Associated natural gas, on the other hand, is a co-product of crude oil and, therefore, its production is determined by the rate of production of the accompanying oil. It has long been considered a waste product and most of it was flared (for safety reasons). With the present energy situation, however, associated gas represents a feedstock and energy source of great potential value. Moreover, utilization, instead of flaring, is in better agreement with environment protection measures. An increasing number of schemes are being developed to utilize such gas.

Natural gas from the well is treated depending on the compounds present in the gas. A dry gas needs little or no treatment except for H_2S and possibly CO_2 removal if the amounts are appreciable (sour gas). Condensable hydrocarbons are removed from a wet gas and part can be sold as liquefied petroleum gas (LPG, propane, butane). The other part, C_5^+ can be blended with gasoline. So, once natural gas has been purified and separated, a large part of the gas is a single chemical compound, methane.

QUESTION:
            Why would associated natural gas always be 'wet'?

2.3.3.2 Crude oil

The composition of crude oil is much more complex than that of natural gas. Crude oil is not a uniform material with a simple molecular formula. It is a complex mixture of gaseous, liquid, and solid hydrocarbon compounds, occurring in sedimentary rock deposits throughout the world. The composition of the mixture depends on its location. Two adjacent wells may produce quite different crudes and even within a well the composition may vary significantly with depth. Nevertheless, the elemental composition of crude oil varies over a rather narrow range as shown in Table 2.4.

Although at first sight these variations seem small, the various crude oils are extremely different. The high fraction of C and H suggests that crude oil consists of hydrocarbons, which indeed has been proven to be the case. From detailed analysis it appears that crude oil contains alkanes, cycloalkanes (naphthenes), aromatics, poly-

Table 2.4   Elemental composition of crude oil

Element	Percentage range (wt%)
C	80–87
H	10–14
N	0.2–3
O	0.05–1.5
S	0.05–6

cyclic aromatics, sulfur-containing compounds, nitrogen-containing compounds, oxygen-containing compounds, etc.

QUESTION:

> *Does the elemental composition agree with the presence of these compounds?*

The larger part of crude oil consists of alkanes, cycloalkanes, and aromatics (see Figure 2.12). Both linear and branched alkanes are present. In gasoline applications the linear alkanes are much less valuable than the branched alkanes, whereas in diesel fuel the linear alkanes are desirable. One of the aims of catalytic reforming is to shift the ratio branched/linear in the desired direction. Cycloalkanes are often called naphthenes. Aromatics have favorable properties for the gasoline pool. However, currently their potential dangerous health effects are receiving increasing attention. The most important binuclear aromatic is naphthalene.

QUESTION:

> *Crude oil does not contain alkenes. Is this surprising?*

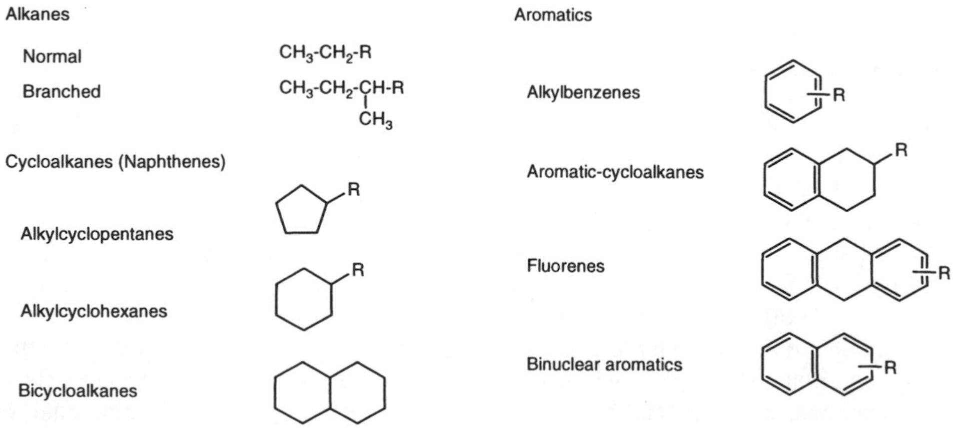

Figure 2.12   Examples of alkanes, cycloalkanes and aromatics present in crude oil.

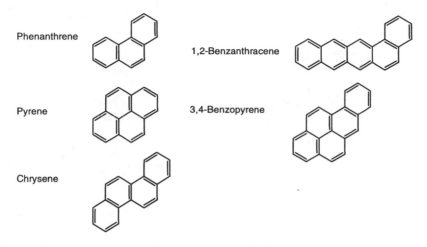

Figure 2.13   Examples of polycyclic, polynuclear aromatics in crude oil.

The heavier the crude the more polycyclic aromatic compounds it will contain. Evidently, heavy crudes render less useful products. Moreover, polycyclic aromatics may lead to carbonaceous deposits referred to as 'coke'. The implications of coke formation will be discussed in Chapter 3. Figure 2.13 shows examples of polycyclic, polynuclear aromatics.

Crudes do not consist exclusively of C and H; minor amounts of so-called hetero-atoms are also present, the major ones being S, N, and O. Although the amount of sulfur at first sight may seem low, its presence in petroleum fractions has many

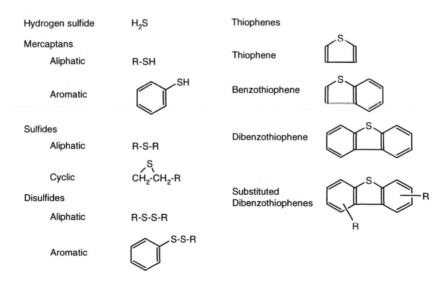

Figure 2.14   The most important sulfur-containing hydrocarbons in crude oil.

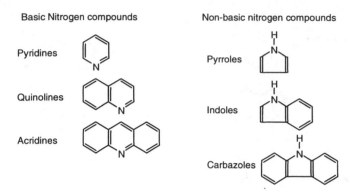

Figure 2.15   The most important nitrogen-containing hydrocarbons present in crude oil.

consequences for processing of these fractions. The presence of sulfur is highly undesirable, because it leads to corrosion, poisons catalysts, and is environmentally harmful. Chapter 3 discusses the technology for sulfur removal.

Figure 2.14 shows examples of sulfur-containing hydrocarbons. They are ranked according to reactivity: sulfur in mercaptans is relatively easy to remove in various chemical reactions, whereas sulfur in thiophene and benzothiophenes has an aromatic character, resulting in high stability. Aromatic sulfur compounds in particular are present in heavy crudes.

The nitrogen content of crude oil is lower than the sulfur content. Nevertheless, nitrogen compounds (Figure 2.15) deserve attention because they disturb major catalytic processes, such as catalytic cracking and hydrocracking. Basic nitrogen compounds react with the acid sites of the cracking catalyst and thus destroy its acidic character. Nitrogen compounds, just as aromatic sulfur compounds, are present in the higher boiling hydrocarbon fractions in particular.

Although the oxygen content of crude oil usually is low, oxygen occurs in many

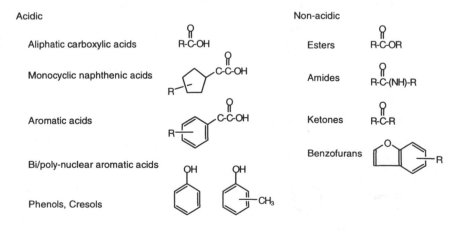

Figure 2.16   Oxygen-containing hydrocarbons present in crude oil.

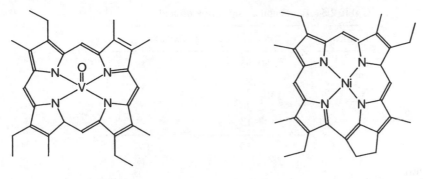

VO-ethioporphyrin I (VO-Etio-I)   Ni-decarboxydeoxophytoporphyrin (Ni-DCDPP)

Figure 2.17 Examples of metal containing hydrocarbons in crude oil.

different compounds (see Figure 2.16). A distinction can be made between acidic and non-acidic compounds. Organic acids and phenols belong to the class of acids.

Metals are present in crude oil only in small amounts. Even so, their occurrence is of considerable interest, because they deposit on and thus deactivate catalysts for upgrading and converting oil products. Part of the metals is present in the water phase of crude oil emulsions and may be removed by physical techniques. The other part is present in oil-soluble organometallic compounds and can only be removed by catalytic processes. Figure 2.17 shows examples of metal containing hydrocarbons.

Figure 2.18 shows metal contents of various crudes. Most of the metal-containing compounds are present in the heavy residue of the crude oil. Clearly, the metal contents vary widely. The most abundant metals are nickel, iron, and vanadium.

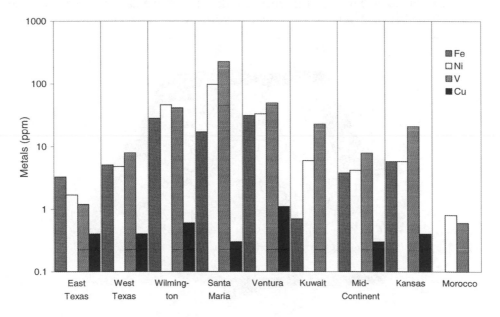

Figure 2.18 Metal content of various crudes.

Table 2.5   Elemental composition of coal

Element	Percentage range (wt%)
C	60–95
H	2–6
N	0.1–2
O	2–30
S	0.3–13

2.3.3.3 Coal

In contrast to crude oil and natural gas, the composition of coal varies over a wide range (see Table 2.5). The elemental composition range is based on only the organic component of coal. In addition, coal contains an appreciable amount of inorganic material (minerals), which forms ash during combustion and gasification (see Chapter 5). The amount ranges from 1 to over 25%. Furthermore, coal contains water: the moisture content of coal ranges from about 2 to nearly 70%.

The high C/H ratio reflects the fact that a major part of the coal is built up of complex polycyclic aromatic rings.

QUESTION:
> *Show that most coals are aromatic.*

Many scientists have been fascinated by the structure of coal, and have been working hard trying to elucidate it. This structure depends on the age of the coal and the conditions under which it has been formed. Figure 2.19 shows a typical structure of a coal particle.

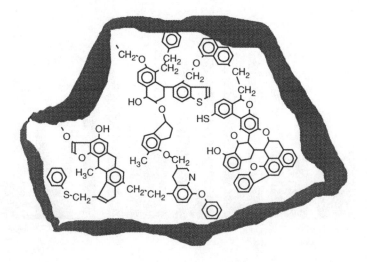

Figure 2.19   Typical coal structure.

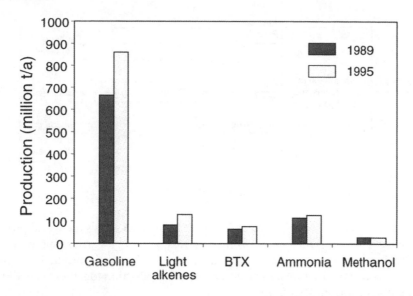

Figure 2.20 Production of base chemicals and gasoline. Based on data from [3,6,9–11].

2.4 BASE CHEMICALS

Petrochemicals, i.e. oil derived chemicals, are the major raw materials for the chemical industry. The products from the chemical industry are of a tremendous variety, although the vast majority, about 85%, is manufactured from a very limited number of simple chemicals, called *base chemicals*. The most important base chemicals are the lower alkenes (ethene, propene, and butadiene), the aromatics (benzene, toluene, and xylene, also referred to as 'BTX'), ammonia, and methanol. Synthesis gas, a mixture of hydrogen and carbon monoxide in varying ratio, can also be considered a base chemical. Most chemicals are produced directly or indirectly from these compounds,

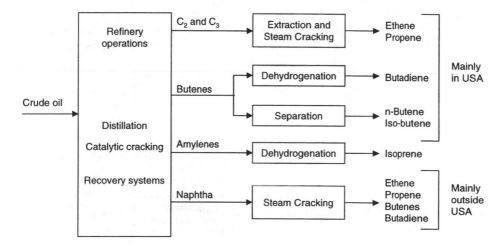

Figure 2.21 Lower alkenes production from oil [12].

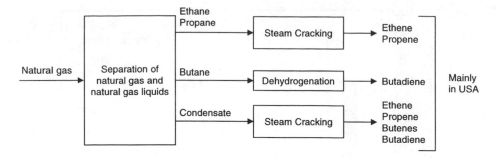

Figure 2.22   Lower alkenes production from natural gas [12].

which can be considered real building blocks. Figure 2.20 shows the world production of the base chemicals in comparison to a major energy carrier, motor gasoline. Although base chemicals account for most of the chemicals production, gasoline production is much larger than the base chemicals production together. This illustrates the huge consumption of gasoline.

The feedstock chosen for the production of the base chemicals will depend on the production unit, the local availability, and the price of the raw materials. For the production of light alkenes, in broad terms, there is a difference between the USA and the rest of the world, as illustrated by Figures 2.21 and 2.22.

One of the reasons for these differences lies in the consumer market: in the USA, the production of gasoline is even more dominant than in the rest of the developed world. As a consequence, lower alkenes are primarily produced from sources that do not yield gasoline. This development was made possible by the discovery of natural gas fields with high contents of hydrocarbons other than methane (see Table 2.3). Natural gas from these fields is very suitable for the production of lower alkenes.

Historically, the production of the simple aromatics has been coal based: coal tar, a by-product of coking ovens, was the main source of benzene, toluene, etc. At present, the aromatic base chemicals for nearly 95% are petroleum-based. Figure 2.23 shows the main processes involved in aromatics production.

The main sources of aromatics, viz. catalytic reforming and steam cracking, also are producers of motor gasoline. Therefore, production of aromatics is closely related to

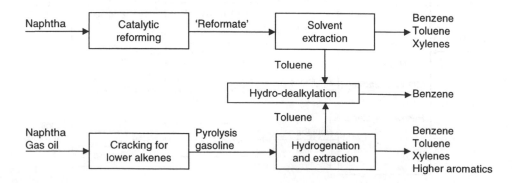

Figure 2.23   Aromatics production [12].

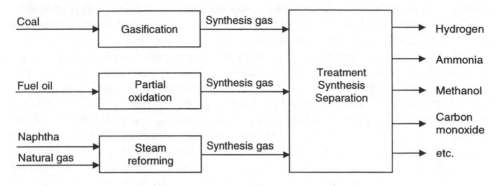

Figure 2.24 Ammonia and methanol production.

fuel production: aromatics and gasoline have to compete for the same raw materials. The demand for benzene is greater than the demand for the other aromatics. Hence, part of the toluene produced is converted to benzene by hydrodealkylation.

Ammonia and methanol are two other important base chemicals. Their production generally involves the conversion of synthesis gas, a mixture of mainly hydrogen and carbon monoxide, which can be produced from either coal, petroleum products, or natural gas, or any other hydrocarbon source. Figure 2.24 shows a schematic diagram, which indicates the processes used for the conversion of the different raw materials.

Most of the processes referred to in Figures 2.21–2.24 will be the topic of subsequent chapters.

As mentioned earlier, a very large part of organic chemicals is produced from base chemicals. Therefore, the production of base chemicals is a useful indicator of the growth of the petrochemical industry. In the 1950s and 1960s essentially exponential growth of the petrochemical industry took place, as illustrated in Figure 2.25.

In Western Europe and to a lesser extent in North America the growth rate has

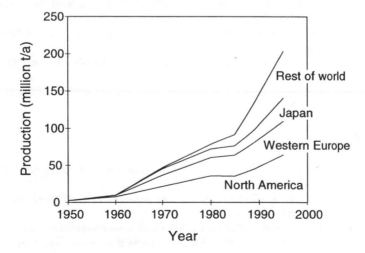

Figure 2.25 Growth of the petrochemical industry, production of six major base chemicals (ethene, propene, butadiene, benzene, toluene and xylenes) [3,6].

declined, in contrast to Japan and some other regions referred to as 'Rest of world'. The latter category includes Singapore, Taiwan, Korea, the Middle East, and China. It is not surprising that growth is taking place in these regions in particular.

In the Western world the industry, in many respects, is mature and investments tend to be stable or even decreasing for two reasons. Firstly, partly due to good maintenance the lifetime of the plants is much larger than anticipated. Secondly, capacities of existing plants have been increased by good engineering ('de-bottlenecking'). For instance, new packings in distillation units, new catalysts, and new control systems can rejuvenate a plant. In addition to the foregoing, new incentives for radically novel plants have been emerging. For instance, environmental concerns call for new technology. It is generally accepted that a sustainable technology is desired. Elements are renewable feedstocks and energy, less polluting routes, and more efficient processes. 'Process Intensification' is an umbrella for many new concepts.

References

1 Vogel H (1992) 'Process development' in: Gerhartz W, et al. (eds.) *Ullmann's Encyclopedia of Industrial Chemistry*, vol. B4, 5th ed., VCH, Weinheim, pp. 438–475.
2 Anon. (1987) 'A guide to chemicals from hydrocarbons – some examples' *Hydroc. Process* 66(11), 60–61.
3 Shell International Chemical Company Limited (1990) *Chemicals Information Handbook 1990.*
4 Smith JM and Van Ness HC (1987) *Introduction to Chemical Engineering Thermodynamics*, 4th ed., McGraw-Hill, New York.
5 Shell International Chemical Company Limited (1995) *Energy in Profile.*
6 The British Petroleum Company (1998) *BP Statistical Review of World Energy.* http://www.bp.com/bpstats/
7 *Our Industry; Petroleum* (1977), 5th ed., London, The British Petroleum Company.
8 Woodcock KE and Gottlieb M (1994) 'Gas, natural' in: Kroschwitz JI and Howe-Grant M (eds.) *Kirk Othmer Encyclopedia of Chemical Technology*, vol. 12, 4th ed., Wiley, New York, pp. 318–340.
9 Crocco JR (1991) 'The outlook for global methanol supply and demand during the short term crisis period', *World Methanol Conference.*
10 Chauvel A and Lefebvre G (1989) *Petrochemical Processes 1. Synthesis Gas Derivatives and Major Hydrocarbons*, Technip, Paris.
11 SRI international (1998) *Chemical Industries Newsletter*, September 19. http://piglet.sri.com/CIN/
12 *The Petroleum Handbook* (1983), 6th ed., Compiled by staff of the Royal Dutch Shell Group of Companies, Elsevier, Amsterdam.

General Literature

Gary JH and Handwerk GE (1994) *Petroleum Refining, Technology and Economics*, 3rd ed., Marcel Dekker, New York.
McKetta JJ and Cunningham WA (eds.) (1976) *Encyclopedia of Chemical Processing and Design*, Marcel Dekker, New York.
Schobert HH (1990) *The Chemistry of Hydrocarbon Fuels*, Buttersworth, London.
Van Krevelen DW (1993) *Coal: Typology – Physics – Chemistry – Constitution*, 3rd ed., Elsevier, Amsterdam.

Chapter 3

Processes in the Oil Refinery

Refining of crude oil is a key activity in the process industry. Over 600 refineries are in operation worldwide with a total annual capacity of over 3500 million metric tons. The goal of oil refining is twofold: production of fuels for transportation, power generation, and heating purposes, and production of raw materials, especially for the chemical industry.

3.1 OIL REFINERY – AN OVERVIEW

Oil refineries are complex plants. They constitute a relatively mature and highly integrated industrial sector. Besides physical processes such as distillation and extraction a large number of chemical processes are applied, many of which are catalytic. Figure 3.1 shows a flow scheme of a complex modern refinery, with the catalytic processes indicated by gray blocks.

After desalting and dehydration (not shown in Figure 3.1), the crude oil is separated in fractions by distillation. The distilled fractions are not directly applicable in the market. Many different processes are carried out in order to produce the required products. This chapter will discuss most of the processes in Figure 3.1. They can be divided in physical and chemical processes, as shown in Table 3.1. Oil refineries show large differences and, besides the processes mentioned here, many others are applied. For instance, Figure 3.1 and Table 3.1 do not show processes for treating refinery off-gases and sulfur recovery (discussed in Section 3.6) and facilities for wastewater treatment, etc. More information on processes not described in this chapter can be found in, for instance Refs. [1,2].

The reason that such a complex set of processes is needed is the difference between the properties of the crude oil delivered and the requirements of the market. Extensive processing is required in order to obtain products with satisfactory performance, especially for fuels in the transport sector. Another reason for the complexity of an oil refinery lies in environmental considerations. Legislation calls for increasingly cleaner processes and products. In fact, at present legislation related to minimizing the environmental impact is the major drive for process improvement and the development of novel processes.

3.2 PHYSICAL PROCESSES

3.2.1 Desalting/Dehydration

Crude oil often contains water, inorganic salts, suspended solids, and water-soluble trace metals. These contaminants can cause corrosion, plugging, and fouling of equip-

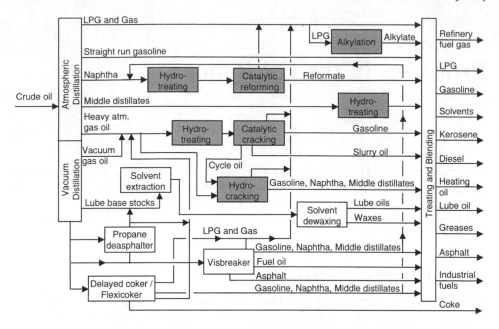

Figure 3.1 Flow scheme of a complex modern oil refinery; catalytic processes are indicated by gray blocks. Adapted from [1].

ment and poisoning of catalysts in catalytic processes. Therefore, the first step in the refining process is desalting and dehydration. This can be accomplished by adding hot water with added surfactant to extract the contaminants from the oil. Upon heating, the salts and other impurities dissolve into the water or attach to it. The oil and water phase are then separated in a tank, where the water phase settles out. Wastewater and contaminants are discharged from the bottom of the settling tank and the desalted crude is drawn from the top and sent to the crude distillation tower.

Table 3.1 Physical and chemical processes in a typical oil refinery

Physical processes	Chemical processes	
	Thermal	Catalytic
Distillation	Visbreaking	Hydrotreating
Solvent extraction[a]	Delayed coking	Catalytic reforming
Propane deasphalting	Flexicoking	Catalytic cracking
Solvent dewaxing[a]		Hydrocracking
Blending[a]		Catalytic dewaxing[a]
		Alkylation
		Polymerization
		Isomerization

[a] Not discussed in this book, see for instance [1,2].

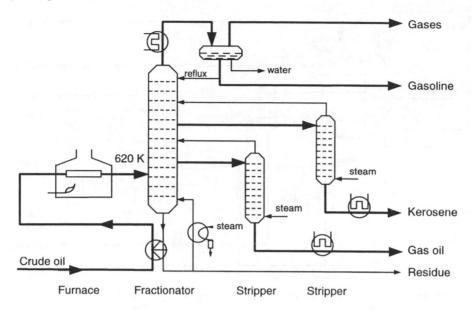

Figure 3.2 Simple crude distillation.

3.2.2 Crude Distillation

The desalted crude has to be separated into products, which each have specific uses. Crude oil consists of thousands of different compounds, and as a consequence, it is impossible and undesirable to separate crude oil in chemically pure fractions. The central separation step in every oil refinery is distillation of the crude, which separates the various fractions according to their *volatility*. Figure 3.2 shows a schematic of a simple crude distillation unit.

After heat exchange with the residue the crude oil feed is heated further to about 620 K in a furnace. It is discharged to the distillation tower as a foaming stream. The vapors flow upward and are fractionated to yield the products given in Figure 3.2. A distillation column typically is 4 m in diameter and 20–30 m in height and contains 15–30 trays. The stripping columns, which may be 3 m high with a diameter of 1 m, serve to remove the more volatile components from the side streams.

With this simple distillation unit a satisfactory set of products is not obtained. The market calls for clean products (no S, N, O, metals, etc.), more gasoline (high octane number), more diesel (high cetane number), specific products (aromatics, alkenes, etc.), and less residue. The logical steps to meet these demands are more sophisticated distillation and other physical separation steps, followed by chemical conversion steps.

The major limitation of atmospheric distillation is the fact that hydrocarbons should not be heated to high temperature. The reason is that thermal decomposition reactions take place above a temperature of about 630 K. Decomposition reactions ('thermal cracking' or 'pyrolysis') are highly undesirable because they result in the deposition of carbonaceous material on the tube walls.

Formation of carbonaceous deposits is not restricted to thermal processing of hydrocarbons. Also in catalytic processes such deposits, referred to as 'coke', are formed. In

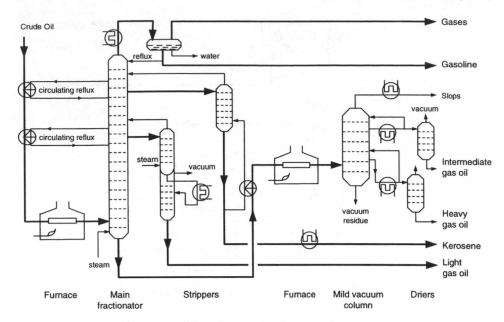

Figure 3.3   Modern crude distillation unit.

the next sections it will become clear that coke formation has tremendous implications for hydrocarbon processing. Note that in coal processing the solid residue, formed in thermal processing, is called 'char' (see Chapter 5).

Figure 3.3 shows a schematic of a typical modern two-column distillation unit, consisting of an atmospheric distillation column and a column operating under reduced pressure (<0.1 bar) for fractionation of the atmospheric residue. This advanced distillation requires substantial energy, and therefore, optimal heat recovery is essential. Preheating of the feed by heat exchange with the residue, as is done in a simple distillation unit, is not convenient because the residue is further distilled. Therefore, the crude oil passes through several heat exchangers, where it recovers heat from circulating reflux streams. This provides an optimal heat recovery and temperature control system.

The feed enters the first distillation column at 570–620 K, depending on the crude and the desired product distribution. As the vapor passes up through the column heat is removed and the vapor is partially condensed by the circulating reflux streams, thus providing a flow of liquid down the column. As a result, the circulating reflux streams also influence (improve) the composition of the fractions that are withdrawn as side-streams by establishing a new vapor-liquid equilibrium. The light gas oil side-stream is stripped with steam to recover more volatile components. The kerosene side-stream is heat exchanged with the column bottom product. In this case no steam for stripping is used to meet the water content specification for use of the kerosene as aviation fuel.

The heavy liquid fraction leaving the bottom of the main fractionator is sent to a mild-vacuum column for further fractionation. Heat is withdrawn from the vacuum column at various levels by circulating reflux streams. The residue from the mild-vacuum column may be further processed in a high-vacuum fractionator to recover additional distillates, which are used as fuel oil or as feedstock for catalytic cracking.

A vacuum column usually has a much larger diameter (up to 15 m) than an atmospheric distillation column because of the large volumetric flow rate due to reduced vapor pressures.

QUESTION:

> *Packed columns allow larger liquid-gas contact areas than tray columns. Two types of packings can be applied, viz. structured and random packings. What would be the preferred packing for a vacuum distillation column?*

3.2.3 Propane Deasphalting

The coke-forming tendencies of heavier distillation products (mainly atmospheric and vacuum residue) can be reduced by removal of 'asphaltenic' materials by means of extraction. Liquid propane appears to be a suitable solvent, but butane and pentane are also commonly applied. Deasphalting is based on the *solubility* of hydrocarbons in propane, i.e. on the type of molecule, rather than on molecular weight as in distillation. Figure 3.4 shows a schematic representation of the propane deasphalting process [2,3].

The heavy feed, usually vacuum residue, is fed to the deasphalting tower and is contacted with liquefied propane countercurrently. Alkanes that are present in the feed dissolve in the propane, whereas the asphaltenic material (aromatic compounds), the 'coke precursors', do not dissolve. Propane evaporates on depressurization and is condensed and recycled to the deasphalting tower. The remaining deasphalted oil is stripped with steam to remove any residual propane. The asphaltenic material that leaves the deasphalting unit undergoes similar treatment. The asphalt residue is then

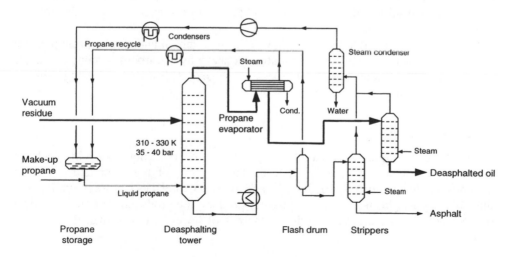

Figure 3.4 Propane deasphalting process.

sent to thermal processes like 'Flexicoking', 'Delayed Coking', and 'Visbreaking'. It may also be processed in a HYCON plant or blended with other asphalts.

QUESTIONS:

> *Why would asphalt be the bottom product?*
> *Typical propane deasphalting temperatures are 310–330 K.*
> *What would be the pressure in the deasphalting tower? What would it be for butane deasphalting (first decide what temperature is best)?*

3.3 THERMAL PROCESSES

Thermal processes are widely applied. When heating a hydrocarbon to a sufficiently high temperature, thermal cracking takes place. This is often referred to as 'pyrolysis' (in particular in processing of coal, see Chapter 5) and, slightly illogically, as 'steam cracking' (pyrolysis of hydrocarbons in the presence of steam, see Chapter 4). In the latter case steam acts as a diluent; cracking of steam does not occur. Nevertheless, because in the reacting mixture a large quantity of steam is present, the term 'steam cracking' is generally used.

QUESTION:

> *Why is steam generally used as diluent?*

Pyrolysis of coal yields a mixture of gases, liquids, and solid residue ('char'). The same process is applied in oil conversion processes. In this section we will discuss two of them, viz. 'visbreaking' and 'delayed coking'. The more recent 'Flexicoking' process will be treated in Section 3.5. Thermal processes are flexible but a disadvantage is that in principle large amounts of low value products are formed. This is the reason for innovations like Flexicoking and processes optimized towards the production of certain types of high-quality coke.

Coal and heavy residue processes have a lot in common and similar principles are applied. An example is the partial oxidation of heavy hydrocarbons. Among the major processes are the Texaco and Shell processes (see, e.g. [2]) with a technology similar to their coal gasification analogues.

3.3.1 Visbreaking

Visbreaking (see Figure 3.5) is a relatively mild thermal cracking process in which the viscosity of vacuum residue is reduced. The severity of the visbreaking process depends on temperature (710–760 K) and reaction time (1–8 min). Usually, less than 10% by weight of gasoline and lighter products are produced. The main product (ca. 80 wt%) is the cracked residue, which is of lower viscosity than the vacuum residue.

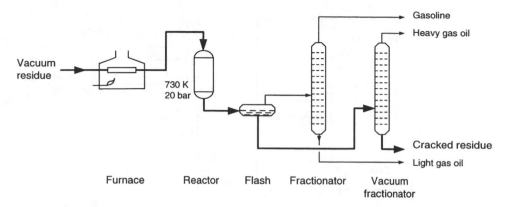

Figure 3.5 Schematic of the visbreaking process.

3.3.2 Delayed Coking

Delayed coking is a thermal cracking process with long residence times, making it more severe than visbreaking. As a consequence, also solid residue ('coke') is formed:

$$\text{heavy feed} \begin{cases} \text{gas} \\ \text{liquid} \\ \text{coke} \end{cases} \tag{1}$$

Originally, the goal of delayed coking was minimization of the output of residual oil and other high-molecular-weight material, but over the years it was found profitable to produce certain types of cokes with desired properties for the application it is produced for. A good example is the production of coke for electrodes, which in fact is the largest non-fuel end-use for petroleum coke. Figure 3.6 shows a typical process.

The heated feed is introduced in a coke drum, where the reaction takes place. Usually, a coke drum is on stream for about 24 h, during which it becomes filled with porous coke. The feed is subsequently introduced in the second coke drum, while coke is removed from the first drum with high-pressure water jets.

QUESTION:
> *Explain the names 'delayed coking' and 'visbreaking'.*

3.4 CATALYTIC PROCESSES

Catalytic processes are the pillars of oil processing. In fact, chemists and chemical engineers from this sector have had tremendous impact on the discipline of catalysis and (chemical) engineering. The most important catalytic processes in the oil refinery, in terms of throughput, are fluid catalytic cracking (FCC), hydrotreating, hydrocracking, catalytic reforming, and alkylation.

The major part of the oil fractions is used in the transport sector, especially as diesel

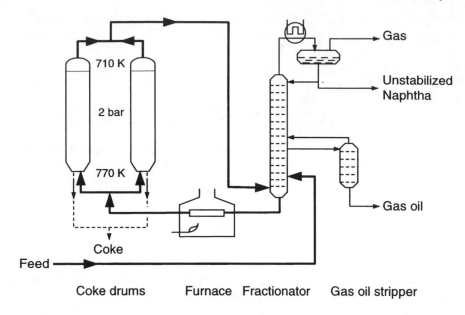

Figure 3.6   Schematic of delayed coking process.

and gasoline. Table 3.2 gives a survey of the octane number of various gasoline range hydrocarbons.

QUESTIONS:

> *Some octane numbers are larger than 100. Explain why this is possible. From the difference between RON and MON for the refinery products can you guess the approximate composition (alkanes, alkenes, aromatics)?*

The octane number is a measure of the quality of the gasoline. Modern gasoline-powered cars require a minimum octane number of 92, 95, or 98 (RON). Gasoline with lower octane number can cause irreversible damage to the engine. The amount of gasoline obtained from simple distillation is too low. Therefore, either high-octane number compounds have to be added or additional conversion steps such as catalytic cracking and hydrocracking, which convert heavy hydrocarbons into lighter ones, are required. Moreover, the octane number of 'straight run' gasoline is much lower than required. Therefore, steps like catalytic reforming, which literary re-forms gasoline range hydrocarbons into hydrocarbons of higher octane number, and alkylation, which combines lower alkenes with isobutane to form high octane gasoline components, are commonly carried out. Moreover, ethers like MTBE (RON 116) are added at a level of a few percent. In addition, hydrotreating is often necessary to remove hetero-atoms (S, N, etc.).

Diesel is another important transportation fuel. The diesel equivalent of the octane

Table 3.2 Octane numbers of various hydrocarbons and typical octane numbers of main refinery gasoline streams

Hydrocarbon	Octane number[a]		Boiling point (K)
	RON	MON	
n-Pentane	62		309
2-Methylbutane	90		301
Cyclopentane	>100		322
n-Hexane	19		342
2-Methylpentane	73		333
2,2-Dimethylbutane	93		323
1-Hexene	76		337
2-Hexene	92		341
Benzene	106		353
Cyclohexane	83		354
***n*-Heptane**	**0**[b]		**362**
2-methylheptane	13		381
2,2,4-Trimethylpentane	**100**[b]		**372**
1-Octene	35		395
2-Octene	56		398
3-Octene	68		396
Xylenes	>100		≈415
Ethylbenzene	108		410
1,2-Dimethylcyclohexane	79		403
Ethylcyclohexane	41		403
Methyl-tertiary-butyl-ether (MTBE)	116		328
Ethyl-tertiary-butyl-ether (ETBE)	118		345
Tertiary-amyl-methyl-ether (TAME)	111		359
Light straight-run gasoline	68	67	
Isomerate	85	82	
FCC light gasoline	93	82	
FCC heavy gasoline	95	85	
Alkylate	95	92	
Reformate (CCR)	99	88	

[a] RON, research octane number; MON, motor octane number; MON is generally lower, depending on the particular compound. The difference is particularly large for alkenes and aromatics.
[b] By definition.

number is the cetane number, which in Europe should at least be 49 [4]. Especially the diesel fraction of the hydrocracking process contributes to the cetane number (55–60). Table 3.3 shows cetane numbers of hydrocarbons in the diesel fuel range.

Cetane numbers are highest for linear alkanes, while naphthalenes, which are very aromatic compounds (see Chapter 2), have the lowest cetane numbers. Note that this situation is the opposite for octane numbers (Table 3.2). In general, the more suitable a hydrocarbon is for combustion in a gasoline (Otto) engine, the less favorable it is for combustion in a diesel engine.

Table 3.3 Cetane numbers of various hydrocarbon classes and of main refinery gas oil
streams (typical ranges)

Hydrocarbon class	Cetane number
n-**Hexadecane (cetane)**	**100**[a]
n-Alkanes	100–110
iso-Alkanes	30–70
Alkenes	40–60
Cyclo-alkanes	40–70
Alkylbenzenes	20–60
Naphthalenes	0–20
α-**Methyl naphthalene**	**0**[a]
Straight-run gas oil	40–50
FCC cycle oil	0–25
Thermal gas oil	30–50
Hydrocracking gas oil	55–60

[a] By definition.

Otto versus Diesel engines

The differences between the required properties of gasoline and diesel fuel result
from the difference between the ignition principles of gasoline and diesel
engines. In gasoline (Otto) engines the premixed air/fuel mixture is ignited by
a spark, and combustion should proceed in a progressing flame front. Hence,
uncontrolled self-ignition of gasoline is highly undesired. In contrast, in diesel
engines the fuel is injected in hot compressed air present in the cylinder and self-
ignition has to occur.

3.4.1 Catalytic Cracking

The main incentive for catalytic cracking [2,5–7] is the need to increase gasoline
production. Originally cracking was performed thermally, but nowadays cracking in
the presence of a catalyst predominates. Feedstocks are heavy oil fractions, typically
vacuum gas oil. Cracking is catalyzed by solid acids, which promote the rupture of
C–C bonds. The crucial intermediates are carbocations (positively charged hydrocarbon ions). They are formed by the action of the acid sites of the catalyst. The nature of
the carbocations is still subject of debate. In the following it will be described that they
are 'classical' carbenium ions and protonated cyclopropane derivatives.

Besides C–C bond cleavage a large number of other reactions occur:

- isomerization;
- protonation, deprotonation;
- alkylation;
- polymerization;
- cyclization, condensation (eventually leading to coke formation).

So, catalytic cracking comprises a complex network of reactions, both intra-molecular and inter-molecular. The formation of coke is an essential feature of the cracking process. It will be shown that although coke deactivates the catalyst, its presence has enabled the development of an elegant practical process.

Catalytic cracking is one of the largest applications of catalysis: worldwide cracking capacity exceeds 500 million t/a. Catalytic cracking was the first large-scale application of fluidized beds. This explains the name 'fluid catalytic cracking' (FCC). Nowadays, entrained-flow reactors are used instead of fluidized beds, but the name 'FCC' is still retained.

3.4.1.1 Cracking mechanism

Catalytic cracking proceeds via a chain reaction mechanism in which organic ions are the key intermediates. The role of the catalyst is to initiate the chain reactions.

Alkenes can abstract a proton from a Brønsted site of the catalyst to form carbenium ions:

$$R-CH=CH_2 + H^+ \longrightarrow R-CH_2-\overset{\overset{\displaystyle H}{|}}{\underset{\underset{\displaystyle H}{|}}{C}}{}^+ \quad or \quad R-\overset{\overset{\displaystyle }{}}{\underset{\underset{\displaystyle H}{|}}{C}}{}^+-CH_3 \tag{2}$$

Here, the alkene is a terminal alkene and either a primary or a secondary carbenium ion is formed. Branched alkenes can also lead to tertiary carbenium ions. The probability for the existence of these ions is not distributed randomly, because (i) their stability differs profoundly and (ii) they are interconverted (e.g. primary to secondary). The relative stability of the carbenium ions decreases in the order:

$$tertiary > secondary > primary > ethyl > methyl$$

In reality these ions are not present as such, but they form surface ethoxy species on the catalyst. For the discussion here it suffices to treat them as adsorbed carbenium ions.

QUESTION:

> *Why are ethyl and methyl carbenium ions less stable than other primary carbenium ions?*

Alkanes are very stable and only react at high temperature in the presence of strong acids. They react via carbonium ions under the formation of H_2:

$$R-CH_2-CH_3 + H^+ \longrightarrow R-\overset{\overset{\displaystyle H}{|}}{\underset{\underset{\displaystyle H\ \ H}{/\ \backslash}}{C}}{}^+-CH_3 \longrightarrow R-\overset{\overset{\displaystyle }{}}{\underset{\underset{\displaystyle H}{|}}{C}}{}^+-CH_3 + H_2 \tag{3}$$

The ion formed upon proton addition contains a pentacoordinated carbon atom and is referred to as a carbonium ion. It decomposes easily into a carbenium ion and H_2.

In a medium where carbeniums ions are present, transfer of a hydride ion to a carbenium ion can be the predominant route:

$$
\begin{array}{c}
\overset{\displaystyle H}{\underset{\displaystyle H}{H_3C-\overset{|}{\underset{|}{C}}-CH_2\text{-}CH_2\text{-}CH_2\text{-}CH_3}} \\[2mm]
+ \\[2mm]
H_3C-\overset{+}{\underset{\displaystyle CH_3}{C}}-CH_2\text{-}CH_2\text{-}CH_3
\end{array}
\quad\longrightarrow\quad
\begin{array}{c}
H_3C-\overset{+}{\underset{\displaystyle H}{C}}-CH_2\text{-}CH_2\text{-}CH_2\text{-}CH_3 \\[2mm]
+ \\[2mm]
H_3C-\underset{\displaystyle CH_3}{CH}-CH_2\text{-}CH_2\text{-}CH_3
\end{array}
\tag{4}
$$

The presence of carbenium ions gives rise to a variety of reactions. Isomerization can take place via protonated cyclopropane intermediates, see Scheme 3.1.

The essential reaction of carbenium ions in catalytic cracking is *scission of C-C bonds*. For example:

$$
H_3C-\overset{+}{\underset{\displaystyle CH_3}{C}}-CH_2-\underset{\displaystyle CH_3}{\overset{\displaystyle CH_3}{C}}-CH_3
\longrightarrow
H_3C-\underset{\displaystyle CH_3}{C}=CH_2 \;+\; \overset{+}{\underset{\displaystyle CH_3}{\overset{\displaystyle CH_3}{C}}}-CH_3
\tag{5}
$$

The reaction generates a smaller carbenium ion and an alkene molecule. The reaction rate depends on the relative stability of the reactant and product carbenium ions. When both are tertiary, as in the above example, then the reaction is fast. However, if a linear carbenium ion would undergo C–C bond breaking, a highly unstable primary carbenium ion would be formed, and the reaction would be very slow. Therefore, the reaction is believed to proceed via a protonated cyclopropane derivative as shown in Scheme 3.1.

A number of hydride shifts and the actual C–C bond breaking result in the formation of a linear alkene and a tertiary carbenium ion. The latter is converted to an iso-alkane by hydride transfer to a neutral molecule. The neutral molecule then becomes a new carbenium ion and the chain continues. Isomerization of linear alkanes into branched alkanes can take place in a similar way.

Of course, cracking of isomerized alkanes, and isomerization of large enough cracking fragments is always possible.

The following rules hold for catalytic cracking:

- cracking occurs on β versus + charge;
- 1-alkenes will be formed;
- shorter chains than C_7 are not or hardly cracked.

QUESTIONS:

> *Explain these rules from the reaction mechanism shown in Scheme 3.1. Why is it improbable that C–C bond breaking of the 'classical' carbenium ion would take place directly?*

This cracking mechanism explains why catalytic cracking is preferred over thermal cracking (see Section 3.3 and Chapter 4) for the production of gasoline; in thermal cracking, bond rupture is random, while in catalytic cracking it is more ordered and, therefore, selective. Figure 3.7 shows the difference in product distribution resulting from thermal and catalytic cracking of *n*-hexadecane.

QUESTIONS:

Explain why the product distributions peak at different carbon numbers. (Hint: compare the reaction mechanisms of thermal (Chapter 4) and catalytic cracking.) Are the units of the y-axis correct? Alkylation and polymerization only play a minor role in catalytic cracking. Why?

In 'hydrocracking' the catalyst performs two catalytic functions; hydrogenation of alkenes into alkanes and cracking reactions are catalyzed simultaneously. Why would the temperature for hydrocracking be lower than for FCC? Why would the reactions be carried out at high hydrogen partial pressure?

There are many pathways leading to coke formation. Important reactions are cyclization of alkynes, and di-, and poly-alkenes, followed by aromatization and condensation. By these reactions poly-aromatic compounds are formed, which are referred to as coke. The coke is deposited on the catalyst, and thus causes deactivation.

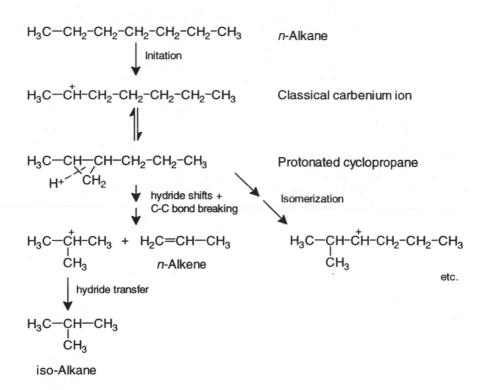

Scheme 3.1 Mechanism for catalytic cracking of alkanes, including isomerization.

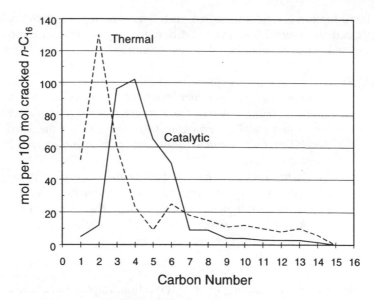

Figure 3.7   Product distribution in thermal and catalytic cracking of *n*-hexadecane [8].

3.4.1.2 Catalysts for catalytic cracking

The types of catalysts used in catalytic cracking have changed dramatically over the years. In the past, $AlCl_3$ solutions were used, which resulted in large technical problems such as corrosion and extensive waste streams. Subsequently, heterogeneous catalysts were applied. From the cracking mechanism described previously it is evident that catalytic cracking calls for acid catalysts. The improvement of cracking catalysts has been crucial in the development of the catalytic cracking process. Initially, acid-treated clays were used. Later, it appeared that synthetic materials, viz. *amorphous silica-alumina*, had superior properties due to their higher thermal and attrition stability, higher activity, and optimal pore structure. A great breakthrough in catalytic cracking was the discovery of *zeolites*, which are even better catalysts: they are more active and more stable (less coke, higher thermal stability).

Figure 3.8 shows that with zeolites the feed throughput can be much lower for the same product yield. With current zeolitic catalysts much less gas oil needs to be processed in an FCC unit than would have been the case with the amorphous catalysts to satisfy the increasing gasoline demand.

Silica-alumina is a mixed oxide that can be produced by 'impregnation' of porous silica with an Al^{3+} solution. Silica-alumina contains Lewis and Brønsted acid sites. It is a remarkably strong acid, much more acidic than silica or alumina alone. This is explained by the fact that in deprotonation of silica-alumina the negative charge is spread over the AlO_4-unit as shown in Figure 3.9.

QUESTION:

> *Just as silica-alumina is more acidic than silica alone, HBr is a stronger acid than HCl. Explain this analogue.*

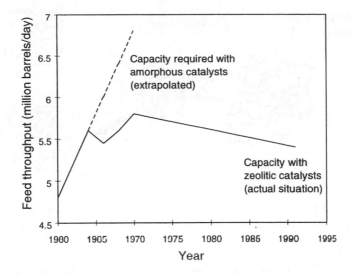

Figure 3.8 Growth of catalytic cracking capacity in United States [8].

The acid zeolites applied in FCC are related to the amorphous silica-aluminates, but zeolites are crystalline materials. A large variety of zeolites have been discovered. In contrast to amorphous silica-alumina, which contains sites with very different acidic strengths, many zeolites contain very well defined (strong) acid sites. The acid zeolites used in FCC are three-dimensional structures constructed of SiO_4 and AlO_4 tetrahedra, which are joined by sharing the oxygen ions [9]. The definitive structure depends on the Si/Al-ratio, which may vary from 1 to >12. Figure 3.10 shows examples of zeolitic structures based on so-called Sodalite cages (composed of SiO_4 and AlO_4 tetraeders).

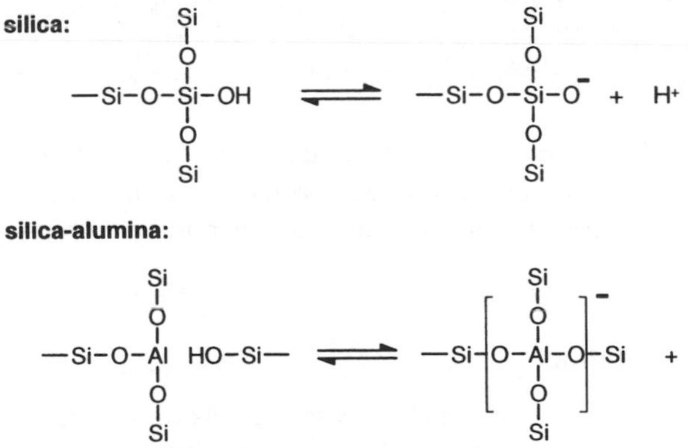

silica:

$$\begin{array}{ccc} & Si & \\ & | & \\ & O & \\ & | & \\ -Si-O-Si-OH & \rightleftharpoons & -Si-O-Si-O^- \ + \ H^+ \\ & | & \\ & O & \\ & | & \\ & Si & \end{array}$$

silica-alumina:

$$\begin{array}{ccc} & Si & \\ & | & \\ & O & \\ & | & \\ -Si-O-Al \quad HO-Si- & \rightleftharpoons & -Si-[O-Al-O]^- Si \ + \ H^+ \\ & | & \\ & O & \\ & | & \\ & Si & \end{array}$$

Figure 3.9 Silica and silica-alumina.

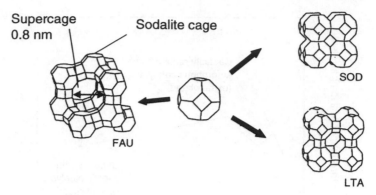

Figure 3.10 Examples of zeolites based on sodalite cages: left, zeolite Y (Faujasite), used in FCC; middle, part of sodalite cage (truncated octahedron); right, sodalite mineral (top), zeolite A (bottom).

QUESTIONS:

> *Why would of the three zeolites in Figure 3.10 only the Y zeolite be applied in FCC? Earlier it was stated that carbenium ions do not occur as such but in the form of ethoxy species. Explain this from the structure of the solid acids used.*

Zeolites have many interesting characteristics (see also Section 3.7). Important for FCC are their high acidity, their low tendency towards coke formation, and their high thermal stability.

Zeolite crystals are not used as such. They are too active in the catalytic cracking reactions (reactivity $\approx 10^6$ mol/(m$^3_{cat}$·s)), which cannot be dealt with in a practical industrial reactor – see Chapter 1). In addition, the pores are too small for a large part of the feed. Moreover, due to the small pore diameters, crystal sizes have to be extremely small to minimize internal mass-transfer limitations. Therefore, the zeolite particles are diluted with a macro/meso-porous matrix material (silica-alumina).

Porosity range	Pore diameter (nm)
Micro pores	<2
Meso pores	2–50
Macro pores	>50

Although the matrix is only moderately active, its function must not be under-estimated: the FCC catalyst matrix provides primary cracking sites, generating inter-mediate feed molecules for further cracking into desirable products by acidic zeolitic sites.

3.4.1.3 Product distribution

The feed to a catalytic cracker is usually a vacuum gas oil but other heavy feeds can also be processed. Figure 3.11 shows the improvement in product selectivity with time. The conversion is usually defined as the yield on gasoline, LPG (C$_3$, C$_4$), light

Synthesis of commercial cracking catalysts

Figure B.3.1 illustrates the production of commercial cracking catalysts. The zeolite crystals are formed in a stirred-tank reactor. After washing they are ion exchanged with rare earth (RE) ions, introduced as the chloride: $RECl_3$ (Zeolites are not acidic in the Na form, after exchanging Na ions with rare earth ions they are). In the other stirred-tank reactor amorphous silica-alumina is produced. The two slurries are mixed and dried in a spray dryer, producing a fine powder of catalyst particles consisting of a silica-alumina 'matrix' with the zeolite crystals dispersed in it.

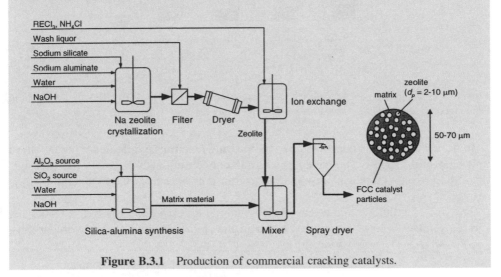

Figure B.3.1 Production of commercial cracking catalysts.

gases, and coke together. Gasoline yields with the present catalysts are between 40 and 50 wt%. (In American literature yields are usually based on vol%, which gives higher values).

Coke formation is a '*mixed blessing*'. The catalyst is poisoned by coke, so regeneration of the catalyst is necessary. Regeneration is carried out by combustion in air at 973 K.

$$C \xrightarrow{O_2} CO/CO_2 \tag{6}$$

This reaction is highly exothermic. When considering that cracking is an endothermic process, an excellent combination of these processes suggests itself: the two are coupled and the regeneration provides the necessary energy for the endothermic cracking operation!

3.4.1.4 Processes

Originally, fixed-bed and moving-bed reactors were applied, but nowadays the predominant process is FCC. The first Fluid Catalytic Cracking processes were based on fluidized beds (see Figure 3.12).

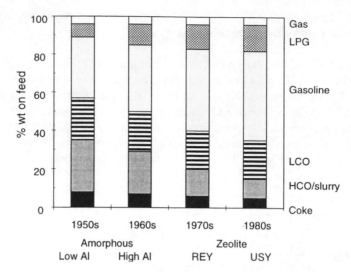

Figure 3.11 Improvement of catalytic cracking product selectivity [10]; REY is a rare earth exchanged Y-zeolite; USY refers to a zeolite that is more stable, 'Ultrastable Y zeolite'.

In the reactor heat is consumed by the cracking reactions, while in the regeneration unit heat is produced. The catalyst is circulated continuously between the two fluidized-bed reactors and acts as a vehicle to transport heat from the regenerator to the reactor.

The use of a fluidized-bed reactor for catalytic cracking has some disadvantages. Firstly, relatively large differences occur in the residence time of the hydrocarbons,

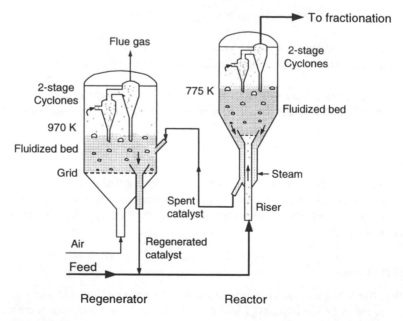

Figure 3.12 Early FCC process based on combination of two fluidized beds.

which results in a far from optimal product distribution. Secondly, the fluidized-bed reactor acts as a CSTR regarding the catalyst particles, causing age distribution. In modern processes, riser reactors are applied, because plug-flow is approximated better in these reactors (see box).

Actual hydrodynamics in riser reactor

Although the flow of catalyst particles in a riser reactor shows much more plug-flow behavior than the flow in a fluidized-bed reactor, a riser is certainly not an ideal plug-flow reactor (see Figure B.3.2). The explanation is as follows: the catalyst flows upward in the core of the reactor, giving a relatively narrow residence time distribution of about 2 s. At the reactor wall, however, the catalyst slides down as a result of friction. This results in a large fraction of the catalyst particles having a residence time in the range of 10–20 s. Current riser designs are fitted with internals to avoid the described behavior.

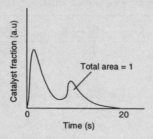

Figure B.3.2 Typical residence time distribution in an industrial FCC riser unit. Adapted from [11].

In the past, riser reactors could not be applied because the catalyst originally was not sufficiently active to apply riser technology. So, the change in technology was driven by a combination of catalyst improvement and chemical reaction engineering insights.

Figure 3.13 shows a flow scheme of a modern fluid catalytic cracking unit employing riser technology. The cracker feed is diluted with steam for better atomization, and fed to the riser reactor together with regenerated catalyst. The mixture flows upward and cracking takes place in a few seconds. The spent catalyst is separated from the reaction mixture in a cyclone. Steam is added to the downcomer in order to strip adsorbed heavy hydrocarbons off the catalyst. It also creates a buffer between the reducing environment in the riser and the oxidizing environment in the regenerator. The catalyst is transported to a fluidized-bed regenerator where coke is removed from the catalyst by combustion with air. Due to the limited life of the commercially applied catalysts, approximately 30 days, up to 5% fresh catalyst is added every day.

Table 3.4 shows typical operating conditions for both the riser reactor and the regenerator.

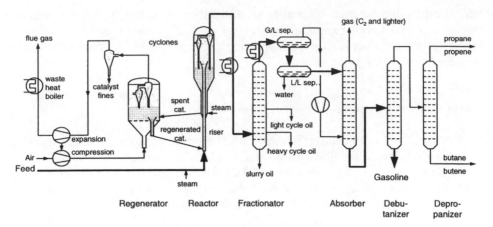

Figure 3.13   Flow scheme of a modern fluid catalytic cracking unit.

Table 3.4   Typical operation conditions in riser FCC

	Reactor	Regenerator
Temperature (K)	775	973
Pressure (bar)	1	2
Residence time	1–5 s	Minutes/half hour

QUESTIONS:

> *Guess the flue gas composition. Schematically draw the temperature versus time profile of a catalyst particle. Why is regeneration not carried out in a riser reactor? Could you design an improved reactor? Why does the regenerator operate at higher pressure than the riser reactor? Instead of a riser reactor a downer could be envisaged. What would be the advantages and disadvantages?*

3.4.1.5 Environment

The FCC unit is one of the most polluting units in the refinery [12]. During catalyst regeneration sulfur oxides referred to as SO_x ($SO_2/SO_3 > 10$) are formed by oxidation of the sulfur present in the coke. In addition to SO_x, NO_x (a mixture of NO and NO_2) is formed. Depending on regulations the SO_x and NO_x emissions in the flue gas might have to be reduced. Catalyst fines in the flue gas from the regenerator are removed in cyclones, but dependent on the separation efficiency a certain amount of particle emission takes place.

Sulfur leaves the FCC unit either in the liquid products, or as SO_x in the flue gas and H_2S in the light gas from the fractionation unit (see Table 3.5). The total sulfur emission of an FCC unit depends on the sulfur content of the feed. This sulfur content may vary from about 0.3 –3 wt%.

The amount of sulfur in the feedstock can be reduced by hydrotreating (see Section 3.4.2), thus providing lower sulfur amounts in all three product streams. If the feed is not hydrotreated or if the SO_x emissions from an FCC unit are still too high, they can

Table 3.5 Example of the distribution of sulfur in FCC products [13]

'Product'	% of sulfur in feed	Ton/day S (SO$_2$)^a
H$_2$S	50 ± 10	83.6
Liquid products	43 ± 5	71.9
Coke → SO$_x$	7 ± 3	11.7 (23.3)

^{a} 50000 barrels/day unit, 500 tons catalyst inventory, 50000 ton/day catalyst circulation rate, catalyst to oil ratio = 6 kg/kg, feedstock 2 wt% sulfur.

be reduced in two ways. A standard technology is flue-gas desulfurization (FGD), as often applied in power generation [14]. Another method of SO$_x$ control is reserved for the FCC process only (see box).

QUESTIONS:

> *In principle, hydrotreating can be applied to the feed of an FCC unit or to the products (gasoline, middle distillates). Which option would you choose and why? Does your choice depend on the type of feedstock and/or on regulations (concerning sulfur content in products, flue gas)?*

With the demand for gasoline increasing, and the market for heavy fuel oil declining, more and more heavier feedstocks with accompanying higher sulfur contents are processed by FCC units. For these heavy feeds certainly, desulfurization is necessary in order to comply with the gasoline and flue gas standards.

3.4.1.6 Novel developments in FCC

Production of lower alkenes

In the future, the role of FCC may shift from gasoline to light alkenes producer (also see Section 3.7). Already, there is an increasing incentive for the production of larger amounts of C$_3$ and C$_4$ alkenes in FCC units, because these are valuable as petrochemical feedstocks (see also Chapter 4). Furthermore, there is an increasing demand on isobutene for the production of octane enhancing oxygenates (e.g. MTBE, ETBE), which are now required constituents of gasoline (see also Chapter 6). In this context it is fair to state that MTBE has raised environmental concerns because it easily dissolves in groundwater and is not biologically degradable, thus presenting a possible health hazard. Therefore, it is doubtful whether MTBE will keep its prominent place as a gasoline additive. For higher analogues such as ETBE this disadvantage might be less or even absent, but still, all ethers will be abandoned in the USA. Another option is the addition of ethanol or methanol, but these compounds have other problems such as lower miscibility with gasoline and higher volatility.

The production of larger amounts of lower alkenes can be achieved by the addition of a small amount of the zeolite ZSM-5 (Zeolite Synthesized by Mobil, see Figure 3.14) to a conventional Y zeolite catalyst [15,16]. ZSM-5 has narrower pores and thus is only accessible to linear or slightly branched alkanes and alkenes and not to the

Reduction of SO$_x$ emissions from FCC units

Scheme B.3.1 shows the principle of the removal of SO$_2$ in an FCC unit. It involves the addition of a metal oxide (e.g. MgO, CeO or Al$_2$O$_3$), which captures the SO$_x$ in the regenerator and releases it as H$_2$S in the reactor and stripper. SO$_x$ removal is based on the difference in stability of sulfates and sulfites. An oxide is selected such that the sulfate (formed by reaction with SO$_x$) is stable under oxidizing conditions (in the regenerator) and unstable under reducing conditions (in the riser).

The metal oxide can either be incorporated in the FCC catalyst or added as a separate solid phase. The latter option has the advantage that the metal oxide supply is flexible and can be adjusted for feeds with different sulfur contents. The commercially available sulfur traps can remove up to 80% of the sulfur from the regenerator.

The H$_2$S released in the reactor and stripper can be removed by absorption and subsequent conversion in a Claus plant (see Section 3.6) together with the normal quantities of H$_2$S formed.

Oxidation of SO$_2$ to SO$_3$ in regenerator and subsequent adsorption on metal oxide (MO):

$$2\,SO_2 + O_2 \longrightarrow 2\,SO_3$$

$$SO_3 + MO \longrightarrow MSO_4 \qquad \text{(stable in regenerator)}$$

Reduction of the metal sulfate in the riser reactor and release of H$_2$S:

$$MSO_4 + H_2 \longrightarrow MSO_3 + H_2O \qquad \text{(unstable in riser)}$$

$$MSO_3 \; \xrightarrow{\;H_2\;} \; \begin{cases} MO + H_2S \\ MS + H_2O \end{cases}$$

Regeneration to form the metal oxide in the stripper with release of H$_2$S:

$$MS + H_2O \longrightarrow MO + H_2S$$

Scheme B.3.1   Removal of SO$_2$ in an FCC unit.

more branched ones and aromatic compounds. Therefore, in particular low octane hydrocarbons are cracked in ZSM-5, which has the additional advantage of improving gasoline quality, although at the expense of gasoline yield. Probably, linear

0.51 - 0.55 nm

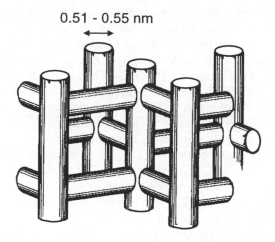

Figure 3.14 Channel system in zeolite ZSM-5.

unbranched and some monobranched alkenes formed in the cracking reactions are converted to lower alkenes in the small pores of the ZSM-5 added. This is an example of the advantageous use of the shape selectivity of zeolite catalysts (see also Section 3.7).

QUESTION:

> *Does the statement that only unbranched and some monobranched alkanes are converted in the ZSM-5 zeolite added to the catalyst conflict with the theory that branched alkanes are cracked most efficiently during FCC?*

Processing of heavier feedstocks

Catalytic cracking of residual feedstocks has become possible as a result of improved reactors and strippers, improved feed injection and gas/solid separation, the use of catalyst cooling, and the application of higher temperatures in the regenerator. However, the possibilities are limited due to much higher coke production, metal deposits on the catalyst (heavy feeds contain relatively large amounts of Ni and V compounds), and the higher amount of sulfur in products and regenerator off-gas. Therefore, conversion of very heavy feeds, such as vacuum residues, remains reserved for thermal and hydro-conversion processes (Flexicoking and hydrogenation of residues, see Section 3.5).

QUESTION:

> *Discuss the pros and cons of catalyst cooling.*

3.4.2 Hydrotreating

Hydrotreating belongs to the class of conversions involving reaction with hydrogen. The term 'hydrotreating' is limited to hydrogenolysis and hydrogenation reactions in which removal of 'hetero-atoms' (especially S, N, and O) and some hydrogenation of double bonds and aromatic rings take place. In contrast to 'hydrocracking' (see Section 3.4.3), in which size reduction is the main objective, in hydrotreating the molecular size is not drastically altered. Hydrogenation and hydrogenolysis both are reactions with hydrogen. The former occurs by addition of hydrogen (e.g. to double or triple bonds), whereas the latter involves breaking of C–S, C–N bonds, etc. Hydrotreating as defined here is sometimes called hydropurification, literally purification with hydrogen.

The major objectives of hydrotreating are protection of downstream catalysts (hetero-compounds often act as poisons), improvement of gasoline properties (odor, color, stability, corrosion), and protection of the environment.

Hydrogenation plays a major role in oil refining. Whether the hydrogenation is non-destructive or destructive depends on the 'severity'. At mild conditions hydrotreating occurs, whereas at more severe conditions cracking of C–C bonds takes place. The latter process is called 'hydrocracking'. When the process is carried out at intermediate severity it is referred to as 'mild hydrocracking'.

In hydrotreating the following terms are used, referring to the hetero-atom removed, i.e. S, N, O, M (metal): hydrodesulfurization (HDS), hydrodenitrogenation (HDN), hydrodeoxygenation (HDO), and hydrodemetallization (HDM).

3.4.2.1 Reactions and thermodynamics

Scheme 3.2 shows typical reactions that occur during hydrotreating. Figure 3.15 gives equilibrium data of selected HDS reactions.

The equilibrium constants are large and positive for all reactions considered under practical reaction conditions (625–700 K). Note that all hydrotreating reactions are exothermic. Although the equilibrium for all reactions is to the desulfurized products at these temperatures, the rates at which the equilibria are achieved are very different for the various sulfur compounds. Substituted benzothiophenes, which are mostly present in heavy oil fractions (see Figure 2.14), in particular show a low reactivity (see also Figure 3.19). Therefore, compared with light oil fractions, hydrotreating of heavy oil fractions requires higher temperature and hydrogen pressure.

QUESTION:

> *It has been proposed to use adsorption technology instead of hydrotreating. Give the pros and cons.*

3.4.2.2 Processes

In naphtha reforming, hydrotreating is always applied as a pre-treatment step to protect the Pt-containing catalyst against S-poisoning [17]. The sulfur content of the feed to the reforming unit should not exceed 1 ppm. The hydrogen used in naphtha hydrotreating is a by-product of catalytic reforming. In contrast, when hydrotreating of heavy residues is performed, separate hydrogen-production units are often required. Unstable by-products, which are formed during steam cracking of naphtha (see Chapter 4), are hydrogenated in order to increase their stability. One of the important reactions occurring is the saturation of alkynes and di-alkenes.

In simple naphtha hydrotreating, the feed is evaporated and led through a fixed-bed reactor. When heavier feedstocks (gas oil, residue) are treated, evaporation is not

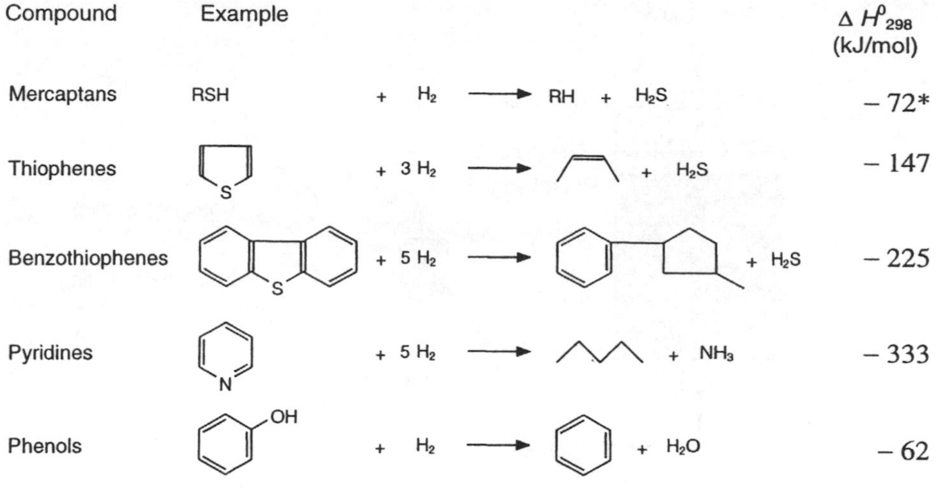

Compound	Example		ΔH^o_{298} (kJ/mol)
Mercaptans	RSH	$+$ H₂ $\longrightarrow$ RH $+$ H₂S	-72^*
Thiophenes		$+$ 3 H₂ $\longrightarrow$ $+$ H₂S	-147
Benzothiophenes		$+$ 5 H₂ $\longrightarrow$ $+$ H₂S	-225
Pyridines		$+$ 5 H₂ $\longrightarrow$ $+$ NH₃	-333
Phenols		$+$ H₂ $\longrightarrow$ $+$ H₂O	-62

Scheme 3.2 Typical hydrotreating reactions; *with R = CH₃.

possible due to the high boiling point of these feeds, and hence 'trickle-flow' operation will be more suitable (see Figure 3.16). In this mode, liquid (heavy part of hydrocarbon feed) and gas (hydrogen, evaporated part of feed) flow downward co-currently.

Most hydrotreating of heavy feedstocks is performed in trickle-bed (or trickle-flow) reactors. The major concern is to prevent maldistribution, i.e. an uneven distribution of the feed over the cross section of the bed. In addition, wetting of the catalyst particles is a point of concern: dry regions might exhibit relatively large rates, resulting in increased conversion or undesired coking because of the increased temperature.

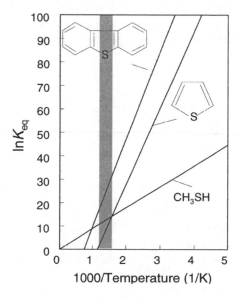

Figure 3.15 Equilibrium data of selected HDS reactions; shaded area: practical reaction conditions (625–700 K).

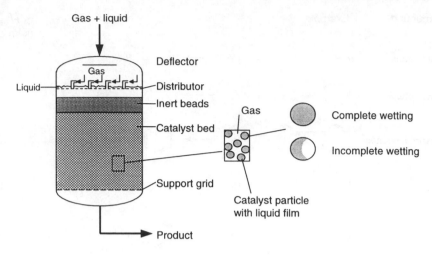

Figure 3.16 Trickle-bed reactor.

Figure 3.17 shows a simplified flow scheme of a hydrotreating process involving a trickle-bed reactor. The feed is mixed with fresh and recycle hydrogen and preheated by heat exchange with the reactor effluent. This reduces the fuel consumption of the furnace. The reactor usually contains several catalyst beds with intermediate cooling by hydrogen injection (quench). In the reactor, the sulfur and nitrogen compounds present in the feed are converted to hydrogen sulfide and ammonia, respectively. The effluent is cooled and sent to a hot high-pressure (HP) separator, where hydrogen is separated for recycle. The liquid from the HP separator is let down in pressure, sent to the low-pressure (LP) separator, and then to the product stripper.

The hydrogen-rich vapor from the hot HP separator is cooled and water is injected to absorb NH_3 and H_2S. The cold HP separator then separates the vapor, liquid water, and liquid naphtha. Further removal of H_2S from the hydrogen recycle stream is done

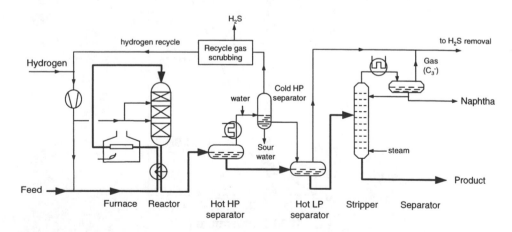

Figure 3.17 Simplified flow scheme of hydrotreating involving trickle-bed reactor. HP, high pressure; LP, low pressure.

Table 3.6 Typical process conditions in hydrotreating of naphtha and gas oils

	Naphtha	Gas oil
Temperature (K)	590–650	600–670
Pressure (bar)	15–40	40–100
H_2/oil (Nm^3/kg)	0.1–0.3	0.15–0.3
WHSV (kg feed/($m_{cat}^3 \cdot$h))	2000–5000	500–3000
Catalyst	Mixed metal sulfides (e.g. CoS/ MoS_2 or NiS/MoS_2) on γ-Al_2O_3	

by amine scrubbing (see also Section 3.6). In the hot LP separator a stream rich in hydrogen sulfide is removed, which is sent to a gas-treating unit. The liquid stream is normally sent to a stripping column, where any remaining gases are removed by stripping with steam.

Table 3.6 shows typical hydrotreating process conditions. The severity of the processing conditions depends on the feed; for light petroleum fractions it will be milder than for vacuum gas oils, while heavy residues require the highest temperatures and hydrogen pressures (see Section 3.5) [18]. Moreover, it is common practice to compensate for deactivation of the catalyst by increasing the temperature of the reactor.

QUESTIONS:

Is hydrogen present in excess? What is the composition of the gas from the last separator? Estimate typical reactor sizes of trickle-bed reactors.

Wetting is an issue in hydrotreating based on trickle-bed reactor technology. In general, partial wetting is considered to be undesired. Is this true in all respects? (Are there also advantages?)

3.4.2.3 Environment

From the viewpoint of clean technology, hydrotreating is gaining increasing interest. A recent example is diesel fuel. Sulfur compounds in diesel fuel cause serious difficulties in catalytic cleaning of the exhaust gases from diesel fuelled cars. Moreover, they contribute to the particulate emissions. Figure 3.18 shows the development of standards for the maximum sulfur level in automotive diesel fuel. Currently, in Europe the maximum sulfur level in diesel fuel is 500 ppm [19]. Undoubtedly, in the near future, novel processes will be needed for so-called 'deep desulfurization' of diesel in order to meet the more stringent demands.

Figure 3.19 shows the activity of conventional and novel hydrotreating catalysts for HDS of a straight-run gas oil that was first hydrotreated to a level of 760 ppm S [19].

Only the most refractory sulfur compounds are still present in this pretreated feed, since the more reactive ones have already been converted in the hydrotreating step. The figure clearly shows that with current catalysts (CoMo and NiMo) the lower sulfur levels required in the future cannot be achieved, except maybe with inconveniently large reactors. Supported noble metal catalysts like Pt and especially Pt/Pd on amorphous silica-alumina can achieve much better conversions and show a lot of promise for deep desulfurization.

QUESTIONS:

Why would sulfur compounds contribute to particulate emissions? European legislation is expected to require S-levels below 50 ppm for transportation diesel by the year 2005 [19]. Estimate the required reactor size (the sulfur level of gas oils may be assumed to be about 1.5 wt%). Currently, the discussion focuses on a further reduction of the S-content, e.g. down to 10 ppm. Again estimate the reactor size needed.

Would you use a Pt catalyst for desulfurization of straight-run gas oil (1.5 wt% S)? (see also Section 3.4.4). Suggest a reactor configuration for deep desulfurization of straight-run gas oil.

3.4.3 Hydrocracking

Hydrocracking [1,2,16,21–23] is a catalytic oil refinery process of growing importance. Heavy gas oils and vacuum gas oils are converted into lighter products, i.e. naphtha, kerosene, and diesel oil. Factors contributing to its growing use are the increasing demand for transportation fuels and the decline in the heavy fuel oil market. More recently, the increasing need for the production of clean fuels has also had a significant impact [21].

As the name implies, hydrocracking involves the cracking of a relatively heavy oil

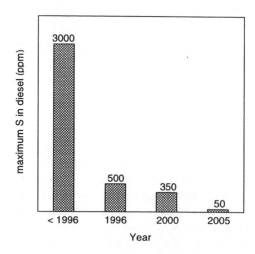

Figure 3.18 Development of maximum sulfur content in automotive diesel in Europe [20].

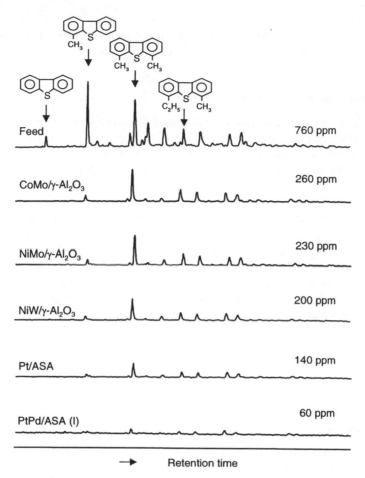

Figure 3.19 Activity of various catalysts for HDS of gas oil that had undergone HDS up to 760 ppm total sulfur; 613 K; 50 bar; ASA, amorphous silica-alumina [19].

fraction into lighter products in the presence of hydrogen. This distinguishes the process from the FCC process (see Section 3.4.1), which does not have hydrogen in the feed, and from the hydrotreating process (see Section 3.4.2), in which virtually no C–C bond breaking takes place.

Hydrocracking is a very versatile and flexible process, which can be aimed at the production of naphtha or at the production of middle distillates, viz. jet and diesel fuel. Although at first sight one might expect that hydrocracking competes with fluid catalytic cracking, this is certainly not the case; both processes work as a team. The catalytic cracker takes the more easily cracked alkane-rich atmospheric and vacuum gas oils as feedstocks, while the hydrocracker mainly uses more aromatic feeds, such as FCC cycle oils and distillates from thermal cracking processes (see Section 3.5). On the other hand, it also takes heavy atmospheric and vacuum gas oils and deasphalted oil.

QUESTION:

> *When comparing gasoline range products from hydrocracking and FCC which one will contain most sulfur?*

3.4.3.1 Reactions and thermodynamics

Hydrocracking can be viewed as a combination of hydrogenation and catalytic cracking. The former reaction is exothermic while the latter reaction is endothermic. Since the heat required for cracking is less than the heat released by the hydrogenation reaction the overall hydrocracking process is exothermic. Scheme 3.3 shows examples of reactions taking place.

Dehydrogenation, hydrogenation, and hetero-atom removal require a hydrogenation catalyst, while the actual cracking reactions proceed via carbenium ions, and therefore require an acidic catalyst. Various catalysts are in use. As in hydrotreating, mixed metal sulfides (mainly NiS/MoS_2 and NiS/WS_2) are used to supply the hydrogenation function. Noble metals are also used. The additional cracking function is fulfilled by a more acidic carrier than used in hydrotreating, for instance silica-alumina, and, increasingly often, zeolites. Figure 3.20 shows a model of a typical hydrocracking catalyst system.

QUESTION:

> *Explain why aromatic feeds are more easily processed in a hydrocracker than in a catalytic cracker.*

3.4.3.2 Processes

Nitrogen compounds (many of them basic) present in the feed play an important role in hydrocracking, because they adsorb on the acidic catalyst and thus inhibit the cracking reactions. Therefore, in most processes, hydrodenitrogenation (HDN) is applied as a first step before the actual hydrocracking.

Various process configurations have been developed, which can be classified as single-stage, two-stage, and series-flow hydrocracking (see Figure 3.21). In hydrocracking process design hydrogenation and cracking reactions have to be balanced carefully.

Single-stage hydrocracking is very similar to hydrotreating (see Figure 3.17), except for the different catalyst and more severe process conditions (see Table 3.7). It is the simplest configuration of the hydrocracking process, with the lowest investment costs. The conversion of the feedstock is not complete, i.e. there is still material present with the same molecular weight range as the feedstock, but this 'unconverted' product is highly saturated and free of feed contaminants (i.e. it has been hydrotreated). The single-stage hydrocracking process is applied for the partial production of naphtha (for instance as feed to a naphtha cracker for ethene production, see Chapter 4) together with an FCC feed, and for the production of lube oils. The process is optimized for hydrogenation rather than for cracking.

Recently, a new technology referred to as mild hydrocracking, has been introduced (see [21] and references therein). The advantages of this process are that it can be

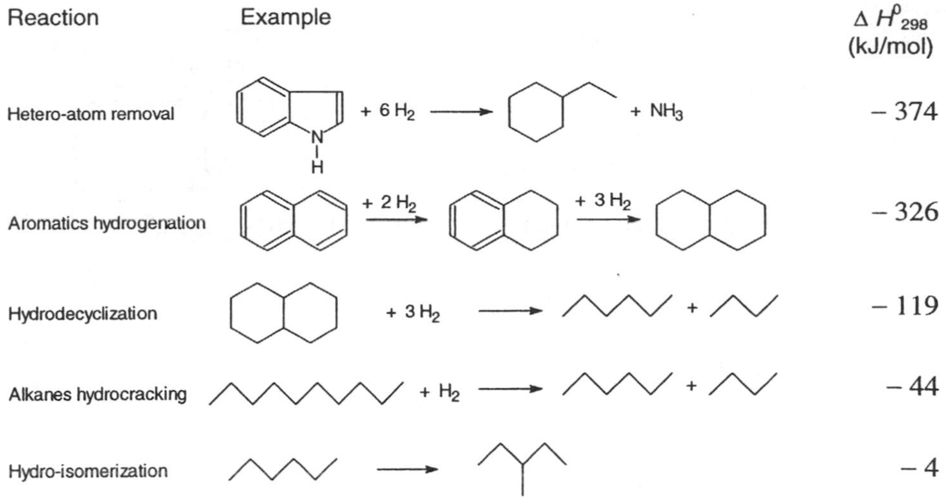

Reaction	Example	ΔH^{ρ}_{298} (kJ/mol)

Hetero-atom removal ⬡(indole) + 6 H₂ ⟶ ⬡(ethylcyclohexane) + NH₃ − 374

Aromatics hydrogenation ⬡⬡(naphthalene) + 2 H₂ ⟶ ⬡⬡(tetralin) + 3 H₂ ⟶ ⬡⬡(decalin) − 326

Hydrodecyclization ⬡⬡(decalin) + 3 H₂ ⟶ ∧∧∨ + ∧∨ − 119

Alkanes hydrocracking ∧∨∧∨∧∨ + H₂ ⟶ ∧∧∨ + ∧∨ − 44

Hydro-isomerization ∧∨∧∨ ⟶ (branched alkane) − 4

Scheme 3.3 Examples of reactions during hydrocracking.

implemented in existing hydrotreaters by increasing the severity of operation. So, without much investment the flexibility of the oil refinery is increased.

QUESTIONS:

> *Explain the term 'mild hydrocracking'. Also explain the term 'hydrowax'.*

In two-stage hydrocrackers (see Figure 3.22) the conversion of nitrogen (and sulfur) compounds and the hydrocracking reactions are carried out in two separate reactors with intermediate removal of ammonia. The effluent from the first reactor, after cooling and removal of hydrogen sulfide and ammonia, is fractionated and the bottom stream of the fractionator is subsequently cracked in the second reactor.

The feed is completely converted into lighter products, in contrast with the single-stage process. The product yield can be tailored towards maximum naphtha production or maximum middle distillates production by either changing the fractionator operation or using alternative catalysts for the second stage [24].

In the series-flow process, the product from the first reactor is directly fed to the second reactor, without prior separation of hydrogen sulfide and ammonia. The heavy

Figure 3.20 Models of hydrocracking catalysts: left, NiS/MoS₂/Silica-alumina; right, Pt/amorphous silica-alumina.

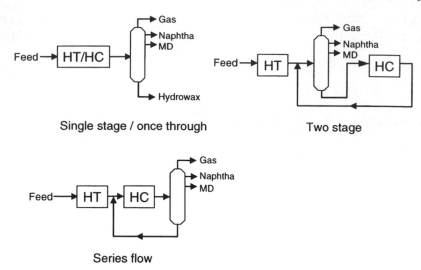

Figure 3.21 Hydrocracking process configurations: HT, hydrotreating; HC, hydrocracking; MD, middle distillates. Adapted from [16].

fraction from the stripper is recycled to the first reactor. So, the catalyst in the hydrocracking reactor has to operate under NH_3 rich conditions (NH_3 is not removed upstream the hydrocracking reactor). This results in a lower activity due to adsorption of NH_3 on the catalyst. Therefore, development of the series-flow process has only become possible due to the development of suitable catalysts.

QUESTION:
> *Hydrocracking catalysts can serve for two years before they have to be regenerated (coke removal). Explain this extreme difference with catalytic cracking catalysts (which only last for a few seconds).*

Table 3.7 summarizes the processing conditions used in the various hydrocracking processes. Regarding the catalyst, it is also possible to place different catalysts in different beds in one reactor. This somewhat blurs the distinction between the processes.

Table 3.7 Summary of processing conditions of hydrocracking processes

	Mild	Single stage/first stage	Second stage
Temperature (K)	670–700	610–710	530–650
H_2 pressure (bar)	50–80	80–130	80–130
Total pressure (bar)	70–100	100–150	100–150
Catalyst	Ni/Mo/S/γ-Al$_2$O$_3$ + P[a]	Ni/Mo/S/γ-Al$_2$O$_3$ + P[a]	Ni/W/S/USY zeolite

[a] Increases activity for nitrogen removal.

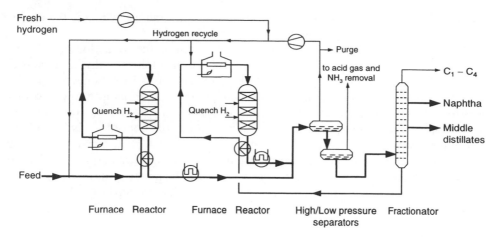

Figure 3.22 Flow scheme of two-stage hydrocracker.

QUESTIONS:

> *Draw a flow scheme of the series-flow hydrocracking process. Give advantages and disadvantage of this process compared to the two-stage process (think of investment costs, energy requirements, process flexibility, catalyst stability, etc.)*
>
> *Compare the FCC process and the hydrocracking process in terms of feedstocks, products, environmental issues, operation costs, etc.*
>
> *Under practical conditions slow catalyst deactivation occurs. This is coped with by increasing the reactor temperature. The undesired consequence is an increase in aromatic contents of the product. Explain. Hint: consider the data on aromatics hydrogenation given in the next section on Catalytic Reforming.*

3.4.4 Catalytic Reforming

Catalytic reforming is a key process in the production of gasoline components with a high octane number. It also plays an important role in the production of aromatics for the chemical industry. Furthermore, catalytic reforming is a major source of hydrogen. Feedstocks are straight-run naphtha and other feeds in the gasoline boiling range (about C_6–C_{11}). During catalytic reforming the change in molecular weight of the feed passed through the process is relatively small, as the process mainly involves rearrangement of hydrocarbon structures.

3.4.4.1 Reactions and thermodynamics

Scheme 3.4 shows examples of major reactions that occur during catalytic reforming. Other, undesired, reactions that occur are hydrocracking of alkanes and naphthenes and hydrodealkylation of aromatics to form butane and lighter alkanes. An important undesired side reaction is thermal cracking, leading to coke formation.

Except for isomerization, which only involves a minor enthalpy change, all reactions that occur during catalytic reforming are moderately to highly endothermic, particularly the dehydrogenation of naphthenes (aromatization), which is the predominant reaction during reforming [25] (see Table 3.8).

An important side reaction that has to be dealt with is the formation of coke. In addition, light alkanes ($<C_5$) are formed by dealkylation of toluene, ethylbenzene, etc. An additional amount of methane is formed by reaction of coke with hydrogen.

Thermodynamically favorable reaction conditions are high temperature and low pressure. Figure 3.23 illustrates this for the conversion of cyclohexane into benzene. At atmospheric pressure the thermodynamically attainable conversion is essentially 100% at temperatures over 620 K. At higher pressure much higher temperatures are required for high conversions. The equilibrium conversions shown in Figure 3.23 were obtained without addition of hydrogen to the feed. In practice, however, product hydrogen is recycled to limit coke formation. This has an adverse effect on the cyclohexane conversion, as shown in Figure 3.24.

QUESTION:

> *Explain the difference between the influence of total pressure (large) and the H_2/cyclohexane feed ratio at constant total pressure (small) on the cyclohexane conversion.*

The dehydrogenation and aromatization reactions are catalyzed by metal catalysts. The isomerization and cracking reactions proceed via carbenium ions, and therefore require an acidic catalyst (see Section 3.4.1). In catalytic reforming, the catalysts consist of platinum metal (which explains why the name *platforming* is also often used), dispersed on alumina. New generation bi-metallic catalysts also contain rhenium ('Rheniforming') or iridium in order to increase catalyst stability. The alumina support has some acidity, which is enhanced by adding chlorine. Figure 3.25 shows a schematic of a catalytic reforming catalyst.

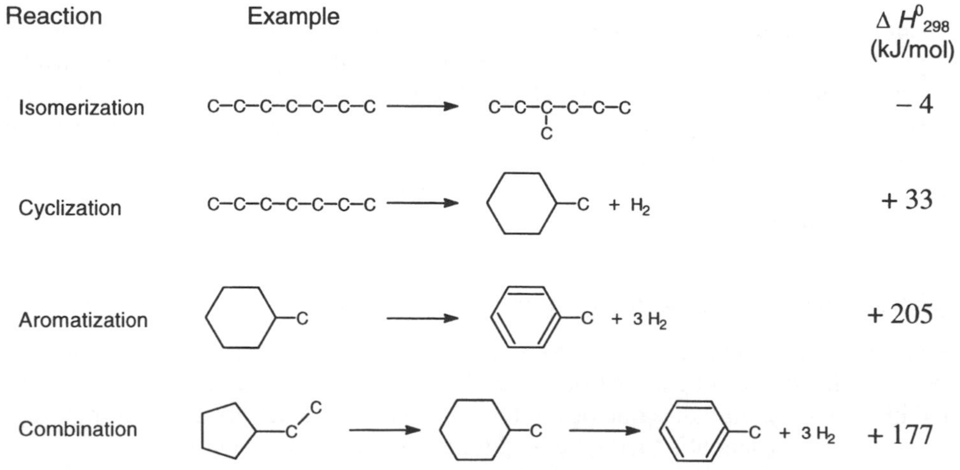

Reaction	Example	ΔH^{ρ}_{298} (kJ/mol)
Isomerization	C–C–C–C–C–C–C $\longrightarrow$ C–C–C–C–C–C (with branch C)	-4
Cyclization	C–C–C–C–C–C–C $\longrightarrow$ ⬡–C + H_2	$+33$
Aromatization	⬡–C $\longrightarrow$ ⬢–C + 3 H_2	$+205$
Combination	⬠–C–C $\longrightarrow$ ⬡–C $\longrightarrow$ ⬢–C + 3 H_2	$+177$

Scheme 3.4   Reactions occurring during catalytic reforming.

Table 3.8 Typical catalytic reforming feedstock and product analysis (vol%) [1]

Component	Feed	Product
Alkanes	45–55	30–50
Alkenes	0–2	0
Naphthenes	30–40	5–10
Aromatics	5–10	45–60

QUESTION:

Explain why alumina becomes strongly acidic by chlorination (hint: compare CH_3COOH and $CH_2ClCOOH$; which is most acidic?)

3.4.4.2 Feed pretreatment

Platinum, the main metal of most catalytic reforming catalysts, is sensitive to deactivation by hydrogen sulfide, ammonia, and organic nitrogen and sulfur compounds [1]. Therefore, naphtha pretreatment by hydrotreating (see Section 3.4.2) is required. The organic sulfur and nitrogen compounds are catalytically converted with hydrogen into H_2S and NH_3, which are then removed from the hydrocarbon stream in a stripper. The off-gas from the stripper is then further treated (see Section 3.6). Hydrogen needed for the hydrotreater is obtained from the catalytic reformer.

3.4.4.3 Reforming processes

The most important issue in the development of a catalytic reforming process is to find a compromise between optimal octane number and product yield (low hydrogen pressure optimal) and optimal catalyst stability, i.e. minimal coke formation (high hydrogen pressure optimal). The two main types of processes applied in catalytic reforming are *semi-regenerative reforming* (SRR) and *continuously-regenerative reforming* (CRR).

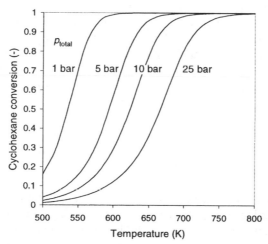

Figure 3.23 Effect of temperature and pressure on the aromatization of cyclohexane without hydrogen addition.

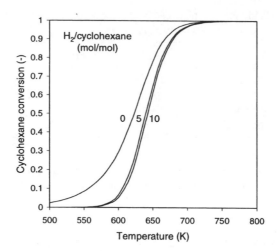

Figure 3.24 Effect of temperature and H_2/cyclohexane feed ratio on the aromatization of cyclohexane at constant total pressure of 10 bar.

Semi-regenerative reforming (SRR)

SRR units consist of several (three to four) adiabatic fixed-bed reactors in series (see Figure 3.26) with intermediate heating. Early SRR units (~1950) were operated at relatively high pressure (25–35 bar) in order to maximize the run time [2]. The development of improved catalysts that had a lower tendency for coke formation enabled operation at lower pressure (15–20 bar), while maintaining the same catalyst life (typically 1 year).

The feed is preheated and sent to the first reactor. In this reactor the kinetically fastest reactions occur. These are the dehydrogenation reactions of naphthenes to aromatics. To maintain high reaction rates the product gases are reheated in a furnace before being sent to a second reactor, etc. The temperatures shown in Figure 3.26 are indicative of industrial practice [26].

The effluent from the last reactor is cooled by heat exchange with the feed. Then hydrogen is separated from the liquid product and partly recycled to the process. The net hydrogen production is used in other parts of the refinery, for instance in the naphtha hydrotreater. The liquid product is fed to a stabilizer in which the light hydrocarbons are separated from the high-octane liquid reformate.

QUESTION:

> *Explain the decreasing temperature difference over the reactors? What can you say about reactor sizes? (Would you choose all reactors of equal size?)*

Some catalyst deactivation, mainly due to coke deposition, is unavoidable. It is normal practice to balance decreased activity by gradually increasing the temperature. This temperature increase, however, results in lower selectivities and higher rates of coke deposition. The catalyst is regenerated after 0.5–1.5 years by carefully burning off the coke deposits in diluted air with the unit taken off-stream. The catalyst is also

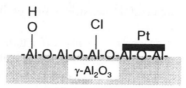

Figure 3.25   Schematic of catalytic reforming catalyst.

treated with Cl_2 in order to replenish the Cl-content of the catalyst, which is reduced by the coke-removal step, and to re-disperse the platinum.

Fully-regenerative reforming

For further reduction of the operating pressure to improve the product octane number and the yield on liquid products and hydrogen, catalyst development is not sufficient. Therefore, process innovations were needed. In the 1960s, *fully-regenerative* or *cyclic regenerative reforming* was developed. This technology still employs fixed-bed reactors, but an additional 'swing' reactor is present. It is thus possible to take one reactor off-line, regenerate the catalyst, and then put the reactor back on-line. In this way regeneration is possible without taking the reforming unit off-stream and losing production [2], which means that a lower operating pressure can be used.

Continuously-regenerative reforming (CRR)

A real innovation was the catalytic reformer with a moving catalyst bed, which enabled continuous regeneration of the catalyst, i.e. the CRR-process (see Figure 3.27). This process, introduced by UOP and IFP, enabled the use of a much lower operating pressure (3–4 bar), since coke deposits were no longer such an issue, as they were continuously being burned off. Over 95% of new catalytic reformers presently designed are reformers with continuous regeneration [2].

Regenerated catalyst enters the first reactor at the top. The catalyst withdrawn from the bottom enters the next reactor, etc. The catalyst from the fourth reactor is sent to the regenerator. A continuous flow of catalyst particles is crucial for this system. In

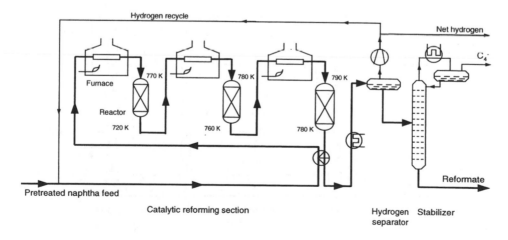

Figure 3.26   Flow scheme of the semi-regenerative catalytic reforming process.

practice, spherical particles are used. Downstream processing of the crude reformate is similar as in the SRR process.

Summarizing, the different processes mainly differ in the manner and frequency of catalyst regeneration, which in turn determines the operating pressure and temperature, and thus the product quality and yield. Table 3.9 shows typical operating conditions used in practice for the three reformer types.

Catalytic reforming is primarily applied for improving the gasoline properties, but it is also used for the production of aromatics. In this case the operating conditions are somewhat more severe (higher temperature, lower pressure) than for the production of high-octane gasoline.

QUESTION:

What would be the most practical way of arranging the four reactors in the CRR process spatially?

Why, despite the fact that hydrogen is produced during reforming reactions, is hydrogen added to the reforming process?

How much is the octane number increased for a conceptual feedstock represented by cyclohexane, going from the semi- to the continuously-regenerative process? (Assume equilibrium.)

3.4.4.4 Reactors

During reaction the atmosphere is reducing, but during regeneration it is oxidizing. In the semi-regenerative and fully-regenerative processes this calls for materials that can withstand these cyclic conditions.

In all catalytic reforming processes that have been discussed adiabatic fixed- or moving-bed reactors are applied. Initially, axial-flow reactors were used, but later often radial-flow reactors have been introduced because of their lower pressure drop (see Figure 3.28) [17].

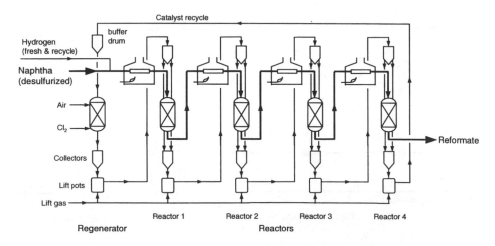

Figure 3.27   Flow scheme of the continuous regenerative reforming process (IFP).

Table 3.9 Practical operating conditions in catalytic reforming

	Semi	Fully	Continuous
Feed H_2/HC (mol/mol)	10	4–8	4–8
Pressure (bar)	15–35	7–15	3–4
Temperature (K)	740–780	740–780	770–800
Catalyst life	0.5–1.5 y	Days–weeks	Days–weeks

QUESTION:

> *Discuss material problems (corrosion, reactor walls, attrition of catalyst particles, etc.) for the three reforming technologies.*

3.4.5 Alkylation

Alkylation is the reaction of alkenes with isobutane to form higher branched alkanes ('alkylate'). The aim is the production of high octane number gasoline components from low molecular weight alkenes (propene, butenes, and pentenes) and isobutane. Alkylation used to be a thermal process (770 K, 200–300 bar), but nowadays a catalytic process (<340 K, 2–20 bar) is applied, using H_2SO_4 or HF as the catalyst. The advantages of alkylation are that gas-phase molecules are eliminated and a valuable liquid product is formed, i.e. gasoline with high octane number (87–98).

QUESTION:

> *Why would the pressure be so high in the thermal process?*

3.4.5.1 Reactions

An example is the alkylation of isobutane with propene yielding the products shown in Scheme 3.5.

The alkylation reaction occurs via carbenium ions, as shown in Scheme 3.6 for the alkylation of propene with isobutane.

Besides alkylation, also undesired side reactions occur, in particular oligomerization/polymerization of the alkene. In that case the cation formed in the initiation

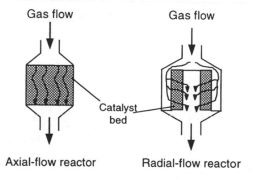

Figure 3.28 Reactors used for catalytic reforming.

reaction may react with an alkene molecule (in this example propene) resulting in oligomers (2 $C_3H_6 \rightarrow C_6H_{12} \rightarrow C_9H_{15}$, etc.). This is both a highly undesired and a relatively easy reaction, which may prevail unless isobutane is present in large excess. Other side reactions are hydrogen transfer, disproportionation, and cracking.

QUESTIONS:

Why is isobutane used and not n-butane? Which alkene will produce the highest octane alkylate?
What would be the source of isobutane and alkenes?

Alkylation is an exothermic reaction. The heat evolved when isobutane is alkylated with propene is about 80 kJ/mol alkylate produced. Alkylation is thus favored thermodynamically by low temperature. It is also favored by high pressure, since the number of moles is reduced in the reaction. The same applies to the undesired side and consecutive reactions. The pressures applied in practice are low.

Excess isobutane is fed to the process to help the alkylation reaction to proceed to the right and to minimize polymerization of propene. Typical isobutane-to-propene feed ratios are in the range of 8–12.

3.4.5.2 Sulfuric acid processes

Two types of catalysts are used, HF and H_2SO_4 [2]. The H_2SO_4 process is the oldest alkylation process. In this process the reactor temperature has to be kept below 293 K to prevent excessive acid consumption as a result of oxidation-reduction reactions, which result in the formation of tars and SO_2. Therefore, cryogenic cooling is required.

There are two main technologies for alkylation with H_2SO_4; the autorefrigeration process licensed by Exxon Research and Engineering, and the effluent refrigeration process, licensed by Stratford Engineering Corporation.

Figure 3.29 shows a flow scheme of the autorefrigeration process. A cascade of reaction stages is combined in one reactor vessel. Every compartment is stirred.

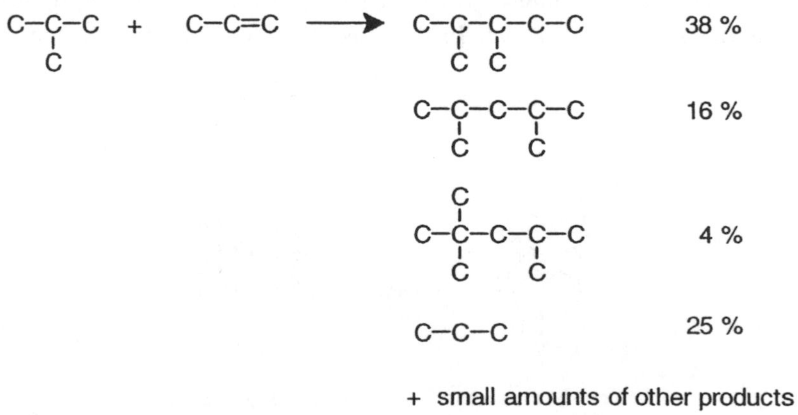

Scheme 3.5 Products from alkylation of isobutane with propene [27].

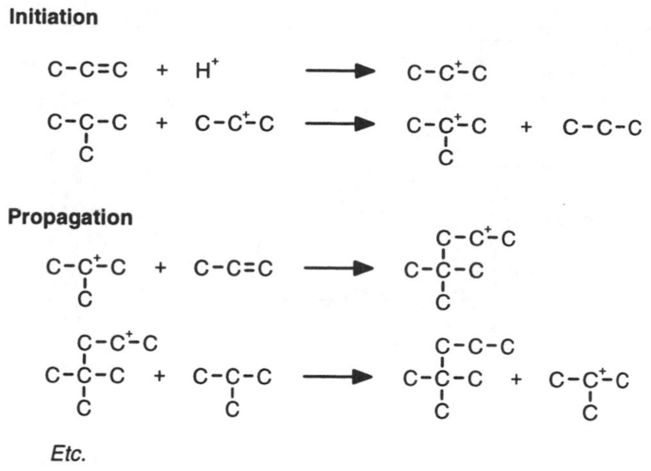

Initiation

$$C-C=C \quad + \quad H^+ \quad \longrightarrow \quad C-C^+_-C$$

$$C-\underset{\underset{C}{|}}{C}-C \quad + \quad C-C^+_-C \quad \longrightarrow \quad C-\underset{\underset{C}{|}}{C^+_-}C \quad + \quad C-C-C$$

Propagation

$$C-C^+_-C \quad + \quad C-C=C \quad \longrightarrow \quad \underset{\underset{C}{|}}{\overset{C-C^+_-C}{\underset{|}{C}}}-C-C$$

$$\underset{\underset{C}{|}}{\overset{C-C^+_-C}{C}}-C-C \quad + \quad C-\underset{\underset{C}{|}}{C}-C \quad \longrightarrow \quad \underset{\underset{C}{|}}{\overset{C-C-C}{C}}-C-C \quad + \quad C-C^+_-C$$

Etc.

Scheme 3.6 Mechanism of alkylation of isobutane with propene.

Isobutane and acid are fed to the first reactor compartment and pass through the reactor. The alkene feed is injected into each of the stages.

The reactor is held at a pressure of approximately 2 bar to keep the temperature at about 278 K. At this low temperature the reaction is rather slow, so relatively long contact times are required (20–30 min).

The hydrocarbon/acid mixture is an emulsion, which is separated in the last compartment of the reactor vessel, the settler. The acid phase is recycled to the reactor. The hydrocarbon phase is neutralized by a caustic wash (e.g. with NaOH) and then fractionated in two distillation columns. First, isobutane is recovered for recycle, and subsequently any *n*-butane present is removed to yield alkylate as the bottom product. Fractionation can also be carried out in one tower, with removal of *n*-butane as a side-stream.

The heat of reaction is removed by evaporation of hydrocarbons, in particular isobutane and some propane, which may be present in the feed and is formed in the reaction. The vapors are compressed and liquefied. A portion of this liquid is vaporized in an economizer, which is an intermediate-pressure flash drum, to cool the alkene feed. The remainder is sent to a depropanizer for removal of propane, which would otherwise accumulate in the system. Isobutane from the bottom of the depropanizer is recycled to the reactor.

QUESTION:

> *In the past a single stirred-tank reactor was used for alkylation with sulfuric acid. The acid was removed in a separate settler. Give the advantages of the scheme in Figure 3.29 over the old process.*

Figure 3.30 shows the reactor section of the effluent refrigeration process, licensed by Stratford Engineering Corporation.

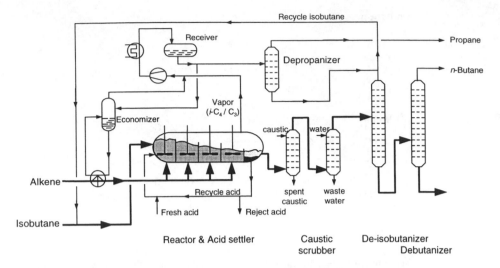

Figure 3.29 Alkylation with H_2SO_4 in cascade of reactors with autorefrigeration (Exxon R&E).

The system pressure is kept relatively high (4 bar) to prevent vaporization of light hydrocarbons in the reactor and settler [28]. In this system, hydrocarbons from the settler are flashed across a control valve into a large number of heat transfer tubes contained within the reactor to provide cooling. The reaction emulsion is pumped across these cooled tubes to keep the reaction temperature lowered. The vapors from the heat transfer tubes are routed to the refrigeration system after separation from the alkylate.

QUESTIONS:

Explain the names for the two sulfuric acid catalyzed processes described.

Compare the energy requirements of the processes shown in Figs. 3.29 and 3.30. Also compare the two processes with respect to concentration and temperature profiles. Which one is the most favorable in this respect?

3.4.5.3 Process comparison

A major problem of the H_2SO_4 alkylation process is the consumption of acid (about 100 kg/t of product!). The alternative HF processes are more favorable in this respect. The acid consumption is two orders of magnitude lower. The process scheme for alkylation with HF is comparable to that of H_2SO_4 alkylation. The most important difference is the installation of a regenerator for HF in the HF process. Another advantage of the HF alkylation process is that reaction can be carried out at higher temperature; because HF is a non-oxidizing gaseous acid, with HF the reaction can be carried out at 310 K, which eliminates the need for cryogenic cooling. Table 3.10 shows typical process conditions for isobutane/butene alkylation.

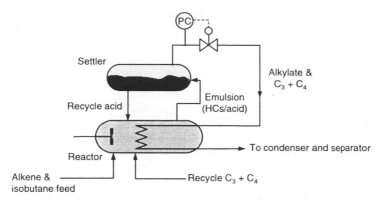

Figure 3.30 Alkylation with H_2SO_4 in Stratco contactor with autorefrigeration (Stratford Engineering Corp.).

Both the HF and H_2SO_4 catalyzed processes suffer disadvantages such as pollution, safety, and corrosion problems. Of the two, sulfuric acid is safer because it stays in the liquid form at ambient conditions. Therefore, a leak of sulfuric acid is much more easily contained than a leak of hydrofluoric acid, which is highly volatile and toxic under ambient conditions. This explains why the sulfuric acid process for alkylation dominates. Indeed, US legislation is directed to restriction, and in some cases elimination, of the use of HF in alkylation units [27].

QUESTION:

> *Why would the sulfuric acid produce much more waste than the HF process?*

3.4.5.4 Novel developments – solid-catalyzed alkylation

Much research is being carried out to replace the liquid acid processes by a more environmentally benign heterogeneously catalyzed process. This is logical when considering the refining catalyst market; in 1991 alkylation accounted for about 5% of the total capacity of catalytic processes in the United States, while alkylation catalysts were responsible for over 30% of catalyst costs [27]. For the largest capacity process, viz. hydrotreating, these figures are 44% of capacity and 12% of catalyst cost.

The development is in harmony with the history of fluid catalytic cracking, which also has progressed from a thermal process via homogeneous catalysis to the present state of the art using zeolite catalysts. The overwhelming success of zeolites in FCC in the 1960s triggered the exploration of the potential of these catalysts for isobutane/alkene alkylation, which can be viewed as the reverse of cracking.

Various zeolites and other strong solid acids, among which so-called solid superacids (with an acid strength higher than that of 100% sulfuric acid), have been tested for alkylation and found to be capable, in principle, of catalyzing the reaction. The progress in the development of solid catalysts for isobutane/butene alkylation has been reviewed [27,29]. Although many solid catalysts yield alkylate gasoline of the same quality as alkylate produced by the traditional liquid acid processes, one very impor-

Table 3.10 Typical process conditions for isobutane/butene alkylation

	H_2SO_4 process	HF process
Temperature (K)	277–283	298–313
Pressure (bar)	2–6	8–20
Residence time (min)	20–30	5–20
Isobutane/butene feed ratio	8–12	10–20
Acid strength (wt%)	88–95	80–95
Acid in emulsion (vol%)	40–60	25–80
Acid consumption per mass of alkylate (kg/t)	70–100	0.4–1

tant drawback has so far become clear: the solid catalysts quickly loose their activity and selectivity as a result of the formation of carbonaceous deposits (coke), by alkene polymerization reactions. An additional disadvantage of super-acids, which are considered to be very promising, is their extreme sensitivity to water, which requires special measures for feed drying.

The precise nature of this loss of activity and selectivity has not yet been elucidated, so here lies a task for chemists. From a chemical engineering viewpoint the deactivation problem presents a challenge in choosing a suitable reactor configuration; the formation of carbonaceous deposits must preferably be limited, while relatively frequent regeneration must be possible. In spite of these difficulties, it is expected that solid-catalyzed alkylation will predominate in the near future.

QUESTIONS:

> *The product of alkylation is used in the gasoline pool. Compare alkylate with the other components of the gasoline pool. Rank them with respect to environmental impact.*
>
> *Why would super-acids be very sensitive to water? (hint: typical super-acids are SO_4^{2-}/ZrO_2 and SO_4^{2-}/TiO_2, etc.).*

3.5 CONVERSION OF HEAVY RESIDUES

The processing of a light crude oil in even a complex refinery, including FCC, hydrocracking, etc. does not yield a satisfactory product distribution (see Figure 3.31). The amounts of fuel oil are too high.

For heavy oil the situation is even worse (see Figure 3.32): about 50% fuel oil is produced even in a complex refinery. It should be noted that fuel oil is worth less than the original crude oil; the value of the products decreases in the order: gasoline > kerosene/gas oil > crude oil > fuel oil.

There are several reasons for an increased incentive to convert fuel oil into lighter products:

- The demand for light products such as gasoline and automotive diesel fuels continues to increase, while the market for heavy fuel oil is declining.
- Environmental restrictions become more important. Fuel oil contains a high amount of impurities such as sulfur, nitrogen, and metals, so measures must be taken to lower emissions.

- With the exception of Western Europe, the quality of crude oil shows a worsening trend. It becomes heavier, with higher amounts of hetero-atoms, so more extensive processing is required to obtain the same amount and quality of products.

In principle, two solutions are feasible for upgrading residual oils and for obtaining a better product distribution. These are *carbon out* and *hydrogen in* processes.

QUESTION:

Discuss the logic of this statement. How do the other thermal processes (visbreaking, delayed coking) fit in this respect?

Examples of carbon rejection processes are the Flexicoking process (Exxon) and the FCC-process (see Section 3.4.1). The LC-fining process (Lummus), and the HYCON process (Shell) are examples of the hydrogen addition route.

3.5.1 Flexicoking

The Flexicoking process has been developed by Exxon to minimize coke production [2,30]. Figure 3.33 shows a flow scheme of this process. A residual oil feed is injected into a reactor, and contacted with a hot fluidized bed of coke. Thermal cracking takes place producing gas, liquids, and more coke. Coke constitutes approximately 30% of the cracking products. The coke particles are transported to a heater, also a fluidized bed. The primary function of the heater is to transfer heat from the gasifier to the reactor. Part of the hot coke particles is recirculated to the reactor, while another part is fed to the gasifier, which also is a fluidized bed. Here the coke is gasified by reaction with steam and oxygen (comparable to coal gasification, Chapter 5):

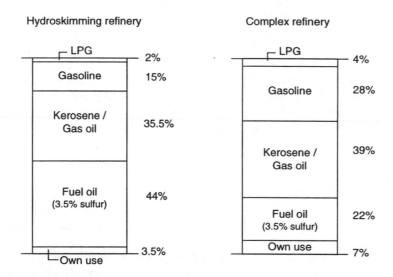

Figure 3.31 Light crude oil product distribution in a simple ('hydroskimming') refinery and a more advanced refinery.

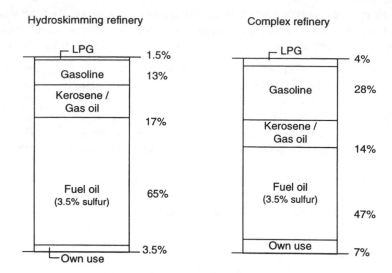

Figure 3.32 Heavy crude oil product distribution.

$$C + O_2/H_2O \rightarrow H_2, CO, CO_2, CH_4, \text{pollutants} \tag{7}$$

QUESTIONS:

   Explain the term 'hydroskimming' refinery.

The gas leaves the top of the gasifier and flows to the heater where it serves to fluidize the coke particles and provide part of the heat required in the reactor. The gas leaving the heater is cooled by steam generation, and then passes through a cyclone and a venturi type gas scrubber for the removal of coke fines. The venturi scrubber is a device in which dust (in this case coke fines) is removed from a gas stream by washing with a liquid, usually water. The turbulence created by the flow of gas and liquid through the narrow part of the venturi tube promotes the contact between the dust particles and liquid droplets. The resulting slurry is separated from the gas in a separator. Finally, H_2S is removed from the gas (see Section 3.6.1).

The recirculating coke particles serve as nuclei for coke deposition and as heat carriers. Of course, also part of the coke is withdrawn to prevent accumulation of metals, etc.

QUESTIONS:

   How is the heat required for thermal cracking produced? Guess the composition of the gases produced. The gases produced are low caloric. What is the reason? Why is a cyclone required in the heater but not in the gasifier?
   When the Flexicoking units were built as part of the Exxon refinery in Rotterdam, The Netherlands, also a hydrogen plant was added. Why?

3.5.2 Catalytic Hydrogenation of Residues

Catalytic hydrogenation is a 'hydrogen-in' route. It serves two purposes: removal of sulfur, nitrogen, and metal compounds, and the production of light products. The reactions taking place are very similar to those occurring during hydrotreating and hydrocracking of gas oils. However, there are two important differences. Firstly, residues contain much higher amounts of sulfur, nitrogen, and polycyclic aromatic compounds. Secondly, the removal of metals, which are concentrated in the residual fraction of the crude oil, is essential. Hence, the operating conditions are much more severe and more hydrogen is required in catalytic hydrogenation of residues than in hydroprocessing of gas oils.

3.5.2.1 Catalyst deactivation

A crucial point in catalytic hydroconversion of residual oils is the deactivation of the catalyst by deposition of metals. Essentially, all metals in the periodic table are present in crude oil, but the major ones are Ni and V. At the applied reaction conditions H_2S is present, and as a consequence, metal sulfides rather than metals are formed. The reaction scheme is complex, see [31], but the overall reactions can be represented by the following simplified equations:

$$Ni - porphyrin + H_2 \rightarrow NiS + hydrocarbons \tag{8}$$

$$V - porphyrin + H_2 \rightarrow V_2S_3 + hydrocarbons \tag{9}$$

The catalyst is poisoned by this process. This may occur even at relatively low levels of deposition. The reason is that for a large part the deposition occurs in the

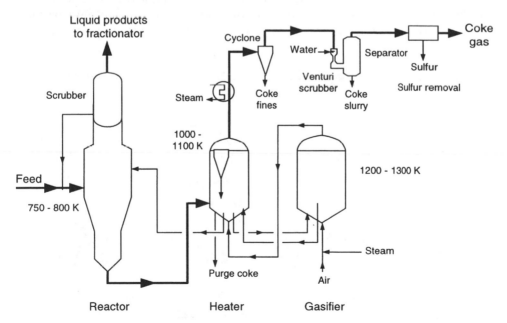

Figure 3.33 Schematic of the Flexicoking process.

outer shell of the catalyst particles. Figure 3.34 illustrates the phenomena occurring during metal deposition. Initially, the catalyst pellets will be poisoned rather homogeneously, although always with some preference for the outer shell. With time, diffusion will be progressively hindered and the metals will be deposited more and more in the outer shell. This leads to so-called pore plugging rendering the catalyst particle inactive although in a chemical sense the inner part of the pellets is still active.

3.5.2.2 Reactors

Many processes for residue hydrotreating are industrially applied or under development, see e.g. [18,30,33–36]. They can be distinguished based on the type of reactor used, i.e. fixed-bed reactors, fluidized-bed reactors (also called ebullated-bed reactors), and slurry reactors. Figure 3.35 shows examples of the three reactor types. In each case three phases are present in the reactor: gas (H_2), liquid (residual oil), and solid (catalyst).

Fixed-bed reactors are operated as so-called trickle-flow reactors. The gas and liquid flow co-currently downward. In fluidized-bed reactors, gas, and liquid flow upward and keep the catalyst particles in suspension. In principle, the catalyst remains in the reactor. In slurry reactors, the catalyst is very finely divided and is carried through the reactor with the liquid fraction. Slurry reactors are usually mechanically stirred, but in slurry reactors for the conversion of heavy petroleum fractions, suspension of the catalyst particles is no problem: the liquid/solid slurry behaves as a homogeneous phase [30].

It is interesting to compare the reactor systems. The most important advantages of the fluidized-bed reactor are the excellent heat-transfer properties, the option to use highly dispersed catalyst systems, and the ease of addition and removal of catalyst particles. Compared to fixed-bed reactors the particle size can and must be much smaller, and as a consequence, the apparent activity and the capacity for metal

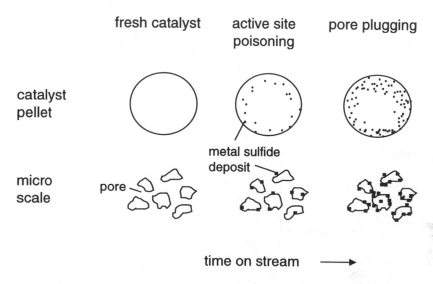

Figure 3.34 Metal sulfide deposition on catalysts. Adapted from [32].

removal are higher. In slurry reactors, the particle size is even smaller, with the accompanying advantages. However, separation of the fine particles from the product is often a problem. In residue processing, recovery of the catalyst is usually not practical, so that it is discarded with the unconverted residue [30].

QUESTIONS:

> *In which reactor type would internal diffusion rates be highest?*
> *Give advantages and disadvantages of the three reactor types.*
> *Discuss safety aspects of these reactors: is a 'runaway' (dangerous temperature rise) possible?*

3.5.2.3 Processes with fluidized-bed reactors

Figure 3.36 shows an example of a process scheme based on fluidized-bed reactors, the LC-fining process developed by Lummus [34].

The feed, a vacuum residue, and hydrogen are preheated in separate heaters and fed to the first reactor. The velocities required for fluidization of the catalyst bed are accomplished by the usual gas recirculation, together with internal recirculation of the liquid phase. In the reactors, removal of hetero-atoms and conversion of the feed take place. The reaction conditions are more severe than in hydrotreating of distillates. The products are separated in a series of high and low pressure gas-liquid separators at high and low temperature. Fresh catalyst is added during operation periodically (e.g. twice weekly) and part of the spent catalyst removed. The most important problem of residue hydrotreating, viz. catalyst deactivation, is solved in this way.

QUESTIONS:

> *Why three reactors in series? What is the purpose of the internals in the reactors?*

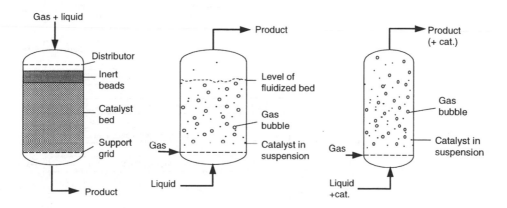

Figure 3.35 Fixed-bed (left), fluidized-bed (middle), and slurry reactor (right).

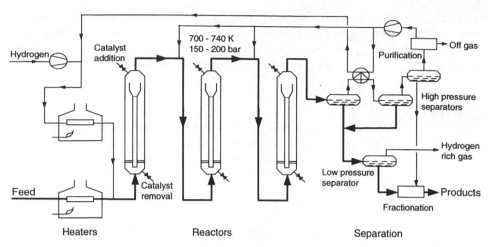

Figure 3.36   Catalytic hydrogenation of residues in fluidized-bed reactors (Lummus).

3.5.2.4 Processes with fixed-bed reactors

Replacement of deactivated catalyst in a conventional fixed-bed reactor is not possible during operation. Therefore, in the processing of heavy residual fractions in a fixed-bed reactor, either a catalyst with a long life (highly metal-resistant with accompanying low activity) is required, or a solution has to be found for easy catalyst replacement. Depending on the metal content of the feedstock various combinations can be applied (see Figure 3.37).

HYCON

The HYCON process (Figure 3.38) [35,36] solves the dilemma that a fixed-bed reactor is most convenient but that an active catalyst deactivates fast, by applying two reactor systems in series. A reactor that enables easy catalyst replacement is used for

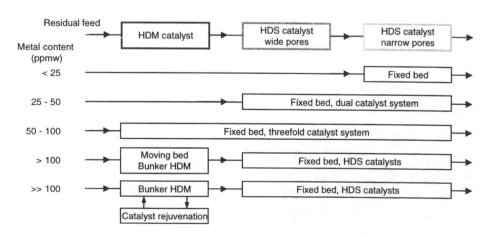

Figure 3.37   Catalyst configuration dependence on metal content in feed. Adapted from [35].

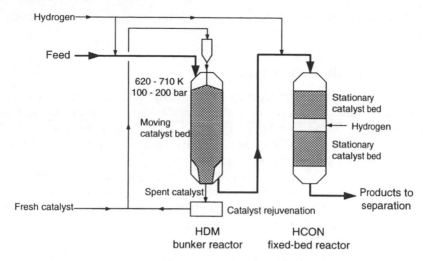

Figure 3.38 Reactor section of HYCON process.

the removal of metals (hydrodemetallization: HDM). Further conversion, i.e. sulfur removal and conversion into lighter products (hydroconversion: HCON) takes place in a conventional fixed-bed reactor. This process is suitable for feedstocks having a metal content $\gg$100 ppmw.

The HDM reactor is of the moving-bed type, or more accurately a 'bunker' reactor. At regular intervals, part of the catalyst loading is replaced. This is referred to as 'bunker flow'. It is crucial that the catalyst particles flow freely. Therefore, in practice spherical particles are used. Both reactors are operated as trickle-flow reactors with the gas and liquid flowing co-currently downward.

The HDM reactor contains a catalyst with relatively large pores. This catalyst is metal resistant, but not very active. The HCON reactor contains a catalyst with smaller pores, leading to higher activity but lower metal-resistance. Actually, two catalyst beds are applied in the HCON reactor, the latter containing a catalyst with smaller pores than the former.

If the vacuum residue feed contains very large amounts of metals, the HDM catalyst consumption would still be too high, and catalyst 'rejuvenation' would be necessary (see Figures 3.37–3.39). Catalyst 'rejuvenation' is achieved by removal of metal sulfides and carbonaceous deposits (essentially by oxidation), and by extraction of the metals.

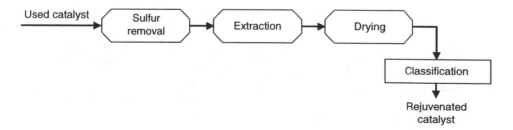

Figure 3.39 HDM catalyst rejuvenation.

The first commercial HYCON process came on stream in 1993 at Pernis, The Netherlands.

3.5.2.5 Processes with slurry reactors

Slurry processes for residue processing are normally designed with the objective of maximizing the residue conversion. Downstream reactors are then used to treat the liquid products for S and N removal. Examples of the slurry process are the Veba Combi-Cracking process [37] and the CANMET process [30]. Figure 3.40 shows a flow scheme of the Combi-Cracking process.

Conversion of the residual feed takes place in the liquid phase in a slurry reactor. After separation of the residue from the products, these products are further hydro-treated in a fixed-bed reactor containing a HDS catalyst. The process uses a cheap once-through catalyst, which ends up in the residue.

3.6 TREATMENT OF REFINERY GAS STREAMS

3.6.1 Removal of H₂S from Refinery Exhaust Gases

Removal of H_2S from H_2S-containing gases in the refinery, such as the exhaust gas of a hydrodesulfurization (hydrotreating) plant, is usually performed by absorption in the liquid phase [38]. The concentrated H_2S is frequently converted to elemental sulfur by the 'Claus' process [38,39]. In this process 95–97% of the H_2S is converted. To obtain greater conversions several technologies are in use. Some are already proven technology, others are under development [38].

QUESTION:

> *S present in crude oil either ends up in the refinery hydrocarbon products (gasoline, diesel, etc.) or in the inorganic products (S, H_2S, SO_2). What would be the preferred situation?*

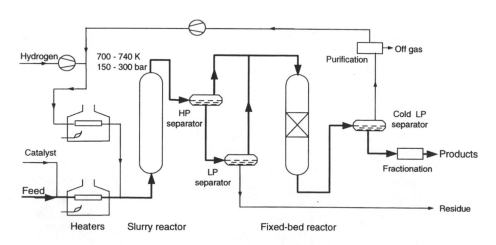

Figure 3.40 Veba Combi-Cracking process.

3.6.1.1 Regenerative absorption of H₂S

H$_2$S is often removed by absorption in the liquid phase with a regenerable solvent. Commonly used absorbents are alkanolamines:

HO−CH$_2$−CH$_2$−NH$_2$ MEA (mono-ethanolamine)

CH$_3$−CH−CH$_2$−N−CH$_2$−CH−CH$_3$
| | |
OH   H   OH
DIPA (di-isopropanolamine)

Advantages of these amines are their high solubility in water (solutions of 15–30 wt% are commonly applied) and low volatility. Furthermore, their reaction with H$_2$S is much faster than with CO$_2$ so that in case the gas to be treated contains both H$_2$S and CO$_2$ the amount of absorbed CO$_2$ can be limited by choosing appropriate conditions.

Figure 3.41 shows a typical process scheme for the removal of H$_2$S from exhaust gases. Raw gas is fed to the bottom of the absorber containing typically 20–24 trays or an equivalent packing. The lean alkanolamine solution is pumped to the top of the absorber. Clean gas leaves the top. The rich solution leaving the bottom is preheated and fed to the regenerator, where the acid gases are stripped with steam. This steam is generated by passing solution of the column bottom through a reboiler, and reintroducing it into the column, upon which steam flashes off. The acid gases are taken from the top and fed to a condenser, where most of the steam condenses and is returned to the regeneration column.

3.6.1.2 Claus process

In the Claus process, elemental sulfur is produced by partial oxidation of H$_2$S in a furnace. The overall reaction may be represented simply by the highly exothermic reaction:

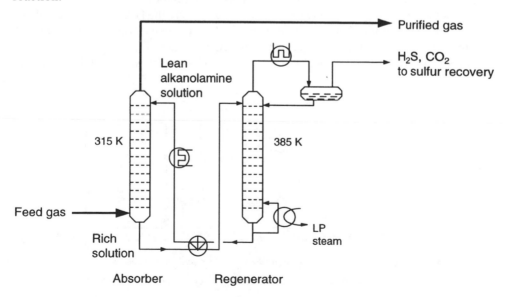

Figure 3.41 Flow scheme for H$_2$S removal by amine absorption.

$$2H_2S + O_2 \rightleftarrows S_2 + 2H_2O \qquad \Delta H_{298}^0 = -444 \text{ kJ/mol} \qquad (10)$$

but in reality the chemistry is much more complex, see [38,39]. The overall reaction is accomplished in two steps. One third of the H_2S is oxidized to SO_2 and H_2O, and then the remaining H_2S reacts with the formed SO_2 into sulfur:

$$2H_2S + 3O_2 \rightarrow 2SO_2 + 2H_2O \qquad \Delta H_{298}^0 = -1038 \text{ kJ/mol} \qquad (10a)$$

$$2H_2S + SO_2 \rightleftarrows 3/2S_2 + 2H_2O \qquad \Delta H_{298}^0 = -147 \text{ kJ/mol} \qquad (10b)$$

Sulfur recovery is limited by the equilibrium. The temperature of the furnace is high: >1300 K. At this temperature, oxidation rates are very high and in agreement with thermodynamics SO_2 is formed. Because only a substoichiometric amount of O_2 is present sulfur can be formed in a consecutive reaction (along with a number of by-products). Sulfur recovery is limited to 50–70% due to the equilibrium, which is to the left at high temperature.

To increase the conversion, catalytic converters (usually two or three in series) are added, which operate at relatively low temperature. In practice about 95% of the H_2S is converted to sulfur. Figure 3.42 shows a flow scheme of a typical Claus process.

The H_2S containing gas is fed to an oxidation chamber where it is partially combusted with air to form SO_2, which then reacts with the remaining H_2S to form elemental sulfur. The heat of reaction is recovered in a waste heat boiler by raising high-pressure steam. Gas leaving the combustion chamber is cooled to about 450 K for sulfur condensation. The gas is then reheated (to 520 K) and passed into the first Claus reactor, where the Claus equilibrium is re-established and additional sulfur is formed. After each reactor sulfur is removed by condensation. In the condensers low-pressure steam is generated. The tail gas, which besides residual H_2S and SO_2 contains small amounts of CS_2 and COS, is either treated in a SCOT plant (see box) or incinerated and emitted to the atmosphere, depending on emission regulations.

In order to ensure optimal conversion in the Claus plant, the $H_2S:SO_2$ ratio should be 2:1. The ratio is measured in the tail gas from the final condenser (QC, quality control), and the air flow to the furnace is adjusted to achieve this ratio.

The sulfur produced is a saleable product. The Claus process is widely applied in

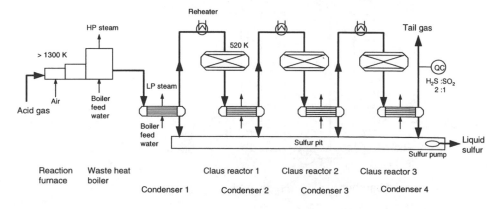

Figure 3.42 Flow scheme of a typical Claus process. QC = quality control.

SCOT Process

In the SCOT process (see Figure B.3.3), all the S-containing components in the Claus off-gas (SO_2, CS_2, COS) are converted to H_2S in the presence of H_2 and a catalyst.

The Claus tail gas is heated in a special burner, which produces reducing gas (H_2 and CO) by incomplete fuel combustion. If available, an external source of reducing gas can also be applied. In the SCOT reactor nearly all sulfur compounds are converted into H_2S by reaction with hydrogen. After cooling of the gas, H_2S can then be removed from the gas stream by absorption in an alkanolamine solution (Section 3.6.1.1).

An appreciable amount of CO_2 is present in the gas stream to the absorber, so H_2S absorption should be selective. This is achieved by choosing the conditions in the absorber such that the bonding of H_2S to the amine is instantaneous, while reaction of CO_2 with the amine hardly takes place.

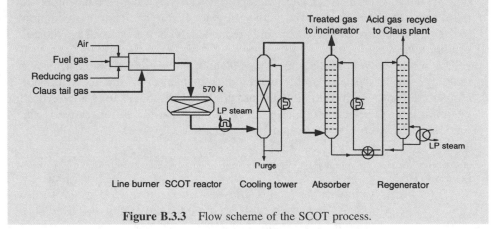

Figure B.3.3   Flow scheme of the SCOT process.

refineries. World wide several hundred plants with a typical capacity of 100 t/d sulfur are in use.

QUESTIONS:

> *What would be the temperature profile in the reaction furnace? In principle, removal of H_2O from the gas mixture could be applied to enhance the conversion of H_2S. How could this be done? Why would the conversion be higher? Why is incineration of the residual gas necessary? What compound is formed? How should an operator adjust the air supply when the $H_2S:SO_2$ ratio measured in the tail gas is lower than 2:1?*

With more stringent regulations, the amount of residual sulfur compounds in the tail gas from a Claus plant is still unacceptable. Recent developments such as the Shell

SuperClaus Process

The SuperClaus process has been developed recently by Stork-Comprimo BV, in co-operation with Gastec NV and The University of Utrecht, The Netherlands. The process is based on the use of a selective oxidation catalyst in the last reactor stage, increasing the sulfur recovery to about 99%. The catalyst catalyzes the direct oxidation of H_2S to elemental sulfur (Eq. (10)). Due to the low operating temperature, the equilibrium is shifted to the right, so a high conversion to S_2 is obtained. It is essential that the catalyst exhibits a high selectivity, viz., it should produce only very small amounts of SO_2.

Figure B.3.4 shows a flow scheme of the SuperClaus Process. The most important differences with the Claus process are the use of a selective oxidation reactor and a less sensitive, more flexible air:acid gas control.

The acid gas feed is burned in the thermal stage in a substoichiometric amount of oxygen so that the $H_2S:SO_2$ ratio is much higher than the conventional value of 2:1. Hence, after the two Claus reactors nearly all SO_2 is converted, while H_2S is present. This stream is mixed with excess air before it enters the oxidation reactor. After the oxidation reactor and the final sulfur condensation step, the tail gas is incinerated.

The sulfur recovery in the SuperClaus plant is not as high as can be achieved in SCOT plants because sulfur compounds such as CS_2 and COS are not converted. However, it is possible to convert these compounds to H_2S by hydrogenation prior to selective oxidation. It is claimed that the sulfur recovery can then be increased to the value achieved in the SCOT process.

QUESTION:

> *Compare the SuperClaus with the combination of the Claus and SCOT processes. What are advantages and disadvantages of both sulfur removal options?*

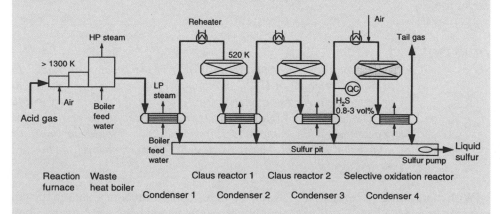

Figure B.3.4   Flow scheme of the SuperClaus process.

Liquid Redox Sulfur Recovery [38,40]

The processes discussed previously are suited for large-scale processes. In small-scale processes (say <10 kt/d) the gas flow rate and composition is usually not constant. This alone makes the above technologies less applicable. Moreover, Claus plants require a large investment, which is not economical for small-scale applications.An alternative is to add catalysts to the liquid used for absorption (see Section 3.6.1.1) and to convert H_2S directly into elemental sulfur:

$$2H_2S + O_2 \rightarrow S_2 + 2H_2O \qquad \Delta H^0_{298} = -444 \text{ kJ/mol} \qquad (11)$$

In this respect, one might speak of a multifunctional reactor. Several processes have been developed based on various catalyst systems. Vanadium-based catalysts appeared to work well and have been used in many plants. Because of environmental and economical concerns new generations of catalysts have been developed, including iron complexes and bio systems.

QUESTION:

Evaluate this type of process for large-scale applications. Give the pros and cons.What technology (absorption + Claus or the process described here) would you choose for the cleaning of H_2S containing off-gases from:
– coal gasifiers;
– hydrotreatment of oil fractions;
– sewer gas?

Claus Off-gas Treating (SCOT) process and the so-called SuperClaus process have resulted in sulfur recoveries of nearly 100%.

QUESTIONS:

What is the origin of the CO_2 in the gas? Why would its co-absorption have to be avoided?

3.6.2 Recovery of Hydrogen from Refinery Gas Streams

Exit gases from units such as fluid catalytic crackers and hydrotreaters contain considerable amounts of H_2. It is often economical to recover this H_2 instead of using it as fuel. H_2 is a valuable commodity because of the increasing need for hydroprocessing of feeds and products of refinery processes.

Several methods are available for gas purification and recovery. The most important are cryogenic distillation, absorption, adsorption, and membrane separation. Cryogenic distillation is amongst others used for the separation of the product stream from naphtha and ethane crackers (see Chapter 4). This process requires a large amount of energy, because very low temperatures are required (ca. 120 K). Absorption

Pressure swing adsorption (PSA)

PSA processes were developed as replacement for adsorption processes involving thermal regeneration. Figure B.3.5 shows the principle of PSA for a hydrogen purification process (for clarity, not all tubing is shown). Adsorption is carried out at elevated pressure (typically 10–40 bar), while regeneration is carried out at low, often atmospheric, pressure. In industry several units (at least two, but usually four or more) are placed in parallel.

High-pressure gas is fed to adsorber 1. Here the adsorbable components (methane, ethane, etc.) are adsorbed. The purified hydrogen gas exits the adsorber. When an adsorber is taken out of the adsorption stage (in this case adsorber 4), it expands the gas (mainly hydrogen, which is contained in the voids of the adsorber packing) into two vessels at lower pressure. The first is the vessel in which adsorption is to take place in the next step, but which is still at low pressure (adsorber 2). The remaining available pressure is expanded in the remaining vessel (3), which is in its regeneration (purging) mode; the feed impurities are desorbed from the bed.

Since the gas from adsorber 4 alone cannot bring adsorber 2 to the required pressure, some of the purified hydrogen is added in order to do so. Now, adsorption can take place in adsorber 2, while adsorber 1 undergoes depressurization, etc. Figure B.3.6 shows the cycle-sequence of the PSA system. In an elegant way the pressures of the four adsorbers are varied in such a way that a quasi-continuous process is realized.

QUESTIONS:

> *What would be the pressure of vessels 4 and 2 in Figure B.3.5 after the depressurization/ pressurization? Draw a schematic of the next step after the situation in the figure.*

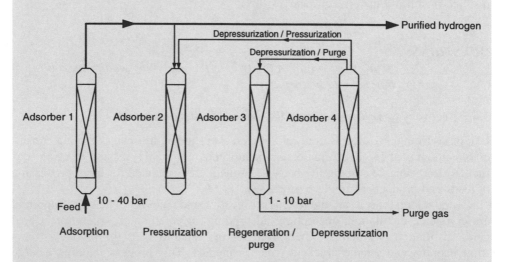

Figure B.3.5 Hydrogen purification using PSA. One step shown, not all piping shown.

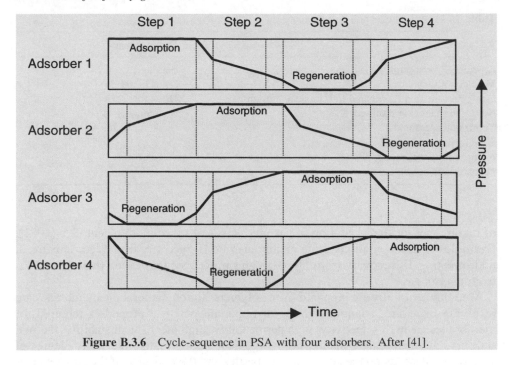

Figure B.3.6 Cycle-sequence in PSA with four adsorbers. After [41].

has been discussed in Section 3.6.1.1 for the removal of H_2S and CO_2 from a gas stream, but it is not applicable for obtaining hydrogen at the required purity (80–95 vol%). At present, adsorption and membrane separation are the preferred technologies.

3.6.2.1 Adsorption

Gas-phase adsorption on solids is widely used in industry for air drying, N_2 production, H_2 purification, etc. It is based on the selective adsorption of one or more components. Hydrogen is practically unadsorbable, and is thus easy to purify by this method. Adsorption processes are usually carried out as a transient process and consist of two steps:

- Impurities are adsorbed yielding a purified gas stream containing the hydrogen.
- The adsorbent bed is regenerated either by raising the temperature ('temperature swing adsorption', TSA) or by reducing the pressure ('pressure swing adsorption', PSA).

3.6.2.2 Membrane separation

A membrane is a selective barrier allowing separation of compounds on the basis of molecular properties such as molecular size, strength of adsorption on, or solubility in the membrane material. Although osmosis through membranes has been known since the 1830s, and has been applied for liquid separation since the 1970s, the application

Table 3.11 Advantages and disadvantages of membranes

Advantages	Disadvantages
Low energy consumption (no phase transfer)	Fouling
Mild conditions	Short lifetime
Low pressure drop (compared to 'classical' filtration)	Often low selectivity
No additional phase required	No economy of scale
Continuous separation	(scale-up factor ≈ 1)
Easy operation	
No moving parts (except recycle compressor in gas separation applications)	

of membranes for the separation of gas mixtures is of much more recent date (1981). Examples of gas separations using membranes are H_2 recovery from bleed streams in hydrotreating, H_2 recovery from bleed streams in NH_3 synthesis, and removal of CO_2 from natural gas.

Membranes usually are prepared from polymers, although inorganic membranes are receiving increasing attention. The most important practical properties required for membranes are high selectivity, high permeability, high mechanical stability, thermal stability, and chemical resistance. For inorganic membranes, the selectivity (degree of separation) depends on the adsorption properties and the pore size of the surface layer that is in contact with the fluid (gas or liquid). For polymer membranes, not adsorption but solubility plays a major role. The permeability (flow rate through the membrane) is also determined by the membrane thickness. Table 3.11 summarizes advantages and disadvantages of membranes in separation processes.

The performance of most pressure-driven membranes is hindered by progressive fouling of the membrane. Solutions are found either by changing the surface characteristics of the membrane or by hydrodynamic changes aimed at reducing the concentration of foulants at the surface of the membrane.

QUESTIONS:

Why would membranes not exhibit a favorable economy of scale?
Membranes are especially applied in the separation of purge streams. Why?

Recovery of hydrogen from refinery streams

The central part of a membrane plant is the module: the technical arrangement of the membranes. An example is the Monsanto module (see box), which is used in many gas separation applications. One of the main applications of the Monsanto module is the recovery of H_2 in gaseous flows in oil refineries. Figure 3.43 shows a flow scheme for application in an oil refinery (the arrows represent the flow of hydrogen).

Hydrogen streams from hydrodesulfurization units, after removal of H_2S, pass through a sequence of membrane units, and are concentrated to about 90%.

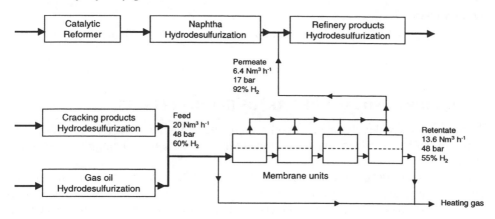

Figure 3.43 Flow scheme of hydrogen recovery from refinery gas. After [42]. For clarity only the H_2-flows are shown. ©John Wiley & Sons Ltd. Reproduced with permission.

Membrane Modules

Membrane modules are produced on an industrial scale. Most modules are so-called three-end modules (see Figure B.3.7). The feed stream enters the module. Because of the action of the membrane, the composition inside the module will change. A 'permeate' stream, which passes through the membrane, and a 'retentate' stream are formed.

Many configurations are in use. They are either of a flat or of a tubular geometry. An example of the former is the plate and frame module, while the latter is often of the shell and tube heat-exchanger type. Modules based on hollow fibers are quite successful. A hollow-fiber module consists of a pressure vessel containing a bundle of fibers. The flow pattern may be co-current, counter-current or cross flow.

A hollow-fiber module equipped with a polysulfone-silicone membrane is applied in many gas-permeation applications. An example of such a module is the Monsanto hollow-fiber module (see Figure B.3.7). The feed is at the shell side and the permeate comes out of the fibers.

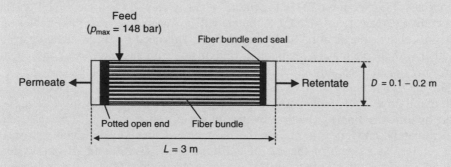

Figure B.3.7 Configuration of the Monsanto hollow-fiber module and some characteristics [42].

QUESTION:

Is the sequence catalytic reforming, HDS in Figure 3.43 correct?

3.7 CURRENT AND FUTURE TRENDS IN OIL REFINING

Current and future developments in oil refining are and will be mainly concerned with upgrading of heavy oil fractions and with environmentally more benign processes and products.

The processes described in Section 3.5 are typical of the 1970s and early 1980s developments, which were aimed at maximizing the conversion of heavy oils to gasoline and middle distillate products. Although this objective is still important, the focus has shifted somewhat since the late 1980s to develop cleaner products [4,43,44]. Ever stricter environmental constraints on both products and refineries have triggered these developments.

3.7.1 Reformulated Gasoline

Over the past decades gasoline production has become much more complicated and it will become even more so. Prior to about 1973 gasoline production involved blending catalytically reformed naphtha with straight-run light naphtha (tops) and other additives (notably lead compounds) to adjust its properties, particularly the octane number, the S content, and vapor pressure.

Environmental regulations, which aim at reducing the toxic and/or polluting components from car exhausts have already had an enormous impact on gasoline production. With the advent of lead-free gasoline from 1973 on, oil refiners had to find other means to enhance the octane number of gasoline. For example, the isomerization of the light straight-run naphtha fraction (mainly C_5 and C_6) [2] has been gaining popularity from then on.

It was expected that the unleaded gasoline market would be essentially 100% in both the USA and Europe by 2000 [45]. Current and future environmental regulations also place upper limits on sulfur, alkenes, benzene, and other lower aromatics content, while the octane number has to be preserved. Moreover, a certain minimum amount of oxygenates is required. In particular MTBE has been added. However, although the octane number is perfect, recently doubts have been expressed on the environmental friendliness (see Section 6.2). In the future, regulations will continue to be tightened, especially in urban areas. Also the S content will be lowered because this enables more efficient catalytic purification of the exhaust gas. This need for reformulated gasoline, together with the decreasing availability of light crude oil has enormous consequences on refinery operations, not in the least from an economic viewpoint.

The environmentally most favorable gasoline consists of highly branched alkanes with mainly five to ten carbon atoms. Such a gasoline is produced by *alkylation*, so for good quality gasoline production in the future its capacity should increase much. Unfortunately, alkylation is one of the most environmentally unfriendly processes in a refinery as a result of the use of the liquid acids H_2SO_4 and HF. In academia and industry large efforts are put in developing stable solid catalysts. It was hoped that solid-catalyzed industrial alkylation process would be available already by the year 2000 [45].

The octane number of gasoline produced by *catalytic reforming* is high owing to the large fraction of aromatics present (see Table 3.7). The process will continue to play an important role as gasoline base stock, and as hydrogen producer (!), but in the long run its importance might diminish if the strict regulations on benzene and aromatics in the USA are extended to other countries. Solutions could be the reduction of the dealkylation activity of the reforming catalyst and the removal of the C_5/C_6 fraction from the reforming feedstock (dehydrogenation of hexanes is the most important source of benzene in reformate), and for instance isomerize this fraction.

The *fluid catalytic cracking* process contributes much to the gasoline pool. However, FCC gasoline, just like reformate, contains a large amount of aromatics, and in addition, a small amount of undesired sulfur. Moreover, the FCC process is a large contributor to refinery emissions. Undoubtedly in the future FCC gasoline will remain important, but sulfur will have to be removed. However, hydrodesulfurization of gasoline will also result in hydrogenation of alkenes that constitute a major fraction of FCC gasoline. As the octane number of an alkene is higher than that of its corresponding alkane, the gasoline quality will be reduced. Additional challenges arise as a result of processing heavier feedstocks. So, although the FCC process was conceived over half a century ago, the technology is far from mature [16].

Gasoline produced by *hydrocracking* has a more favorable composition than FCC gasoline with respect to environmental issues. It does not contain sulfur and is low in aromatics as a result.

3.7.2 Diesel

The main sources of diesel fuel in a refinery are straight-run gas oil, and gas oil from the hydrocracking and FCC processes. Regulations for diesel fuel mainly cover maximum sulfur (see also Section 3.4.2), and aromatics content, and minimum cetane number. The latter might increase from about 50 now to 52 or even 55 beyond 2000 [45], which will imply drastic reduction of the aromatics content as well. Table 3.12 shows these properties for a number of diesel base stocks.

Clearly, of the refinery products, gas oil (diesel fuel) from the hydrocracking process shows the most favorable properties. It is very low in sulfur and has a high cetane number, which is caused by the low amount of aromatics and a high content of only slightly branched alkanes. Hydrocracking will increasingly prevail as a diesel oil

Table 3.12 Properties of diesel base stocks [4,45]

Source	Sulfur (wt%)[a]	Aromatics (wt%)	Cetane number
Straight-run gas oil	1–1.5	20–40	40–50
Light cycle oil from FCC	2–2.8	>70	<25
Gas oil from thermal processes	2–3	40–70	30–50
Gas oil from hydrocracking	<0.01	<10	50–55
Fischer–Tropsch gas oil	0	≈0	>70

[a] Before hydrodesulfurization.

producer. Diesel from other sources will only meet future sulfur standards (see Section 3.4.2) when hydrotreating is applied. A lot of research has been carried out to develop more active and stable desulfurization catalysts. In addition, recently so-called deep desulfurization has received much attention [19,43].

Similarly, to comply with lower aromatics specifications, new hydrogenation catalysts have been introduced [19,43]. In particular upgrading of light cycle oil is highly desired. Hydrogenation and ring opening are the target reactions. Up to now no satisfactory catalysts are available.

QUESTIONS:

How will current and future regulations concerning diesel fuel affect the hydrogen balance in refineries? Suggest means for producing hydrogen.

Fischer–Tropsch synthesis, which converts synthesis gas into long-chain alkanes (see also Chapter 6) produces a nearly ideal diesel oil and might well become the process of the future.

3.7.3 The Use of Zeolites for Shape Selectivity in the Oil Refinery

3.7.3.1 Introduction

Most solid catalysts are only 'class selective'. For example, a classical hydrogenation catalyst will hydrogenate double bonds in all reactant molecules irrespective of their size or of the position of the double bond [46]. Tremendous progress has been made in developing catalysts that show selectivity with respect to the position of the unsaturated bond (regioselectivity). Also stereosectivity is often possible leading to the desired enantiomeric compound. In this respect a breakthrough has been the synthesis of solid catalysts showing shape selectivity as a result of spatial constraints mostly. They can be tuned to selectively convert only particular molecules or produce only a certain product. The most important examples are enzymes and molecular sieves of which zeolites are a special group.

A remarkable property of zeolites is their porosity (up to 0.4 ml/g), although they are crystalline compounds. The pore-size distribution is well defined and the pores are in the nanometer range, corresponding to molecular dimensions.

The pore sizes of most zeolites are between 0.3 and 0.9 nm (see Figure 3.44). For comparison the dimensions of some molecules which are interesting for process technology are also shown. The fact that the dimensions of the pores are in the range of molecular sizes makes so-called shape selectivity possible. Figure 3.44 illustrates the potential selectivities of some common zeolites based on the dimensions. Often three types of shape selectivity are distinguished: reactant, transition-state, and product selectivity (see Figure 3.45).

Reactant selectivity occurs if in a mixture of feed molecules only those with dimensions smaller than the pores will enter and react. An example is hydrocracking of a mixture of linear and branched heptanes: the linear molecules are cracked whereas the branched alkanes cannot enter the pores, and consequently are not transformed.

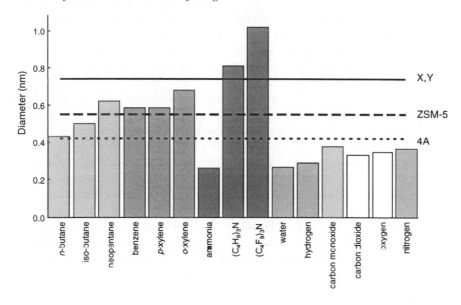

Figure 3.44 Pore sizes of zeolites and dimensions of some molecules [47].

In transition-state selectivity only pathways involving certain transition geometries will be operational. An example is the disproportionation of *m*-xylene (*m*-xylene → trimethylbenzene + benzene) where only one of the three trimethylbenzene isomers is formed, because the others require a coupling geometry that is sterically hindered.

In producrt selectivity, only products that are small enough to diffuse from the

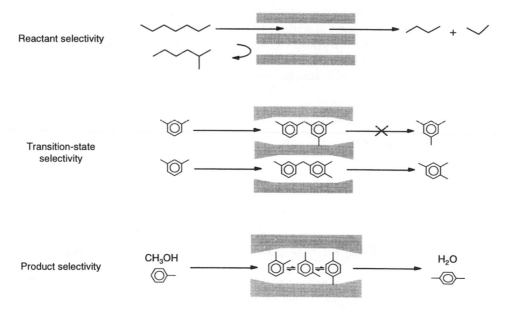

Figure 3.45 Three types of shape selectivity.

interior of the zeolite to the outside can be formed. In the example of Figure 3.45 xylene is formed from toluene. Of the possible products (*o*-, *m*-, and *p*-xylene) only the para-isomer, which is the preferred product, is found in large quantities.

In the examples of Figure 3.45 shape selectivity has provided a means to selectively convert one type of reactant or selectively produce the desired product, where otherwise the thermodynamically determined product distribution would result. Of course, in zeolites, as in other catalysts, intrinsic reaction kinetics also play a role with respect to selectivity; if the desired reaction is fast compared to the undesired reaction, a high selectivity can also be obtained, even if no shape selectivity occurs.

There are also examples where shape selectivity is a disadvantage. An example is processing of heavier feeds in the FCC process discussed in Section 3.4.1. Due to the increasing gasoline demands and decreasing supplies of light crude oils, there is a trend towards processing of heavier FCC feeds. In FCC, zeolites are applied because of their high activity for cracking. However, the shape selectivity of the catalyst is an additional property. In the processing of heavy feeds, this shape selectivity will prevent the larger molecules from entering the pores and hence prevent their conversion.

QUESTION:

   Suggest a solution for the problem of processing heavy feeds in FCC.

The next two sections give examples of the potential of shape selectivity in the production of gasoline components.

3.7.3.2 Production of isobutene

The capacity of plants for the production of oxygenates such as MTBE and ETBE (Scheme 3.7) has increased significantly over the past two decades as a result of government regulations regarding oxygen content in gasoline (see also Chapter 6 for MTBE production). This has led to a shortage of isobutene.

C_4 hydrocarbons are produced by steam crackers (see Chapter 4) and catalytic

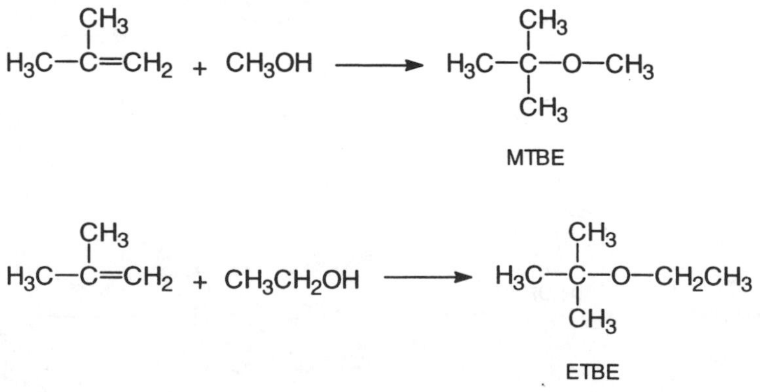

Scheme 3.7   Routes to MTBE and ETBE.

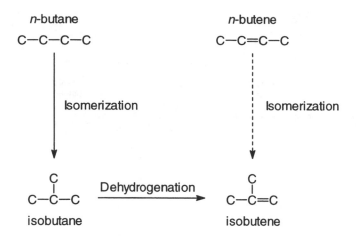

Scheme 3.8 Routes to isobutene.

Butene to Isobutene Isomerization using Ferrierite

The isomerization reaction on the zeolite Ferrierite (zeolite of medium pore size) is believed to proceed via a dimerization step to form a C_8 intermediate, which then undergoes skeletal isomerization followed by selective cracking to produce two isobutene molecules. The Ferrierite catalyst was found to produce isobutene at much higher selectivity than other catalysts due to its shape selectivity: the C_8 isomer responsible for isobutene formation cannot leave the zeolite pores, while less branched C_8 isomers can (see Scheme B.3.2) [48,49]. At first sight one might have expected that single branched C_8 would be the major product. However, since tertiary carbenium ions are more stable than secondary or primary carbenium ions, formation of the highly branched C_8 isomer is favored. Therefore, isobutene is a major product. In addition to the higher selectivity, Ferrierite has a much greater stability than most other catalysts for this reaction [50].

QUESTIONS:

What type of shape selectivity is this? Would direct isomerization of n-butene as shown on the right of the figure be likely?

The Ferrierite-based butene isomerization process has recently been commercialized [51]. Figure B.3.8 shows a simplified flow scheme of the process. A C_4 hydrocarbon feed containing normal butenes (and butanes) is preheated and fed to a fixed-bed reactor, where conversion to isobutene takes place (typical product distribution: 35 wt% isobutene, 40 wt% n-butene, 5 wt% C_1–C_3, 20 wt% C_{5+}) [49]. The reaction takes place in the vapor phase. The reactor effluent is cooled in the feed-effluent heat exchanger prior to compression. The C_{5+} material, which is suitable for blending into the gasoline pool, is then separated from the isomerate by distillation.

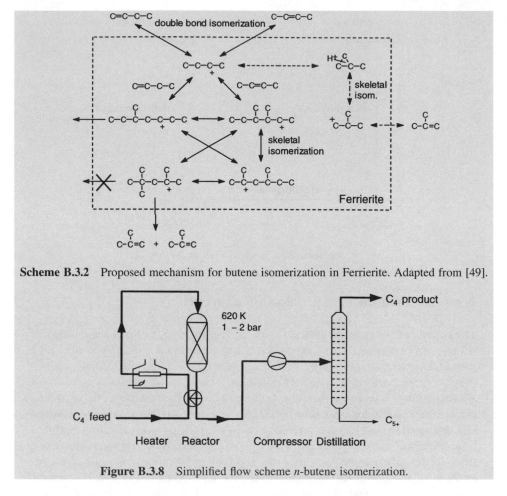

Scheme B.3.2 Proposed mechanism for butene isomerization in Ferrierite. Adapted from [49].

Figure B.3.8 Simplified flow scheme *n*-butene isomerization.

crackers. Isobutene can be produced from these streams by isomerization of *n*-butene or from *n*-butane when a dehydrogenation step is added (see Scheme 3.8).

However, until recently the seemingly simple butene-to-isobutene isomerization was not very attractive due to the lack of a selective catalyst. The reaction was hampered by the dimerization and oligomerization of butenes, which led to large amounts of C_{5+} hydrocarbons and thus poor yields of isobutene. Another problem was rapid coking of the catalyst.

Dimerization reactions (exothermic) can be suppressed by operating at high temperature, but this also leads to low isobutene yields as a result of the unfavorable thermodynamic equilibrium (compare Figure 3.46, which shows the equilibrium composition for hexanes). Furthermore, under these conditions, coke formation results in rapid catalyst deactivation. An alternative, which has recently been commercialized, is the application of highly selective skeletal isomerization catalysts (Ferrierite zeolites, see box: Butene to Isobutene Isomerization using Ferrierite) [48–52].

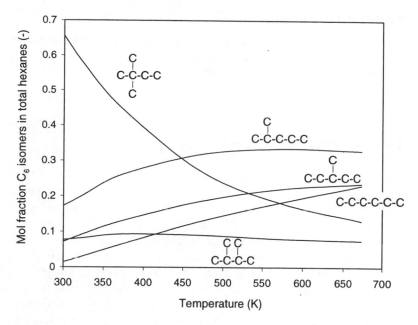

Figure 3.46   Gas-phase equilibrium composition of hexanes as a function of temperature.

3.7.3.3 Isomerization of pentanes and hexanes

Isomerization of C_5 and C_6 alkanes is not a new process (it was developed in the 1960s) but it has become of much greater importance in oil refineries as a result of the requirements of reformulated gasoline. From Table 3.2 it is clear that isomerization of the linear pentanes and hexanes present in a straight-run naphtha will significantly boost its octane number.

Figure 3.46 shows the gas-phase equilibrium composition of hexanes at a pressure of 1 bar as a function of temperature. Clearly, a low temperature is favorable for the formation of branched hexanes.

QUESTION:

How would the equilibrium composition depend on pressure?

Many isomerization processes have been developed, see for instance [2,53] and box: The Total Isomerization Package Process. Two types of catalysts are used, a Pt/Cl/Al$_2$O$_3$ catalyst and Pt on the zeolite H-Mordenite. The former catalyst is more active but the latter has a much longer lifetime and is less susceptible to deactivation by, for instance, water. Therefore, water need not be removed. In addition, problems as a result of the presence of Cl are avoided, which makes zeolite-based processes much simpler.

QUESTION:

Why would Pt/Cl/Al$_2$O$_3$ be sensitive to water?

3.7.4 Alternative Technology and Fuels

Although crude oil will remain the most important source of fuels production for some time still, at one point the resources will dry up. Therefore, either other means of producing gasoline, diesel, etc, or alternative fuels will have to be used.

The Total Isomerization Package Process

The Total Isomerization Package (TIP) process is a combination of the Shell Hysomer process and the Isosiv process of Union Carbide. This process is interesting because it combines catalytic isomerization of linear C_5 and C_6 alkanes over a zeolitic catalyst (Pt/H-Mordenite) with separation by means of shape selectivity in another type of zeolite (molsieve Linde 5A) in one process.

In the Hysomer process linear alkanes are isomerized at about 500–600 K and 10–30 bar in the presence of hydrogen. The reactor is an adiabatic fixed-bed reactor, usually divided in two catalyst beds with intermediate cooling by hydrogen injection. Heat production is predominantly caused by hydrogenation of aromatics that are present and by some hydrocracking (the isomerization reaction itself is nearly thermoneutral). The process flow scheme is very similar to that of the catalytic reforming process (see Section 3.4.4). Only about 40–50% of the normal alkanes in the feed are converted as a result of thermodynamic limitations at the relatively high temperature applied.

In the Isosiv process, the unconverted normal alkanes are separated from the branched alkanes by means of adsorption: the normal alkanes are selectively adsorbed, since iso-alkanes have too large a molecular diameter to enter the pores. Usually, the Isosiv process operates by a pressure difference between adsorption and desorption, i.e. pressure-swing adsorption (see Section 3.6.2.1). However, when the Isosiv process is integrated with the Hysomer process, desorption takes place by purging with hydrogen instead of by pressure reduction [53]. Close integration of both processes is possible since they operate at similar temperatures, and at the same pressure (10 bar in this case).

As a result of recycle of the normal alkane fraction, the TIP process yields a product with a much higher octane number than the Hysomer process only. Figure B.3.9 shows this for a feed consisting of 60% pentanes, 30% hexanes and 10% cyclic compounds.

Depending on the amount of normal alkanes in the feedstock, the Hysomer reactor is situated upstream or downstream the Isosiv adsorbers (see Figure B.3.10).

QUESTIONS:

Why does the RON decrease with temperature?
Draw schematic flow sheets of the Hysomer and Isosiv process.
When would you use the Hysomer lead configuration and when the Isosiv lead configuration?

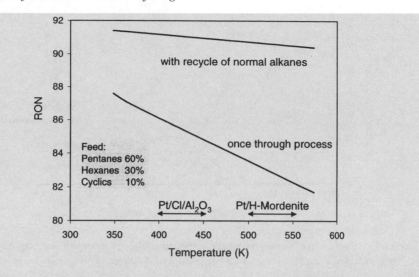

Figure B.3.9 Effect of temperature on product octane number in isomerization without and with recycle of normal alkanes. Adapted from [51].

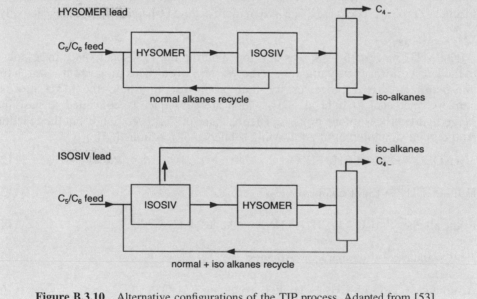

Figure B.3.10 Alternative configurations of the TIP process. Adapted from [53].

3.7.4.1 Alternative technology for the production of gasoline and diesel fuels

The reserves of coal and natural gas are much larger than those of crude oil (see Chapter 2). Currently, technologies are already available for the conversion of natural gas and coal into transportation fuels. Most of these technologies are based on the production of synthesis gas (a mixture of mainly CO and H_2) as an intermediate (see Chapters 5 and 6). Although technologically feasible, in general these processes are not economically attractive (yet). Table 3.13 summarizes processes for the production of transportation fuels from synthesis gas.

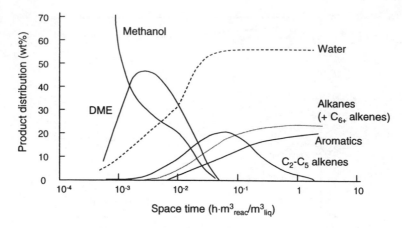

Figure 3.47 Formation of various products from methanol as a function of space time. $T = 643$ K; $p = 1$ bar. Adapted from [56]. Used with permission of Academic Press, Orlando, Florida.

Here we will discuss the MTG (methanol to gasoline) and the MOGD (Mobil olefins to gasoline and distillates) processes. The MTO (methanol to olefins) process and the Fischer–Tropsch process are discussed in more detail in Chapters 4 and 6, respectively.

MTG process

The MTG process [54–58] was developed during the 1970s and 1980s in response to the latest critical energy situation in the Western world. This triggered the search for non-petroleum based processes for fuel production. In short, the MTG process converts methanol to light alkenes, which are subsequently converted to gasoline range hydrocarbons in one process. The mechanism [54] is complex, but the reaction path can be simplified by the following sequence of reactions [55]:

$$2CH_3OH \rightleftarrows H_3C{-}O{-}CH_3 + H_2O \qquad (12)$$

$$H_3C{-}O{-}CH_3 \rightarrow \text{Light alkenes} + H_2O \qquad (13)$$

$$\text{Light alkenes} + H_3C{-}O{-}CH_3 \rightarrow \text{Heavy alkenes} + H_2O \qquad (14)$$

$$\text{Heavy alkenes} \rightarrow \text{Aromatics} + \text{Alkanes} \qquad (15)$$

$$\text{Aromatics} + H_3C{-}O{-}CH_3 \rightarrow \text{Higher aromatics} + H_2O \qquad (16)$$

The initial step is the dehydration of methanol to form dimethyl ether (DME). This is an equilibrium-limited reaction, but the subsequent conversion of DME into light

Table 3.13 From synthesis gas to gasoline and diesel

Process	Via	Main product(s)
MTG	Methanol	Gasoline
MTO	Methanol	Alkenes for MOGD process
MOGD	Alkenes	Distillates/gasoline
Fischer–Tropsch	Synthesis gas	Distillates

alkenes, and then other products drives the dehydration reaction to the right. Figure 3.47 shows a typical product distribution as a function of reaction time, which clearly indicates the sequential character of the reactions.

QUESTION:

> *Are the data in agreement with the sequential character of the reactions.*

Various companies are active in the methanol to fuel conversion area, but Mobil has been the initiator with the discovery that ZSM-5 is a very effective catalyst in converting methanol into high-octane gasoline and still dominates this field [55].

Important features of the MTG process from a technological point of view are the high exothermicity of the overall reaction and the relatively fast deactivation of the catalyst (cycle time of a few months), mainly due to sintering (hydrothermal deactivation) as a result of the presence of a large amount of water.

The MOGD process

In the MTG process methanol is converted according to a sequential scheme via alkenes into gasoline components. It is not surprising that also a direct alkene to gasoline process is feasible. In the MOGD process [57] light alkenes are oligomerized to produce hydrocarbons in the distillate range. With conventional acidic catalysts this would lead to a highly branched product (compare alkylation, Section 3.4.5), which is not desired in diesel fuels. In the MOGD process the same catalyst is used as in the MTO process. Due to the shape selectivity only linear and slightly branched molecules in the distillate range (C_{11}–C_{20}) can escape. Depending on the operating conditions the MOGD process can be used for maximum distillate or maximum gasoline production.

Alternative Diesel

For diesel fuel production there is another alternative. Vegetable oils have approximately the same cetane number as petroleum-based diesel fuel. A great advantage of the use of vegetable oils is their lower sulfur content. Moreover, they do not add to the production of CO_2, which is very important in this time of growing concern regarding the greenhouse effect. Of course, it remains to be seen that this option is realistic in view of the land area needed for a fuel used on such a large scale. A problem in practical applications is that vegetable oils have a much higher viscosity than petroleum-based diesel, which requires a special engine. Another solution is re-esterification of vegetable oils, which leads to a diesel fuel with comparable viscosity as petroleum-based diesel. Rapeseed methyl ester (RME) is the best known product of this kind.

3.7.4.2 Alternative fuels

Alternatives for gasoline and diesel, in principle, include liquefied petroleum gas (LPG), alcohols and ethers (both as such or mixed with gasoline or diesel), compressed natural gas (CNG), hydrogen, and electricity. Table 3.14 gives a brief summary of the advantages and disadvantages of some of these alternative fuels.

Liquefied petroleum gas, a mixture of C_3 and C_4 hydrocarbons, is already common

Mobil MTG process

Figure B.3.11 shows a simplified flow scheme of the Mobil fixed-bed MTG process, which has been in operation in New Zealand since 1985. In the DME reactor crude methanol (containing about 17% water) is converted to an equilibrium methanol/DME/water mixture over a catalyst. This mixture is combined with recycle gas and fed to the gasoline synthesis reactors containing the ZSM-5 catalyst. Recycling of cold product gas is required to remove the heat of reaction from the DME reactor and to limit the temperature rise.

As a result of deactivation of the ZSM-5 catalyst by coke deposition, the composition of the reactor product is not constant. Therefore, and to satisfy the need for frequent regeneration without interrupting the process, the gasoline synthesis section consists of five reactors in parallel, of which one is being regenerated (swing reactor). Regeneration takes place by burning off deposited coke.

QUESTIONS:

> *What alternative reactor would you recommend for the highly exothermic gasoline synthesis reaction and why? (See, e.g. [55,57])*
> *Why would New Zealand have been chosen to build the MTG plant? (Hint: what would be the source of methanol?)*

In the fixed-bed MTG process about 80 wt% of the carbon in methanol is converted to gasoline range hydrocarbons, the other 20 wt% are light alkenes and alkanes and coke. Essentially no hydrocarbons above C_{11}, which corresponds to the end point of gasoline, are produced. This illustrates the advantage of the shape selectivity of ZSM-5 for gasoline manufacture; hydrocarbons with carbon numbers outside the gasoline range cannot escape the pores [58].

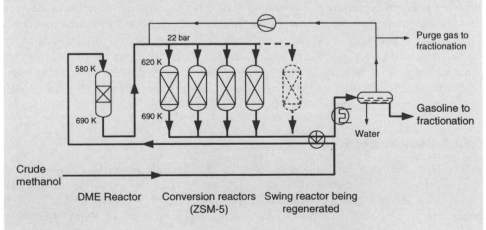

Figure B.3.11   Simplified flow scheme of the Mobil fixed-bed MTG process.

as fuel in many parts of the world. It is the most widely available clean fuel today. It is cheaper than both gasoline and diesel, but with increasing demand the cost of LPG might rise.

CNG is used in some fleet vehicles (e.g. urban buses) as an alternative for diesel fuel. Its major disadvantages are the relatively low energy density and the fact that high pressure storage is unavoidable leading to safety risks.

MOGD process – maximum distillate mode

Figure B.3.12 shows a simplified flow scheme of the MOGD process in the maximum distillate mode. A gasoline recycle stream is mixed with the light alkene feed and preheated. Conversion takes place in three adiabatic reactors in series with intermediate cooling. The product distillate, which is rich in linear and slightly branched alkenes is sent to a hydrotreating section for saturation of the double bonds. This operation increases the cetane number from about 33 to 52 [59].

Catalyst deactivation by coke deposition plays an important role in the MOGD process. Usually a total of four reactors are used. In normal operation, one reactor is being regenerated (by coke burn-off in air), while the other three reactors are positioned such that the first contains the most aged catalyst and the third reactor the freshest catalyst. When a newly regenerated reactor comes on stream, it takes the third position, while the reactor in position three moves to two, etc.

QUESTIONS:

Just as the MTG process, the MOGD process design is largely determined by control of the exothermic heat of reaction. How is this done in the MOGD process (name at least two items)? How would you change the process for maximum gasoline production?

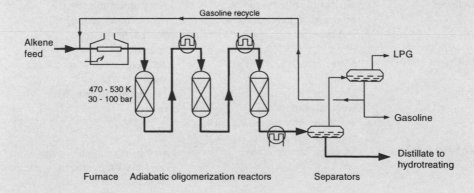

Figure B.3.12 Simplified flow scheme of the MOGD process (maximum distillate mode) [60].

Ethanol produced from sugarcane, is the primary automotive fuel in Brazil, and ethanol/gasoline blends have been used in the USA for many years (ethanol produced from corn). The production of ethanol from crops instead of fossil fuels reduces the accumulation of CO_2 in the atmosphere. However, with current technology and price structures, ethanol is more expensive than gasoline, and hence cannot be expected to be used on a large scale in the near future. On the other hand, specific, subsidized applications in urban areas should not be ruled out [61].

In principle, methanol can also be produced from renewable sources. All major car manufacturers have produced cars that run on 'M85', a blend of 85% methanol and 15% gasoline. Many car manufacturers have developed advanced prototypes of cars running on 'M100', i.e. pure methanol.

Electric vehicles and hybrids

The main advantage of battery-electric vehicles (BEVs) is that they produce virtually no emissions. Of course, the production of electricity in power plants, does lead to emissions, but these emissions can be controlled more easily on the relatively few individual power plants than on millions of cars. Furthermore, electricity can also be produced from other sources such as natural gas, nuclear energy, and hydro, solar, and wind energy. At present, electric cars suffer important disadvantages such as a higher vehicle cost, lower driving range, and long battery recharging time.

Table 3.14 Advantages and disadvantages of alternative fuels

Fuel	Advantages	Disadvantages
LPG	Clean, cheap, widely available	Limited supply, high vehicle cost
CNG	Relatively abundant, clean, can be produced from a variety of sources, including biomass, about same price as gasoline	High vehicle cost (storage under pressure in heavy tanks), less convenient, frequent refueling; safety is a concern
Ethanol	High performance, can be produced from crops, low level of emissions	Expensive
Methanol	High performance, can be produced from, e.g. natural gas and biomass	Low temperature starting difficult modified engine required, hygroscopic, expensive
Electricity (BEV)[a]	Virtually no emissions	High vehicle cost, low driving range
Hybrid	Combines advantage of BEV with onboard means of electricity generation (e.g. in conventional gasoline engine)	More emissions than BEV, expensive
Fuel-cell hybrid	Virtually no emissions	Large hydrogen tank or small-scale onboard hydrogen production plant, under development, expensive

[a] BEV, battery electric vehicle.

Still, recent developments in electric vehicle technology show some promise for the future.

The idea of the hybrid-electric vehicle (HEV) naturally evolves from the inherent limitations of the storage battery. A HEV combines an electrical energy storage system with an onboard means of generating electricity. This can, for example, be a conventional gasoline or diesel engine, or a gas turbine generator.

One of the most widely suggested sources of electricity for a HEV is a fuel cell. The advantage of fuel cells is that they produce electricity instead of heat through reaction of hydrogen with oxygen, with water vapor as the only by-product. For automotive applications the proton-exchange-membrane (PEM) fuel cell (see box) is considered the most attractive (also see Section 5.2.5).

Hydrogen can either be stored on board the car (as in spacecraft), or be produced from natural gas, ethanol, or methanol by reaction with steam (with CO_2 formed as by-product). The main problem with on-board storage of hydrogen, either as a liquid or a compressed gas, is that the required storage space is very large. For example, for the same energy content as gasoline six to ten times more storage space is needed. The *in situ* production of hydrogen from methanol (or other fuels) suffers other problems. Firstly, the system becomes more complicated, because a miniature hydrogen production plant has to be incorporated in the car. Secondly, the catalyst present in the fuel cell is poisoned by carbon monoxide, which is inevitably formed through the water-gas shift reaction. However, the advantage of this method is that the existing infrastructure for gasoline can easily be made suitable for methanol distribution.

Proton-exchange-membrane (PEM) fuel cell

Figure B.3.13 shows a schematic of a PEM fuel cell [62]. In the fuel cell, hydrogen and oxygen are separated by an electrolytic membrane (0.1 mm thick), consisting of a polymer foil coated with platinum catalyst. The Pt catalyst enables the ionization of hydrogen at the anode. The hydrogen protons then pass through the membrane and form water with the oxygen at the other side. This leaves electrons on the hydrogen side, which travel to the cathode through an external circuit producing electricity.

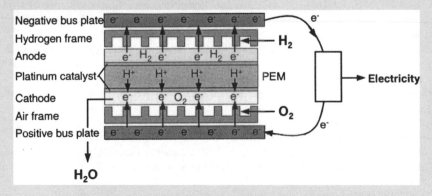

Figure B.3.13   Schematic of a PEM fuel cell. Adapted from [62].

Hybrid technology is progressing rapidly, and most experts agree that the car of the future will be an HEV of some kind. At present, this technology is expensive and not yet proven. In the future this might change, because a tremendous amount of research is carried out in order to develop commercially applicable systems.

Future outlook for alternative fuels

The application of alternative fuels generally requires more or less extensive modifications to the engine and/or fuel distribution system. Since transportation fuels represent a significant part of total chemicals and fuels production, refineries and chemical industries will also have to be reorganized. Furthermore, the infrastructure for the distribution of alternative fuels usually is not sufficient. Therefore, application on a large scale in the very near future is not to be expected. However, some of the alternatives have good prospects for the long term, if large production capacities will become available [62].

References

1 Gary JH and Handwerk GE (1994) *Petroleum Refining, Technology and Economics*, 3rd ed., Marcel Dekker, New York.
2 Meyers RA (ed.) (1997) *Handbook of Petroleum Refining Processes*, 2nd ed., McGraw-Hill, New York.
3 Ditman JG and Godino RL (1965) 'Propane extraction: a way to handle residue' *Hydroc. Process.* 44(9) 175–178.
4 Bousquet J and Valais M (1996) 'Trends and constraints of the European refining industry' *Appl. Catal. A: General* 134 N8–N18.
5 Sadeghbeigi R (1995) *Fluid Catalytic Cracking Handbook; Design, Operation, and Trouble-shooting of FCC Facilities*, Gulf Publishing, Houston, TX.
6 Wilson JW (1997) *Fluid Catalytic Cracking: Technology and Operation*, PennWell Publishing, Tulsa, OK.
7 Magee JS and Mitchell Jr. MM (eds.) (1993) 'Fluid catalytic cracking: science and technology' *Studies in Surface Science and Catalysis*, vol. 76, Elsevier, Amsterdam.
8 Sie ST (1992) *Petroleum Conversie*, lecture course, Delft Technical University (in Dutch).
9 Rollman LD and Valyocisk EW (1981) *Inorganic Syntheses*, vol. 22, Wiley, New York, p. 61.
10 Biswas J and Maxwell IE (1990) 'Recent process- and catalysis-related developments in fluid catalytic cracking' *Appl. Catal.* 63, 197–258.
11 Ambler PA, Mine BJ, Berruti F and Scott DS (1990) 'Residence time distribution of solids in a circulating fluidized bed: experimental and modelling studies' *Chem. Eng. Sci.* 45(8) 2179–2186.
12 Martino G, Courty P and Marcilly C (1997) 'Perspectives in oil refining', in: Ertl G, Knözinger H and Weitkamp J (eds.) *Handbook of Heterogeneous Catalysis*, VCH, Weinheim, pp. 1801–1818.
13 Rheaume L and Ritter RE (1988) 'Use of catalysts to reduce SO_x emissions from fluid catalytic cracking units' in: Occelli ML (ed.) *Fluid Catalytic Cracking, Role in Modern Refining: ACS Symposium Series*, vol. 375, pp. 146–161.
14 Soud H (1995) *Suppliers of FGD and NO_x control systems*, IEA Coal Research, London.
15 Von Ballmoos R, Harris DH and Magee JS (1997) 'Catalytic cracking' in: Ertl G, Knözinger H and Weitkamp J (eds.) *Handbook of Heterogeneous Catalysis*, VCH, Weinheim, pp. 1955–1986.
16 Sie ST (1994) 'Past, present and future role of microporous catalysts in the petroleum industry' in: Jansen JC, Stöcker M, Karge HG and Weitkamp J (eds.) *Advanced Zeolite Science and*

Applications' Studies in Surface Science and Catalysis, vol. 85, Elsevier, Amsterdam, pp. 587–631.

17 Little DM (1985) *Catalytic Reforming*, PennWell Publishing, Tulsa, OK.

18 Speight JG (1981) *The Desulfurization of Heavy Oils and Residua*, Marcel Dekker, New York.

19 Reinhoudt HR, Troost R, Van Langeveld AD, Sie ST, Van Veen JAR and Moulijn JA (1999) 'Catalysts for second-stage deep hydrodesulfurisation of gas oils' *Fuel Process. Tech.* 61 133–147.

20 Reinhoudt HR (1999) *The Development of Novel Catalysts for Deep Hydrodesulfurisation of Diesel Fuel* PhD Thesis, Delft University of Technology.

21 Maxwell IE, Minderhoud JK, Stork WHJ and Van Veen JAR (1997) 'Hydrocracking and catalytic dewaxing', in: Ertl G, Knözinger H and Weitkamp J (eds.) *Handbook of Heterogeneous Catalysis*, VCH, Weinheim, pp. 2017–2038.

22 Mohanty S, Kunzru D and Saraf, DN (1990) 'Hydrocracking: a review' *FUEL* 69, 1467–1473.

23 Gosselink JW and Van Veen JAR (1999) 'Coping with catalyst deactivation in hydrocarbon processing' in: Delmon B and Froment GF (eds.) *Catalyst Deactivation 1999, Studies in Surface Science and Catalysis*, vol. 126, Elsevier, Amsterdam, pp. 3–16.

24 Bridge AG, Jaffe J, Powell BE and Sullivan RF (1993) 'Isocracking heavy feeds for maximum middle distillate production' *API Meeting*, Los Angeles, CA, May 1993.

25 '1970 Refining Process Handbook' (1970) *Hydroc. Process.* 49(9) 183.

26 Chauvel A and Lefebvre G (1989) *Petrochemical Processes. Part 1. Synthesis-gas Derivatives and Major Hydrocarbons*, (Chapter 2) Éditions Technip, Paris.

27 Corma A and Martínez A (1993) 'Chemistry, catalysts, and processes for isoparaffin-olefin alkylation: actual situation and future trends' *Catal. Rev. Sci. Eng.* 35(4) 483–570.

28 Lerner H and Citarella VA (1991) 'Improve alkylation efficiency' *Hydroc. Process.* 70(11) 89–94.

29 Weitkamp J and Traa Y (1997) 'Alkylation of isobutane with alkenes on solid acids' in G. Ertl, H. Knözinger and J. Weitkamp (eds.) *Handbook of Heterogeneous Catalysis*, VCH, Weinheim, pp. 2039–2069.

30 Gray MR (1994) *Upgrading Petroleum Residues and Heavy Oils*, Marcel Dekker, New York.

31 Bonné RLC, Van Steenderen P, Van Langeveld AD and Moulijn JA (1995) 'Hydrodemetalization kinetics of nickel tetraphenylporphyrin over Mo/Al_2O_3 catalysts' *I&EC Research* 34 3801–3807.

32 Janssens JP, Bezemer BJ, Van Langeveld AD, Sie ST and Moulijn JA (1994) 'Catalyst deactivation in hydrodemetallisation of model compound vanadyl-tetraphenylporphyrin' in: Delmon B and Froment GF (eds.) *Catalyst Deactivation 1994, Studies in Surface Science and Catalysis*, vol. 88, Elsevier, Amsterdam, pp. 335–342.

33 Quann RJ, Ware RA, Hung C and Wei J (1988) 'Catalytic hydrodemetallation of petroleum' *Adv. Chem. Eng.* 14 95–259.

34 Van Driesen RP and Fornoff LL (1985) 'Upgrade resids with LC-fining' *Hydroc. Process* 64(9) 91–95

35 Kwant PB and Van Zijll Langhout WC (1985) 'The development of Shell's residue hydroconversion process', Paper presented at the Conference on the Complete Upgrading of Crude Oil, Siofok, 25–27 September 1985.

36 Scheffer B, van Koten MA, Röbschläger KW and De Boks, FC (1998) 'The Shell residue conversion process: development and achievements' *Catal. Today* 43 217-224.

37 Wenzel FW (1992) 'VEBA-COMBI-Cracking, a commercial route for bottom of the barrel upgrading' in: Han C and His C (eds.) *Proceedings of the International Symposium on Heavy Oil and Residue Upgrading and Utilization*, International Academic, Beijing, pp. 185–201.

38 Nehr W and Vydra K (1994) 'Sulfur' in: Gerhartz W, et al. (eds.) *Ullman's Encyclopedia of Industrial Chemistry*, vol. A25, 5th ed., VCH, Weinheim, pp. 507–567.

39 Piéplu A, Saur O and Lavalley JC (1998) 'Claus catalysis and H₂S selective oxidation' *Catal.-Rev.-Sci. Eng.* 40(4) 409–450.

40 Parkinson G, Ondrey G and Moore S (1994) 'Sulfur production continues to rise' *Chem. Eng.* June 30–35.

41 Watson AM (1983) 'Use pressure swing adsorption for lowest cost hydrogen' *Hydroc. Proc.* March, 91–95.

42 Rautenbach R and Albrecht R (1989) *Membrane Processes*, Wiley, New York, pp. 422–454.

43 Absi-Halabi M, Stanislaus A and Qabazard H (1997) 'Trends in catalysis research to meet future refining needs' *Hydroc. Process.* 76(2) 45–55.

44 Courty PR and Chauvel A (1996) 'Catalysis, the turntable for a clean future' *Catal. Today* 29 3–15.

45 Martino G, Courty PR and Marcilly C (1997) 'Perspectives in oil refining' in: Ertl G, Knözinger H and Weitkamp J (eds.) *Handbook of Heterogeneous Catalysis*, VCH, Weinheim, pp. 1801–1818.

46 Weisz PB (1980) 'Molecular shape selective catalysis' *Appl. Chem.* 52, 2091–2103.

47 Van de Graaf, JM (1999) *Permeation and Separation Properties of Supported Silicalite-1 Membranes – A modeling Approach*, Phd Thesis, Delft University of Technology.

48 Maxwell IE, Naber JE and De Jong KP (1994) 'The pivotal role of catalysis in energy related environmental technology', *Appl. Catal. A: General* 113 153–173.

49 Mooiweer HH, De Jong KP, Kraushaar-Czarnetzki B, Stork WHJ and Krutzen BCH (1994) 'Skeletal isomerisation of olefins with the zeolite Ferrierite as catalyst' in: Weitkamp J, Karge HG, Pfeifer H and Hölderick W (eds.) *Zeolites and Related Microporous Materials: State of the Art 1994, Studies in Surface Science and Catalysis*, vol. 84, Elsevier, Amsterdam, pp. 2327–2334.

50 Houzvicka J and Ponec V (1997) 'Skeletal isomerization of *n*-butene' *Catal. Rev.-Sci. Eng.* 39(4) 319–344.

51 Powers DH (1993) *Presentation O-1 at the 1993 DEWITT Petrochemical Review*, Houston, TX, March 23–25.

52 Janssen MJG, Mortier WJ, Van Oorschot CWM and Makkee M (Exxon Chemicals) (1991) 'Process for catalytic conversion of olefins' WO 91/18851.

53 Sie ST (1997) 'Isomerization reactions' in: Ertl G, Knözinger H and Weitkamp J (eds.) *Handbook of Heterogeneous Catalysis*, VCH, Weinheim, pp. 1998–2017.

54 Derouane EG (1984) 'Conversion of methanol to gasoline over zeolite catalysts. I. Reaction mechanisms' in: Ribeiro FR, Rodrigues AE, Rollmann LD and Naccache C (eds.) *Zeolites: Science and Technology*, Martinus Nijhoff, The Hague, pp. 515–528.

55 MacDougall LV (1991) 'Methanol to fuel routes – the achievements and remaining problems' *Catal. Today* 8 337–369.

56 Chang CD and Silvestri AJ (1977) 'The conversion of methanol and other O-compounds to hydrocarbons over zeolite catalysts' *J. Catal.* 47 249–259.

57 Blauwhoff PMM, Gosselink JW, Kieffer EP, Sie ST and Stork WHJ (1999) 'Zeolites as Catalysts in Industrial Processes' in: Weitkamp J and Puppe L (eds.) *Catalysis and Zeolites: Fundamentals and Applications*, Springer-Verlag, Berlin, pp. 437–537.

58 Gabelica Z (1984) 'Conversion of methanol to gasoline over zeolite catalysts. II. Industrial processes' in: Ribeiro FR, Rodrigues AE, Rollmann LD and Naccache C (eds.) *Zeolites: Science and Technology*, Martinus Nijhoff, The Hague, pp. 529–544.

59 Garwood WE (1983) 'Conversion of C₂-C₁₀ to higher olefins over synthetic zeolite ZSM-5' in: Comstock MJ (ed.) *Intrazeolite Chemistry, ACS Symposium Series* 218, pp. 383–396.

60 Tabak SA and Krambeck FJ (1985) 'Shaping process makes fuels' *Hydroc. Process.* 64(9) 72–74.

61 Neeft JPA, Makkee M and Moulijn JA (1996) 'Diesel particulate emission control' *Fuel Process. Technol.* 47 1–69.

62 Van Diepen AE, Makkee M and Moulijn JA (1999) 'Emission control from mobile sources: Otto and diesel engines' in: Janssen FJJG and Van Santen RA (eds.) *Environmental Catalysis*, Imperial College Press, London, pp. 257–291.

Chapter 4

Steam Cracking: Production of Lower Alkenes

The discovery that alkenes can be produced in high yields from both alkanes present in natural gas and from petroleum fractions has laid the foundation for what is now known as the petrochemical industry. The principal process used to convert the relatively unreactive alkanes into much more reactive alkenes is thermal cracking, often referred to as 'steam cracking'. This process produces mainly ethene, but valuable co-products such as propene, butadiene, and pyrolysis gasoline (*pygas*), with benzene as a major constituent, are also produced.

Since the late 1930s, when the petrochemical industry started to take shape, ethene has almost completely replaced coal-derived ethyne and now is the largest volume building block for the petrochemical industry. Its main outlet is the production of polyethenes (>50%), followed by vinyl chloride, ethene oxide, and ethylbenzene.

4.1 INTRODUCTION

In steam cracking a hydrocarbon stream is thermally cracked in the presence of steam, yielding a complex product mixture. The name steam cracking is slightly illogical: under the reaction conditions steam is not cracked but it functions primarily as a diluent, allowing higher conversion. A more accurate description of the process might be 'pyrolysis', which stems from Greek and means bond breaking by heat (also see Sections 3.3 and 5.3). The main product of steam cracking is ethene, which is a very important base chemical.

Feedstocks range from light saturated hydrocarbons such as ethane and propane to naphtha and light and heavy gas oils [1]. In the USA ethane (from natural gas) is the primary feedstock for the production of ethene. In Europe and Japan, by contrast, naphtha is the main feedstock (see Figure 4.1). The latter explains why steam cracking is frequently referred to as naphtha cracking.

From Figure 4.1 it can be seen that most steam cracking raw materials are also feedstocks for fuel production, e.g. naphtha is also converted into gasoline in the catalytic reforming process (see Section 3.4). Furthermore, by-products from steam cracking, such as 'pyrolysis gasoline', usually find their destination in liquid fuels. Therefore, steam cracking is intimately connected with oil refinery operations.

Thermal cracking processes such as steam cracking are key processes in the oil and petrochemical industry. Analogous processes have been discussed in Chapter 3, viz. Visbreaking, Delayed Coking, and Flexicoking. In all these processes the reactions occurring are radical reactions leading to cracking and dehydrogenation. Steam crack-

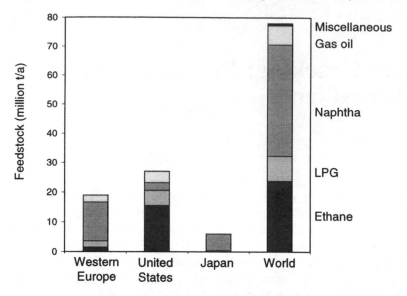

Figure 4.1 Breakdown of steam cracker feedstocks in 1997; miscellaneous: ethanol (Brazil, India) and coal derived gas (Poland, South Africa).

ing, however, occurs at a much higher temperature than the other processes, and in the presence of large amounts of steam.

4.2 CRACKING REACTIONS

Steam cracking yields a large variety of products, ranging from hydrogen to fuel oil. The product distribution depends on the feedstock and on the processing conditions. These conditions are determined by thermodynamic and kinetic factors.

4.2.1 Thermodynamics

In general, lower alkenes, especially ethene, propene, and butadiene are the desired products of steam cracking. Treating of lower alkanes such as ethane, propane, and butane by steam cracking, results in dehydrogenation of the alkanes to form the corresponding alkenes and hydrogen. Figure 4.2 shows the equilibria for the dehydrogenation of lower alkanes.

Figure 4.2 indicates that from a thermodynamic viewpoint the reaction temperature should be high for sufficient conversion. The forward reaction is also favored by a low partial pressure of the alkanes, because for every molecule converted two molecules are formed. A process under vacuum would be desirable in this respect. In practice, it is more convenient to apply dilution with steam, which has essentially the same effect.

Figure 4.2 also shows that the smaller the alkane the higher the temperature has to be for a given conversion.

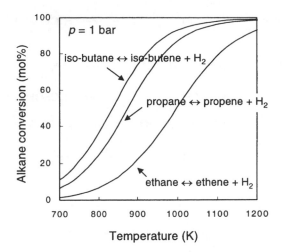

Figure 4.2 Equilibrium conversion in the dehydrogenation of the most important lower alkanes at 1 bar as a function of temperature; corresponding reaction enthalpies at standard conditions are 137.1, 124.4, and 117.7 kJ/mol for ethane, propane, and isobutane dehydrogenation, respectively.

QUESTIONS:

What can you conclude about optimal reaction conditions?
Why is the process not carried out as a catalytic process? Compare steam cracking with FCC.
Compare the reaction conditions used in the various thermal cracking processes discussed in this book. Explain the differences.

4.2.2 Mechanism

Cracking occurs by free-radical reactions (see Scheme 4.1). The simplest feedstock is ethane. The reaction is initiated by cleavage of the C–C bond in an ethane molecule resulting in the formation of two methyl radicals. Propagation proceeds by reaction of a methyl radical with an ethane molecule resulting in the production of methane and an ethyl radical. The ethyl radical subsequently decomposes into ethene and a hydrogen radical, which then attacks another ethane molecule, etc. These latter two reactions dominate in the cracking of ethane, which explains why ethene can be obtained in high yields. Termination results from the reaction of two radicals forming either a saturated molecule or a saturated molecule and an unsaturated molecule.

Scheme 4.1 shows that pyrolysis of ethane besides ethene as the main product also produces methane and hydrogen. Small quantities of heavier hydrocarbons can also be formed as result of the reaction of two radicals.

Similar, although more complex, networks apply to thermal cracking of higher alkanes (Scheme 4.2). The primary cracking products may undergo secondary reactions like further pyrolysis, dehydrogenation, and condensation (combination of two or more smaller fragments to produce aromatics). These secondary reactions may also lead to the formation of coke. Coke is always formed, also when lighter

Initiation: H_3C-CH_3 $\longrightarrow$ $H_3\overset{\cdot}{C}$ + $H_3\overset{\cdot}{C}$

Propagation: $H_3\overset{\cdot}{C}$ + H_3C-CH_3 $\longrightarrow$ CH_4 + $H_3C-\overset{\cdot}{C}H_2$

                     $H_3C-\overset{\cdot}{C}H_2$ $\longrightarrow$ $H_2C=CH_2$ + $H^{\cdot}$

                     $H^{\cdot}$ + H_3C-CH_3 $\longrightarrow$ H_2 + $H_3C-\overset{\cdot}{C}H_2$

                     $H_3C-\overset{\cdot}{C}H_2$ $\longrightarrow$ etc.

Termination: $H^{\cdot}$ + $H^{\cdot}$ $\longrightarrow$ H_2

                     $H_3C-\overset{\cdot}{C}H_2$ + $H_3\overset{\cdot}{C}$ $\longrightarrow$ $H_2C=CH_2$ + CH_4

                     etc.

Scheme 4.1   Mechanism of ethane dehydrogenation/cracking.

alkanes are used as feedstock. However, the heavier the feedstock, the more coke is formed.

4.2.3 Kinetics

The first-order kinetics implies the rate of reaction of alkanes. Figure 4.3 shows rate constants for the cracking of a number of alkanes as a function of temperature. The reactivity increases with chain length. Ethane clearly shows the lowest reactivity.

The first-order kinetics imply that the rate of reaction (but not the conversion) increases with increasing partial pressure of the reactants. However, at higher partial pressures of the reactants, and thus also of the products, unfavorable secondary reac-

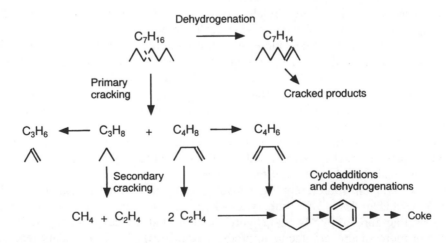

Scheme 4.2   Examples of reactions occurring during thermal cracking of higher alkanes. After [2].

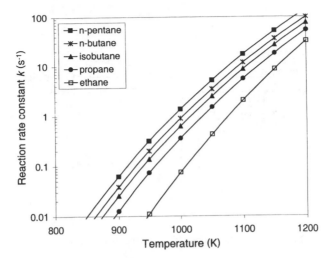

Figure 4.3 Reaction rate constants of various alkanes. Based on data from [3].

tions, such as condensation reactions and the formation of coke, occur more frequently. Hence, the partial pressure of the hydrocarbons must be kept low. For the same reason, conversions should not be too high.

QUESTIONS:

> *Estimate the optimal reaction conditions for the cracking of the alkanes referred to in Figure 4.3. Choose a conversion of 60% and assume isothermal plug flow. Do you expect that the reaction order for the secondary reactions is also 1? Explain your answer.*

From the calculation in the question above it appears that a given conversion corresponds to an infinite number of time–temperature (rate constant) combinations. However, thermodynamics determine the required temperature, and hence the residence time. In particular the ethane–ethene equilibrium calls for a temperature as high as possible.

4.3 INDUSTRIAL PROCESS

From the previous discussion a number of requirements concerning the steam cracking process can be derived:

- considerable heat input at a high temperature level;
- limitation of hydrocarbon partial pressure;
- very short residence times (<1 s);
- rapid quench of the reaction product to preserve the composition.

QUESTIONS:

> *Explain these requirements. Select the optimal temperature.*

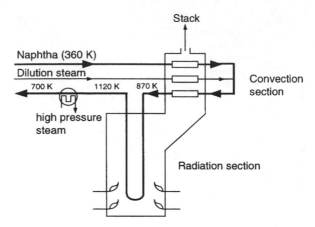

Figure 4.4 Simplified flow scheme of a steam cracker.

In industrial practice these requirements are met in the following way: a mixture of hydrocarbons and steam is passed through tubes placed inside furnaces heated by combustion of natural gas, LPG, or fuel oil. The furnaces consist of a convection section in which the hydrocarbon feed and the steam are preheated and a radiation section in which the reactions take place. The hydrocarbons undergo pyrolysis, and subsequently the products are rapidly quenched to prevent further reaction, and thus preserve the composition. The residence time in the tubes is very short (<1 s). The temperature is chosen as high as possible in the sense that the material properties determine the ceiling temperature. Figure 4.4 shows a simplified flow scheme of a steam cracker for the cracking of naphtha.

4.3.1 Influence of Feedstock on Steam Cracker Operation and Products

Steam cracking yields a complex product mixture. In practice it is crucial to choose the reaction conditions in such a way that the product distribution is optimal. Often, this means that the amount of ethene produced should be as high as possible. Depending on the market developments and the local situation, however, other products might be more desirable.

The cracking process will now be discussed in more detail starting from product distributions calculated for a commercial cracking unit, operating at various conditions (conversion, steam ratio) and with two different feedstocks, viz. ethane and naphtha.

The product distribution in ethane cracking is determined solely by the process parameters, such as temperature, pressure, and steam-to-ethane ratio. For naphtha no general product distribution can be given because 'naphtha' is not a single compound and its composition varies with source and refinery operating conditions. A naphtha with the composition given in Table 4.1, yields the product distribution shown in Table 4.2. The outlet pressure in all cases was 2.05 bar. The inlet temperature for the two feedstocks was chosen differently. Table 4.2 shows product yields, obtained at three different sets of conditions (A, B, C for ethane, and D, E, and F for naphtha). Table 4.3 gives the associated process parameters.

Table 4.1 Composition of naphtha feed for yield patterns in Table 4.2. Adapted from [3]

Component	Analysis (wt% of total)	ASTM D 86 distillation	
		vol%	K
C_3, C_4	8.0	bp_i[a]	306
C_5	22.4	5	320
C_6	19.9	10	325
C_7	17.2	20	334
C_8	12.4	30	344
C_9	11.5	40	355
C_{10} - C_{15}	8.6	50	369
Total	100.0	60	381
		70	397
n-Alkanes	30.5	80	418
iso-Alkanes	39.9	90	460
Naphthenes	17.7	bp_f[b]	480
Aromatics	11.9		
Total	100.0		

[a] Initial boiling point.
[b] Final boiling point.

Table 4.2 Example of commercial ethane (A, B, C) and naphtha (D, E, F) cracking yield patterns. Adapted from [3]

	Ethane A	Ethane B	Ethane C	Naphtha D	Naphtha E	Naphtha F
Conversion (wt%)	53.9	69.0	69.3	81.47[a]	94.19[a]	92.04[a]
Outlet temperature (K)	1073	1106	1107	1053	1073	1073
Steam ratio (kg steam/ kg feed)	0.30	0.30	0.45	0.50	0.50	0.75
Yield (wt%)						
Hydrogen	3.35	4.21	4.27	0.54	0.72	0.71
Methane	3.08	6.21	5.64	11.98	15.26	14.21
Ethyne	0.14	0.32	0.38	0.09	0.25	0.28
Ethene	42.50	50.10	51.45	19.46	23.52	24.00
Ethane	46.00	30.93	30.60	3.97	3.95	3.40
C_3H_4[b]	0.01	0.02	0.02	0.31	0.64	0.68
Propene	1.41	1.67	1.55	16.15	16.15	15.50
Propane	0.16	0.22	0.20	0.56	0.50	0.45
Butadiene	0.89	1.41	1.47	3.73	3.95	4.28
Butenes	0.23	0.24	0.23	7.93	5.44	6.05
Butanes	0.33	0.25	0.24	2.63	1.37	1.65
C_5 – 478 K[c]	1.82	3.94	3.57	30.19	26.04	25.78
Fuel oil	0.08	0.48	0.38	2.46	3.40	2.95
Total	100.00	100.00	100.00	100.00	100.00	100.00

[a] Conversion calculations based on total pentanes.
[b] Methyl-ethyene and propadiene.
[c] Fraction boiling from 301 K (isopentane) to 478 K, pyrolysis gasoline.

Table 4.3 Process parameters for the yield patterns in Table 4.2. Adapted from [3]

	Ethane A	Ethane B	Ethane C	Naphtha D	Naphtha E	Naphtha F
Time in coil (s)	0.737	0.704	0.627	0.537	0.519	0.432
Time to quench (s)	0.050	0.046	0.041	0.030	0.027	0.022
HC partial pressure						
Coil outlet (bar)	1.54	1.57	1.41	1.04	1.10	0.89
Quench point (bar)	1.49	1.52	1.36	1.01	1.06	0.85
Temperature (K)						
Tube wall, max.	1213	1268	1270	1174	1219	1223
Gas inlet	922	922	922	863	863	863

The naphtha composition expressed as the percentages of alkanes, naphthenes, and aromatics present has a pronounced effect on the pyrolysis yield. This can be explained from their different reactivities and hydrogen-to-carbon ratios. Alkanes are relatively easy to pyrolyze and when decomposed produce high yields of light products such as ethene and propene as a result of their high hydrogen content. Aromatics, on the other hand, are very stable and they have a low hydrogen-to-carbon ratio so their yield of light cracking products is negligible. Naphthenes are intermediate between alkanes and aromatics with regard to pyrolysis behavior.

QUESTIONS:

> *Why were the inlet temperatures for the two feedstocks chosen differently? Interpret the data in the tables: compare the reaction conditions and compare ethane and naphtha.*

The data in the example clearly indicate that reaction conditions are very critical and that it pays to operate the plant as close as possible to optimal conditions. These conditions depend on the feedstock and on the desired products. It is not surprising that in practice extensive simulation programs are used. Commercially available simulation packages contain several hundreds of chemical reactions.

An obvious difference between ethane and naphtha cracking is the much wider product distribution of the latter. From ethane, ethene can be produced with a selectivity of over 70 wt%, with hydrogen and methane the second most important products (note: selectivity is defined as mole product/mole feed converted, yield as mole product/mole feed fed). In naphtha cracking the selectivity towards ethene is only about 25 wt%, while also large proportions of methane and propene are formed.

As a result of its stability ethane requires higher temperatures and longer residence times than naphtha to achieve acceptable pyrolysis conversion. The simulation shows that the reactor is far from isothermal (compare calculated inlet temperature with outlet temperature). The feed inlet temperature ranges from about 850 to 950 K depending on the feedstock, while the outlet temperature is between 1050 and 1200 K. Figure 4.5 shows typical temperature profiles.

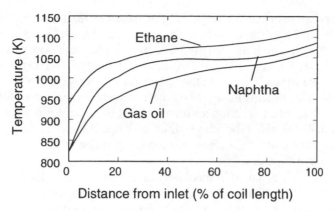

Figure 4.5 Typical reactor temperature profiles in modern steam crackers. Adapted from [3]. Used with permission of Wiley-VCH.

QUESTION:

> *Explain the logic of the temperature profiles in Figure 4.5.*

4.3.2 Cracking Furnace

Figure 4.6 shows a cross section of a typical cracking furnace. The convection section

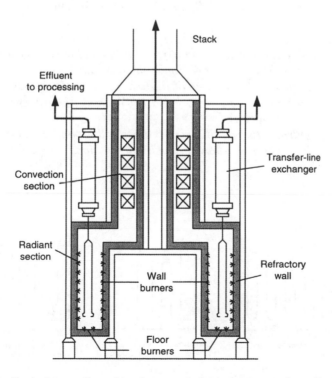

Figure 4.6 Typical two-cell cracking furnace configuration (top outlet coil). After [3,4].

contains various zones used for preheating of the steam and the feed. In the radiant section, which can be as high as 15 m, the actual reactions take place.

The reactor tubes, which are usually hung in a single plane down the center of the furnace, vary widely in diameter (30–200 mm) and length (10–100 m), depending on the production rate for each tube and the rate of coke deposition. For a given total feed rate a furnace either contains many short small-diameter tubes or a smaller number of long large-diameter tubes. The longer tubes consist of straight tubes connected by U-bends. Depending on the tube dimensions and the desired furnace capacity, the number of tubes (or coils) in a single furnace may range from 2 to 180. Figure 4.7 shows some typical tube designs.

The product gas is quenched immediately after leaving the radiant zone in order to retain its composition. The heat exchangers, called transfer-line exchangers, are mounted directly on the outlet line from the furnaces and usually serve several coils.

A typical two-cell cracking furnace (two furnaces sharing one stack) has a capacity of 50,000 to over 100,000 t/a of ethene. With the present plant sizes of about 500,000–800,000 t/a, a plant based on naphtha typically consists of ten naphtha crackers and one or two ethane crackers, in which recycled ethane is processed.

Figure 4.8 shows an artist's impression of a commercial steam cracker, which indicates the size of the furnace.

During past decades, significant improvements have been made in the design and operation of cracking furnaces. The use of better alloys for the cracking coils has enabled operation at higher temperatures (up to 1300 K), which permits operation at shorter residence times and higher capacity [4]. For instance, the residence time has decreased from 0.5–0.8 s in the 1960s to 0.1–0.15 s in the late 1980s [6].

QUESTIONS:

> *Why are the tubes hanging from the ceiling? Compare the configurations in Figure 4.7; give the pros and cons.*
>
> *Estimate the number of burners per furnace (capacity of a burner normally is between 0.1 and 1 MW, modern burners might have a 10 MW capacity).*
>
> *During start-up the reactor tubes will expand. Estimate the increase in the coil length. Are special precautions necessary?*

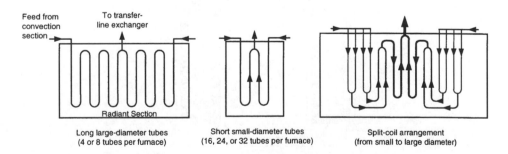

Figure 4.7 Cracking furnace tube designs.

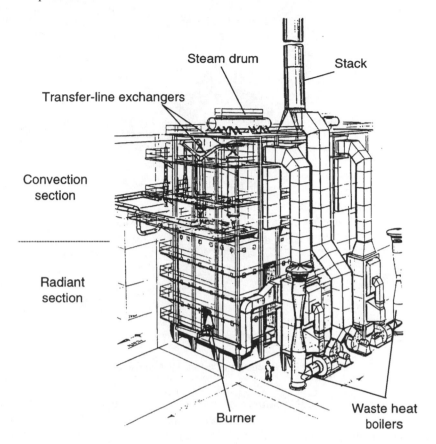

Figure 4.8 Artist's impression of a commercial steam cracker [5].

4.3.3 Heat Exchanger

The cracked gases usually leave the reactor at a temperature exceeding 1070 K. They should be instantaneously cooled to prevent consecutive reactions. Quenching can be direct or indirect or a combination of both. Direct quenching involves injection of a liquid spray, usually water or oil, and cooling can be very fast. However, indirect cooling by *transfer-line exchangers (TLE)*, see Figure 4.9, is common practice, because it allows the generation of valuable high-pressure steam. The generated steam is used in the refrigeration compressors and the turbines driving the cracked gas (see Section 4.4).

In designing transfer-line exchangers the following points are of concern:

- minimum residence time in the section between furnace outlet and TLE;
- low pressure drop;
- high heat recovery. The outlet temperature for liquid cracking should not be too low (condensation of heavy components);
- acceptable run times (fouling occurs by coke deposition and condensation of heavy components).

Usually a transfer-line exchanger is used, followed by direct quenching. In ethane

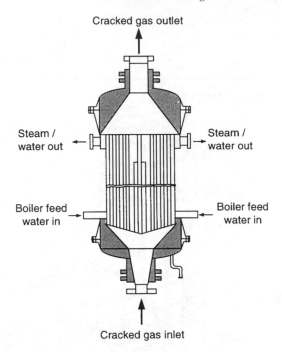

Figure 4.9 Schematic of a transfer-line exchanger.

cracking, where only small amounts of fuel oil are produced, cooling is mainly accomplished in the TLE. During steam cracking of heavy naphtha or gas oil, in which large amounts of fuel oil are produced, the heat exchange area of the TLE is reduced and most of the cooling is performed by direct quench. In the extreme case of vacuum gas oils the TLEs are totally eliminated and supplanted by direct quench.

QUESTIONS:

> *Minimization of pressure drop is important for energy saving and for product quality. Why?*
> *Discuss advantages and disadvantages of direct cooling by water injection.*
> *Why are no TLEs used when the feedstock is a vacuum gas oil?*

4.3.4 Coke

During pyrolysis coke is deposited on the walls of the cracking coils and the heat exchanger tubes. This causes very undesirable phenomena such as reduced heat-transfer rates, increased pressure drop, lowered yields, and reduced selectivity towards alkenes.

QUESTIONS:

> *Explain these phenomena. Does the temperature inside the reactor increase or decrease by coke deposition?*
> *When comparing coke deposited in different processes what do you*

> *expect about the reactivity? (compare coke deposition in FCC, hydrotreating, and ethane cracking).*

As a consequence of these issues frequent interruption of furnace operation is necessary to remove the deposited coke. The interval between decoking operations in the majority of furnaces is 20–60 days for gaseous feedstocks and 20–40 days for liquid feedstocks, depending on the feedstock, the cracking severity (conversion), and special measures taken to reduce coke formation [7,8]. In industrial practice, this means that at least one furnace is always undergoing decoking.

Decoking of the cracking coils takes place by gasification in an air/steam mixture or in steam only. This operation takes from about 5 to 30 h [2]. During decoking of the reactor coils, the TLE is also partly decoked. The procedure for complete decoking of the TLE is usually shutting down the furnace, disconnecting the TLE from the coil and removing the coke mechanically or hydraulically.

Catalytic gasification of coke during production

It is very attractive to avoid coke formation during the pyrolysis reactions. In principle this would be possible when a catalytic coating could be applied that reacts the coke away, for example by steam gasification. Recently, it has been claimed that a certain ceramic coating performs this highly desired catalytic function. When this claim would be justified, it would mean a large cost saving, in particular due to reduced down-time.

4.4 PRODUCT PROCESSING

Figures 4.10 and 4.11 show typical conventional (simplified) flow schemes for downstream processing of the cracked gas. For simplicity, cracked gas processing is described either as gas or liquid feedstock based. Modifications of these schemes are possible and amongst others depend on the type of feedstock and the degree of recovery desired for the different products.

Downstream processing is based on both normal and cryogenic distillation. The cracked gas from gaseous feedstocks (ethane, propane, butane) begins with its entry into the transfer-line exchanger followed by direct quench with water, and multistage compression, typically in four to six stages with intermediate cooling. Before the last compressor stages, acid gas (mainly hydrogen sulfide and carbon dioxide) is removed. After the last compressor stage, water removal takes place by chilling and drying over zeolites. Subsequent fractionation of the cracked gases is based on cryogenic (temperature <273 K) and conventional distillation under pressure (15–35 bar). Cryogenic distillation requires a lot of energy due to the need for a refrigeration system.

In Figures 4.10 and 4.11, ethyne and C_3H_4 compounds are converted to ethene and propene, respectively, by selective hydrogenation in catalytic fixed-bed reactors. Occasionaly, ethyne is produced and sold for welding. The separation of ethene/ethane and propene/propane is difficult: in both cases over 100 trays are needed (often the trays are placed in two towers in series).

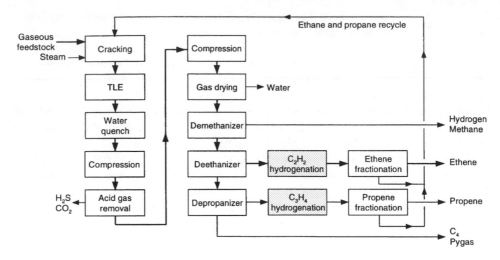

Figure 4.10   Simplified process flow diagram for producing ethene via gas cracking [9].

Downstream processing of cracked gas from liquid feedstocks (naphtha and gas oil) is somewhat more complex because heavier components are present. In this case a primary fractionator is installed upstream the compressor stages in order to remove fuel oil, which is produced in considerable amounts when cracking naphtha or gas oil. This fuel oil is partly used in the direct quench operation, while the remainder may be sold or used as fuel for the cracking furnaces. The primary fractionator also produces a gasoline sidestream, called *pyrolysis gasoline* or *pygas*. C_4 hydrocarbons are produced in sufficient quantities to justify their separate recovery, and therefore usually a debutanizer is installed.

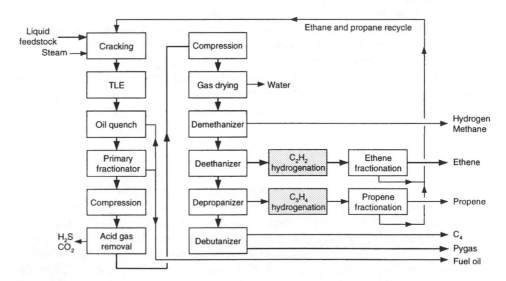

Figure 4.11   Simplified process flow diagram for producing ethene via liquid cracking [9].

Recycled ethane and propane are normally cracked in separate furnaces, rather than including them in liquid cracking furnaces.

QUESTIONS:

Think of alternatives for cryogenic distillation. What are advantages and disadvantages of these alternatives?

Why would separate furnaces be used for naphtha and recycle ethane?

Why is acid gas removal carried out prior to the final compressor stage?

Ethyne and propyne are reduced into the corresponding alkenes in a selective catalytic reduction process. Pyrolysis gasoline is also processed in a selective hydrogenation reactor in order to improve the product quality. What would be the reason? Why is this process carried out in a multi-phase process (gas-liquid) whereas the selective reduction of ethyne and propyne takes place as a gas-phase process (also in the presence of a solid catalyst)?

4.5 CURRENT AND FUTURE DEVELOPMENTS

Steam cracking plants have developed in an evolutionary way. Capital costs dominate and, as a consequence, plant capacities have increased over the years. The capacity of the plants built in the 1970s and 1980s have doubled, compared to the 1960s. De-bottlenecking has also played a role. For example, tower capacities were increased by the application of alternative internals, viz. random or structured packings.

In addition, fundamentally different processes have received attention. The most obvious disadvantage of the steam cracking process is its relatively low selectivity for the desired products. A wide range of products is produced with limited flexibility. This inheres in the non-catalytic nature of the process. In this section alternative feedstocks and processes are presented for the production of the lower alkenes.

4.5.1 Selective Dehydrogenation

In recent years the demand for propene and butenes has increased much more rapidly than the demand for ethene and this situation is expected to continue [10]. In addition, the growth of refinery sources (e.g. from FCC, Section 3.4.1) for these products have not kept pace with this growing demand. Moreover, cracking furnaces are very capital intensive and it is expected that in the next decades many have to be replaced because of their age. Therefore, direct production methods for these specific alkenes receive increasingly more attention. A logical choice of process is the catalytic selective dehydrogenation of the corresponding alkanes.

Alkane dehydrogenation is highly endothermic and high temperatures are required (see Figure 4.2). However, at these high temperatures secondary reactions such as cracking and coke formation become appreciable. Deposition of coke on the catalyst makes regeneration necessary. Furthermore, the thermodynamic equilibrium limits the conversion per pass so that a substantial recycle stream is required.

Several catalytic dehydrogenation processes are commercially available, which

Commercial Alkane Dehydrogenation Processes

Figure B.4.1 shows four alternative alkane dehydrogenation processes and Table B.4.1 compares their performance. The *Oleflex* process [11] uses radial-flow moving-bed reactors in series with interstage heaters. The process is operated in the same way as the UOP continuously- regenerative reforming process (see Section 3.4.4).

A key feature of the *Catofin* process is the principle of storing the needed reaction heat in the catalyst bed. Reactors (usually three to eight in parallel) are alternately on stream for reaction and off stream for regeneration (see also Section 3.7, MTG process). During regeneration by burning the coke off the catalyst, the bed is heated, while this heat is consumed in the endothermic dehydrogenation reaction during the on-stream period. Historically this technology is not new. It has been applied extensively in coal gasification, where coal beds were fed alternatively with air (exothermic) and steam (endothermic).

The *Steam Active Reforming* (STAR) process uses multiple tubular reactors in a firebox similar to the steam reformer used for the production of synthesis gas (see Section 5.2). In this process a similar reaction-regeneration procedure is followed as in the Catofin process.

The *FBD-4* design is similar to older FCC units (see Section 3.4.1) with continuous catalyst circulation between a fluidized-bed reactor and regenerator.

QUESTION:

Compare the four processes. Give pros and cons.

Table B.4.1 Performance of propane and isobutane dehydrogenation processes [11–13]; P = propane dehydrogenation, IB = isobutane dehydrogenation

Process (licensor)	Reactor	Catalyst life (a)	T (K)/p (bar)	Conversion (%)	Selectivity (%)
Oleflex (UOP)	Adiabatic moving beds in series with intermediate heating	Pt/Al$_2$O$_3$, 1–3	820–890 2	P: 25 IB: 35	P: 89–91 IB: 91–93
Catofin (ABB Lummus crest)	Parallel adiabatic fixed beds with swing reactor	Cr/Al$_2$O$_3$, 1–2	860–920 0.5	P: 48–65 IB: 60–65	P: 82–87 IB: 93
STAR (Phillips Petroleum)	Tubular reactors in furnace	Pt-Sn/Zn-Al$_2$O$_3$, 1–2	750–890 5	P: 30–40 IB: 40–55	P: 80–90 IB: 92–98
FBD-4 (Snamprogetti)	Fluidized bed	Cr/Al$_2$O$_3$	820–870 1.3	P: 40 IB: 50	P: 89 IB: 91

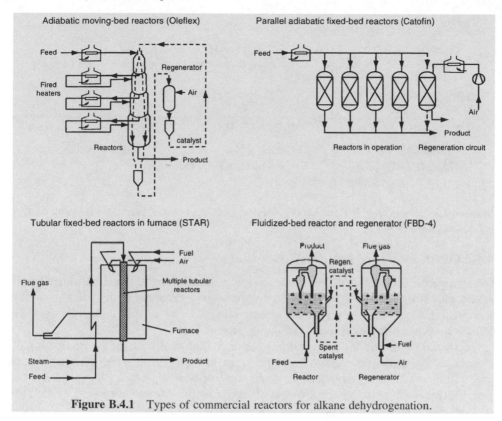

Figure B.4.1 Types of commercial reactors for alkane dehydrogenation.

differ in the type of catalyst used, the reactor design, the method of heat supply, and the method used for catalyst regeneration (see box).

The selectivities in the commercial dehydrogenation processes are much higher than in steam cracking of propane or butane, but the conversions are necessarily lower to prevent excessive side reactions. Still, single-pass yields of 20–40 wt% can be achieved, which is considerably higher than the yields obtained in steam cracking (ca. 15 wt% propene is produced in naphtha cracking (see Table 4.2); propane and butane cracking produce similar yields [3]).

Product processing from dehydrogenation reactors is similar to product processing in a steam cracker plant although less complex as a result of the more concentrated alkene streams. However, as a result of the large recycle streams required, gas compression and purification account for nearly 85% of total capital costs [12]. Understandably, here the largest savings can be made.

In principle, catalytic dehydrogenation of ethane is also possible, but up till now a commercially attractive process has not been realized [6].

QUESTIONS:

> *Why are the conversions low compared to steam cracking?*
> *What properties would the ideal catalyst have?*
> *Why is ethane a more difficult feed than higher alkanes?*

Fischer-Tropsch:

General: $n\,CO + 2n\,H_2 \rightarrow \quad C_nH_{2n} + n\,H_2O$ $\Delta H^o_{298} \ll 0$

Ethene: $2\,CO + 4\,H_2 \rightarrow \quad C_2H_4 + 2\,H_2O$ $\Delta H^o_{298} = -211\ kJ/mol$

Methanol-to-Olefins:

$2\,CH_3OH \qquad \rightleftarrows \quad H_3C\text{-}O\text{-}CH_3 + H_2O$ $\Delta H^o_{298} = -23.6\ kJ/mol$

$H_3C\text{-}O\text{-}CH_3 \qquad \rightarrow \quad H_2C=CH_2 + H_2O$ $\Delta H^o_{298} = -5.5\ kJ/mol$

$H_2C=CH_2 + CH_3OH \rightarrow \quad CH_2=CH\text{-}CH_3 + H_2O,\ etc.$

$\{\rightarrow C_{5}+ \rightarrow aromatics\ (+ alkanes)\}$

Scheme 4.3   Reactions in F-T and MTO processes; reactions between brackets refer to the Methanol-to-Gasoline process.

4.5.2 Other Sources of Lower Alkenes

Currently, the predominant feedstock for the production of ethene (and other lower alkenes) is still naphtha, although in the future natural gas or even coal again may take over this role, as their reserves are greater than the oil reserves (see Chapter 2). Biomass conversion might also gain importance.

The usual procedure for using natural gas or coal as feedstock for the production of chemicals is conversion into synthesis gas, a mixture of hydrogen and carbon monoxide, which can then be converted to a wide range of chemicals (see Chapters 5 and 6). For the production of ethene from synthesis gas the Methanol-to-Olefins (MTO) process and a modification of the Fischer–Tropsch (F-T) process are potential candidates (see Scheme 4.3).

The F-T process [14] (see Chapter 6) is capable of producing a wide range of hydrocarbons and oxygenates, depending on the catalyst, reactor, and reaction conditions. It has mainly been used for the production of liquid fuels (gasoline, diesel) but in principle, the process can also be tuned for the production of light alkenes by using different catalysts. The problem with current catalysts is the low selectivity towards ethene and other lower alkenes, while methane formation is high. In addition, catalyst activity and life are still insufficient [15].

The direct conversion of methane (the largest constituent of natural gas) into valuable chemicals seems very attractive. The direct conversion of methane into ethene is possible by applying so-called 'oxidative coupling', i.e. the catalytic reaction of

Desired reactions:

$2\,CH_4 + 1/2\,O_2 \qquad \rightarrow \quad CH_3CH_3 \quad + H_2O$ $\Delta H^o_{298} = -177\ kJ/mol$

$CH_3CH_3 + 1/2\,O_2 \qquad \rightarrow \quad CH_2=CH_2 \quad + H_2O$ $\Delta H^o_{298} = -105\ kJ/mol$

Undesired reactions:

$CH_4,\ CH_3CH_3,\ etc. + O_2 \quad \rightarrow \quad CO_2,\ CO$ $\Delta H^o_{298} \ll 0$

Scheme 4.4   Reactions in the oxidative coupling of methane

MTO Process

The MTO process [16–19] has evolved from Mobil's Methanol-to-Gasoline (MTG) process [20], which converts methanol into gasoline with alkenes as intermediates. An important issue for the production of lower alkenes is the suppression of the formation of aromatics. This can be accomplished by using a zeolite (ZSM-5, ZSM-34 [16,17]) or another type of molecular sieve catalyst (SAPO-34 [18,19], Ni-SAPO [19]). The keyword here is shape selectivity, in this case product selectivity, i.e. only small molecules can leave the pores of the zeolite, and hence aromatics cannot be formed (see Section 3.7 and Chapter 9).

Coke formation plays an important role in the MTO process. Deactivation of the catalyst by coke makes frequent regeneration necessary. Furthermore, the reaction is highly exothermic, which requires good temperature control. Therefore, it has been proposed to carry out the process in a system consisting of two fluidized-bed reactors, one for reaction and one for catalyst regeneration [20,21]. The heat of reaction (and regeneration) is removed by steam raising in cooling coils.

In 1995 a demonstration plant of the MTO process came on stream. The process uses a proprietary molecular sieve catalyst and operates at nearly 100% methanol conversion. Table B.4.2 compares the product gas composition with the composition from a typical naphtha cracker.

The ratio of ethene to propene can be adjusted by altering the reaction conditions, which makes the process flexible. Whether the MTO process is attractive from an economical viewpoint depends very much on the local conditions and transportation distance, but at least the catalyst activity and obtainable selectivities are favorable for this process [17,21].

Table B.4.2 Comparison of product gas compositions from naphtha cracker and UOP/NorskHydro MTO process [21]

	Naphtha cracker (vol%)	MTO (vol%)
Ethene	31.27	57.90
Propene	10.63	26.60
Butenes	2.33	5.80
Light alkanes	32.54	6.25
C_5+	3.80	0.70
Acetylenes and dienes	3.16	0.25
Hydrogen	16.27	1.90
CO and CO_2	n.a.	0.60
Total	100.0	100.00

methane with oxygen (see Scheme 4.4). The reaction produces ethane, which is converted to ethene *in situ*, while through sequential reactions higher hydrocarbons are also formed in small amounts. The main problem with oxidative methane coupling is the formation of large amounts of CO_2 and CO by oxidation reactions. This not only reduces the selectivity towards ethene, the high exothermicity also presents formid-able heat removal problems.

In the past two decades oxidative methane coupling has received much attention [22,23]. Since the first publication in 1982 [24] considerable progress has been made in the development of catalysts for the oxidative coupling reaction. However, the best combinations of conversion and selectivity (ca. 30 and 80%, respectively) thus far achieved in laboratory fixed-bed reactors are still well below those required for economic feasibility. Accordingly, the emphasis in oxidative coupling research has shifted somewhat to innovative reactor designs, such as staged oxygen addition and membrane reactors for combined reaction and ethene removal [23] (see also Chapter 9).

QUESTION:

> *Can you explain why it is not surprising that a commercially attractive process has not been developed yet for the oxidative coupling of methane, despite the enormous amount of research that has been carried out?*

At present, of the three processes described in this section, the MTO process has the best credentials, and in some cases it already can compete economically with the steam cracking process.

The incentive for the development of the processes discussed in this section and other natural gas conversion processes (see Chapter 5) has been the discovery of large amounts of natural gas located in remote areas. This remote gas is a potentially cheap source of energy and raw materials for the petrochemical industry, but the problem is that it is not marketable due to lack of transportation systems. Conversion to liquid products such as methanol and liquid fuels is a potential route for economical utilization of remote natural gas. For a process to be economical the product must have a high added value and have a large market.

QUESTIONS:

> *When utilizing remote natural gas for ethene production, which of the three processes discussed is favored from a logistics viewpoint? As an ethene producer using the MTO process would you purchase your methanol on the market or build your own methanol plant with remote natural gas as feedstock? If you choose the latter, will you locate your methanol and MTO plants near each other or not.*
> *(Hint: compare total ethene and methanol production and plant sizes).*

References

1 Lee AKK and Aitani AM (1990) 'Saudi ethylene plants move toward more feed flexibility' *Oil and Gas J.* Sept 10 60.

2 Chauvel A and Lefebvre G (1989) *Petrochemical Processes 1. Synthesis Gas Derivatives and Major Hydrocarbons*, Technip, Paris.

3 Grantom RL and Royer DJ (1987) 'Ethylene' in: Gerhartz W, et al. (eds.) *Ullmann's Encyclopedia of Industrial Chemistry* A10, 5th ed., VCH, Weinheim, pp. 45–93.

4 Kniel L, Winter O and Stork K (1980) 'Ethylene: keystone to the petrochemical industry' *Chemical Industries* 2, Marcel Dekker, New York.

5 Van Schijndel J (1992) personal communication.

6 Sundaram KM, Shreehan MM and Olszewski EF (1994) 'Ethylene' in: Kroschwitz JI and Howe-Grant M (eds.) *Kirk-Othmer Encyclopedia of Chemical Technology* vol 9, 4th ed., Wiley, New York, pp. 877–915.

7 Sundaram KM, Van Damme PS and Froment GF (1981) 'Coke deposition in the thermal cracking of ethane' *AIChE J.* 27 946–951.

8 Wysiekierski AG, Fisher G and Schillmoller CM (1999) 'Control coking for olefins plants' *Hydroc. Process.* 78(1) 97–100.

9 Zdonik SB (1983) 'Recovery of ethylene, other olefins, and aromatics' in: Albright LF, Crynes BL and Corcoran WH (eds.) *Pyrolysis, Theory and Industrial Practice*, Academic Press, New York, pp. 402–409.

10 Zehnder S (1998) 'What are Western Europe's petrochemical feedstock options?' *Hydroc. Process.* 77(2) 59–65.

11 Pujado PR and Vora BV (1990) 'Make C_3-C_4 olefins selectively' *Hydroc. Process.* 69(3) 65–70.

12 Calamur N and Carrera M (1996) 'Propylene' in: Kroschwitz JI, Howe-Grant M (eds.) *Kirk-Othmer Encyclopedia of Chemical Technology*, vol 20, 4th ed., Wiley, New York, pp. 249–271.

13 Buonomo F, Sanfilippo D and Trifirò F (1997) 'Dehydrogenation of alkanes' in: Ertl G, Knözinger H and Weitkamp J (eds.) *Handbook of Heterogeneous Catalysis*, VCH, Weinheim, pp. 2140–2151.

14 Fischer F and Tropsch H (1923) 'The preparation of synthetic oil mixtures (Synthol) from carbon monoxide and hydrogen' *Brennst. Chem.* 4 (1923) 276–285.

15 Janardanarao M (1990) 'Direct catalytic conversion of synthesis gas to lower olefins' *Ind. Eng. Chem. Res.* 29 1735–1753.

16 Chang CD (1984) 'Methanol conversion to light olefins' *Catal. Rev.-Sci. Eng.* 26 323–345.

17 Inui T and Takegami T (1982) 'Olefins from methanol by modified zeolites' *Hydroc. Process.* 6(11) 117–120.

18 Kaiser SW (1985) 'Methanol conversion to light olefins over silico-aluminophosphate molecular sieves', *Arab. J. Sci. Eng.* 10 (4) 361–366.

19 Inui T, Phatanasri S and Matsuda H (1990) 'Highly selective synthesis of ethene from methanol on a novel nickel-silicoaluminophosphate catalyst' *J. Chem. Soc. Chem. Commun.* 205-206.

20 Bos ANR and Tromp PJJ (1995) 'Conversion of methanol to lower olefins. Kinetic modeling, reactor simulation, and selection' *Ind. Eng. Chem. Res.* 34 3808-3816.

21 Nilsen, HR (1997) 'The UOP/Hydro Methanol to Olefin process: its potential as opposed to the present application of natural gas as feedstock', *Proceedings of the 20th World Gas Conference*, Copenhagen, Denmark, 10–13 June, FRT 1-05.

22 Lunsford JH (1997) 'Oxidative coupling of methane and related reactions' in: Ertl G, Knözinger H and Weitkamp J (eds.) *Handbook of Heterogeneous Catalysis*, VCH, Weinheim, pp. 1843–1856 and references therein.

23 Androulakis IP and Reyes SC (1999) 'Role of distributed oxygen addition and product removal in the oxidative coupling of methane', *AIChE J.* 45 860–868 and references therein.

24 Keller GE and Bhasin MM (1982) 'Synthesis of ethylene via oxidative coupling of methane. I. Determination of active catalysts' *J. Catal.* 73 9–19.

Chapter 5

Production of Synthesis Gas

5.1 INTRODUCTION

Synthesis gas (or syngas) is a general term used to describe mixtures of hydrogen and carbon monoxide in various ratios. These mixtures are used as such and are also sources of pure hydrogen and pure carbon monoxide. Table 5.1 shows some applications of synthesis gas. From the examples one can note that the term 'synthesis gas' is used more generally than stated above: also the N_2/H_2 mixture for the production of ammonia is referred to as synthesis gas.

Syngas may be produced from a variety of raw materials ranging from natural gas to coal. The choice for a particular raw material depends on cost and availability of the feedstock, and on downstream use of the syngas. Syngas is generally produced by one of three processes, which are distinguished based on the feedstock used.

- Steam reforming of natural gas or light hydrocarbons, optionally in the presence of oxygen or carbon dioxide.
- Partial oxidation of (heavy) hydrocarbons with steam and oxygen.
- Partial oxidation of coal (gasification) with steam and oxygen.

The names of the processes may be somewhat confusing. The term *steam reforming* is used to describe the reaction of hydrocarbons with steam in the presence of a catalyst. This process should not be confused with the *catalytic reforming* process for the improvement of the gasoline octane number (see Section 3.4). In the gas industry, *reforming* is commonly used for the conversion of a hydrocarbon by reacting it with O-containing molecules, usually H_2O, CO_2, and/or O_2. *Partial oxidation* (also called steam/oxygen reforming) is the non-catalytic reaction of hydrocarbons with oxygen and usually also steam (catalysis is possible and the process in that case is referred to as Catalytic Partial Oxidation). This process may be carried out in an autothermic or allothermic way. Coal gasification is the more common term to describe partial oxidation of coal. A combination of steam reforming and partial oxidation, in which endothermic and exothermic reactions are coupled, is often referred to as *autothermic reforming*.

Most syngas today is produced by steam reforming of natural gas or light hydrocarbons up to naphtha. For light feedstocks partial oxidation is usually not an economical option, because of the high investment costs as a result of the required cryogenic air separation. However, partial oxidation processes are applied where steam-reformable feeds are not available, or in special situations where local conditions exist to provide favorable economics. Such conditions could be the availability of relatively

Table 5.1   Synthesis gas applications

Mixtures	Main uses	Chapter of this book
H_2	Refinery hydrotreating and hydrocracking	3
3 H_2:1 N_2[a]	Ammonia	6
2 H_2:1 CO	Alkenes (Fischer–Tropsch reaction)	6
2 H_2:1 CO	Methanol, higher alcohols	6, 8
1 H_2:1 CO	Aldehydes (hydroformylation)	8
CO	Acids (formic and acetic)	8

[a] With N_2 from air.

low-priced heavy feedstocks or the need for a synthesis gas with high carbon monoxide content.

Coal gasification is not very common for the production of synthesis gas for chemical use, but it is increasingly used for power generation.

Figure 5.1 shows general flow schemes of the main processes for syngas production. The steam reforming feed usually has to be desulfurized. Sulfur is a poison for metal catalysts because it can block active sites by the formation of rather stable surface sulfides. In steam reforming and in downstream uses in many reactors transition-metal based catalysts are used. When sulfur is present as H_2S, adsorption (for instance on activated carbon), reaction with an oxide (for instance ZnO), or scrubbing with a solvent may be applied. If the feed contains more stable sulfur compounds hydrotreatment (see Chapter 3) may be required.

Coal and heavy oil based processes for the production of syngas require removal of

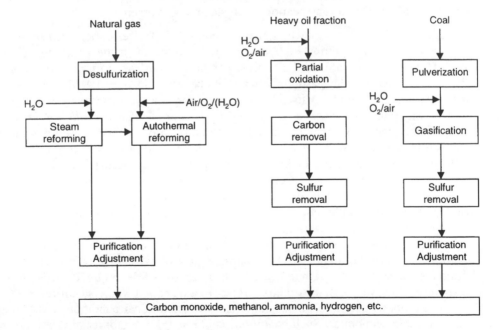

Figure 5.1   General flow schemes for the production of syngas.

sulfur compounds (mainly H_2S) from the syngas. Feed purification is not possible with these raw materials, although attempts have been made using biotechnological approaches.

Depending on the downstream use, the raw syngas may be treated in several ways. Syngas treatment includes such processing steps as CO_2 removal, water-gas shift reaction, methanation, H_2/CO separation, etc.

Partial oxidation of heavy hydrocarbons is very similar to coal gasification and will not be discussed further. This chapter deals with the conversion of natural gas and coal into synthesis gas.

QUESTIONS:

> *Where in the process of producing syngas would you consider scrubbers? What solvents would you use?*

5.2 SYNTHESIS GAS FROM NATURAL GAS

5.2.1 Reactions and Thermodynamics

Although natural gas does not solely consist of methane (see Section 2.3.3) for simplicity we will assume that this is the case. Table 5.2 shows the main reactions during methane conversion.

When converting methane in the presence of steam the most important reactions are the steam reforming reaction (1) and the water-gas shift reaction (2).

Some processes, such as the reduction of iron ore [1] and the hydroformylation reaction (see Table 5.1 and Chapter 8) require synthesis gas with a high CO content, which might be produced from methane and CO_2 in a reaction known as *CO_2 reforming* (3). The latter reaction is also referred to as 'dry reforming', obviously because of the absence of steam.

The main reactions may be accompanied by coke formation, which leads to deactivation of the catalyst. Coke may be formed by decomposition of methane (4) or by disproportionation of CO, the Boudouard reaction (5).

In the presence of oxygen methane undergoes partial oxidation to produce CO and H_2 (6). Side reactions such as the complete oxidation of methane to CO_2 and H_2O (7), and oxidation of the formed CO and H_2 (8) and (9) may also occur.

Table 5.2 Reactions during methane conversion with steam and/or oxygen

Reaction	ΔH^0_{298} (kJ/mol)	
$CH_4 + H_2O \rightleftarrows CO + 3H_2$	206	(1)
$CO + H_2O \rightleftarrows CO_2 + H_2$	-41	(2)
$CH_4 + CO_2 \rightleftarrows 2CO + 2H_2$	247	(3)
$CH_4 \rightleftarrows C + 2H_2$	75	(4)
$2CO \rightleftarrows C + CO_2$	-173	(5)
$CH_4 + \frac{1}{2}O_2 \rightarrow CO + 2H_2$	-36	(6)
$CH_4 + 2O_2 \rightarrow CO_2 + 2H_2O$	-803	(7)
$CO + \frac{1}{2}O_2 \rightarrow CO_2$	-284	(8)
$H_2 + \frac{1}{2}O_2 \rightarrow H_2O$	-242	(9)

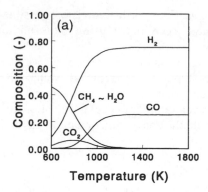
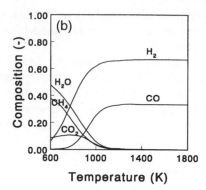

Figure 5.2 Equilibrium gas composition at 1 bar as a function of temperature (a) steam reforming of methane; $H_2O/CH_4 = 1$ mol/mol; H_2O curve coincides with CH_4 curve, and (b) partial oxidation of methane; $O_2/CH_4 = 0.5$ mol/mol.

The reaction of methane with steam is highly endothermic while the reactions with oxygen are moderately to extremely exothermic. Operation can hence be allothermic (steam and no or little oxygen added, required heat generated outside the reactor), or autothermic (steam and oxygen added, heat generated by reaction with oxygen within the reactor), depending on the steam/oxygen ratio.

Figure 5.2 compares the equilibrium gas compositions of the reaction of methane with steam and with oxygen at 1 bar as a function of temperature. Stoichiometric amounts of steam and oxygen were added for reactions (1) and (6), respectively.

Figure 5.2a and b show essentially the same composition pattern as a function of temperature, although the product ratios arc different. The former observation is logical since apart from the irreversible oxidation reactions, the same reactions occur in both steam reforming and partial oxidation. The latter observation is also easily explained, namely by the fact that the ratio of the elements C, O, and H in the feed is different for the two processes. This is most obvious at high temperatures where only H_2 and CO are present. The H_2/CO ratios resulting from steam reforming and partial oxidation are 3 and 2, respectively, at these high temperatures.

The hydrogen and carbon monoxide contents of the equilibrium gas increase with temperature, which is explained by the fact that the reforming reactions (1) and (3) are endothermic.

The carbon dioxide content goes through a maximum. This can be explained as follows: CO_2 is formed in exothermic reactions only, while it is a reactant in the endothermic reactions. Hence, at low temperature CO_2 is formed, while with increasing temperature the endothermic reactions in which CO_2 is converted (mainly reaction (3)) become more important.

QUESTION:

>*Draw a qualitative profile of the heat consumed or produced (as a function of temperature) for both situations in Figure 5.2.*
>*In autothermic gasification/reforming both steam and oxygen are added. Draw a rough composition profile resulting from autothermic operation.*

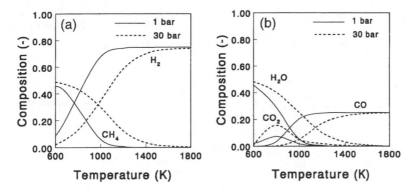

Figure 5.3 Effect of temperature and pressure on equilibrium gas composition in steam reforming of methane; $H_2O/CH_4 = 1$ mol/mol; (a) CH_4 and H_2; (b) CO, CO_2 and H_2O.

Both steam reforming and partial oxidation of methane are hindered by elevated pressure because the number of molecules increases due to these reactions. Figure 5.3 illustrates the effect of pressure on steam reforming of methane at a steam/methane ratio of 1 mol/mol.

At a pressure of 30 bar, the equilibrium conversion to H_2 and CO is only complete at a temperature of over 1400 K. However, in industrial practice, temperatures much in excess of 1200 K cannot be applied as a result of metallurgical constraints. It will be shown below that this point is of great practical significance.

5.2.2 Steam Reforming Process

It is interesting that even though steam reforming is carried out at high temperature (>1000 K) a catalyst (supported nickel) is still required to accelerate the reaction due to the very high stability of methane. The catalyst is contained in tubes, which are placed inside a furnace that is heated by combustion of fuel. The steam reformer (see Figure 5.4) consists of two sections. In the convection section, heat recovered from the hot flue gases is used for preheating of the natural gas feed and process steam, and for the generation of superheated steam. In the radiant section the reforming reactions take place.

After sulfur removal, the natural gas feed is mixed with steam (and optionally CO_2) and preheated to approximately 780 K before entering the reformer tubes. The heat for the endothermic reforming reaction is supplied by combustion of fuel in the reformer furnace (allothermic operation). The hot reformer exit gas is used to raise steam. In a 'knock-out drum' (vessel in which liquid is separated from vapor by gravitation), water is removed and the raw syngas can be treated further, depending on its use.

QUESTION:

> *For which type of syngas applications would you expect CO_2 to be added? What would be the source of this CO_2?*

Figure 5.5 shows the general arrangement of a typical steam reformer. The preheated process stream enters the catalyst tubes through manifolds and pigtails

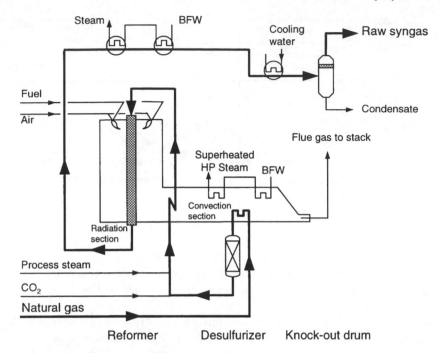

Figure 5.4   Simplified flow scheme of the steam reforming process.

and flows downwards. The product gas is collected in an outlet manifold from which it is passed upwards to the effluent chamber from which it will be further processed [2]. The tubes are carefully charged with the catalyst particles: an even distribution over the tubes is essential and the catalyst bed should be dust free. A furnace may contain 500–600 tubes with a length of 7–12 m and an inside diameter of between 70 and 130 mm.

Table 5.3 shows typical reformer conditions used in practice for steam reforming of

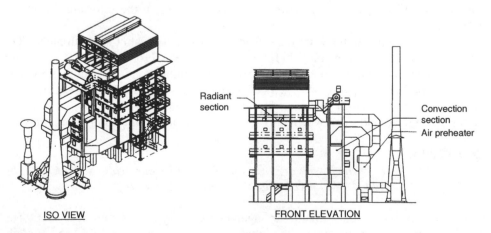

Figure 5.5   General arrangement of a steam reformer (Courtesy of Technip Benelux B.V.).

Table 5.3 Typical reformer conditions for industrial syngas based processes. Based on data from [1,3]

Process	H₂O/C (mol/mol)	T_{exit} (K)	p_{exit} (bar)	Composition (vol%)[a]			
				H₂	CO	CO₂	CH₄
Hydrogen	2.5	1123	27	48.6	9.2	5.2	5.9
Hydrogen[b]	4.5	1073	27	34.6	5.3	8.0	2.4
Ammonia	3.7	1073	33	39.1	5.0	6.0	5.5
Methanol	3.0	1123	17	50.3	9.5	5.4	2.6
Aldehydes/alcohols	1.8	1138	17	28.0	25.9	19.7	1.1
Reducing gas	1.15	1223	5	70.9	22.4	0.9	1.5

[a] Rest is H_2O.
[b] From naphtha.

methane. At first sight the operating conditions seem illogical when considering the reaction stoichiometry and thermodynamics. However, in order to determine optimal conditions for steam reforming one also has to consider the suppression of coke formation and the subsequent use of the syngas.

5.2.2.1 Carbon formation

Carbon formation in steam reformers must be prevented for two main reasons. Firstly, carbon deposition on the active sites of the catalyst leads to deactivation. Secondly, carbon deposits grow that large that they can cause total blockage of the reformer tubes, resulting in the development of 'hot spots'. Hence, the reforming conditions must be chosen such that carbon formation is strictly limited.

QUESTION:

> *Carbon formation changes catalytic activity and, as a consequence, the temperature profile in the reactor changes. Explain that extended carbon deposition can lead to hot spots.*

Carbon forming reactions can be suppressed by adding excess steam. Therefore, it is common practice to operate the reformer at steam-to-carbon ratios of 2.5–4.5 mol H_2O per mol C, the higher limit applying to higher hydrocarbons such as naphtha. Compared to methane, higher hydrocarbons exhibit a greater tendency to form carbonaceous deposits. At high temperature (>920 K) steam cracking (see Chapter 4) may occur to form alkenes which may easily form carbon through reaction (10), which is irreversible:

$$C_nH_{2n} \rightarrow \text{carbon} + H_2 \qquad \Delta H_{298}^0 < 0 \text{ kJ/mol} \qquad (10)$$

An additional advantage of using a larger than stoichiometric quantity of steam is that it enhances hydrocarbon conversion due to the lower partial pressure of hydrocarbons.

5.2.2.2 Syngas use

Most applications require syngas at elevated pressure (e.g., methanol synthesis occurs at 50–100 bar and ammonia synthesis at even higher pressure – see Chapter 6). Therefore, most modern steam reformers operate at pressures far above atmospheric, despite the fact that this is thermodynamically unfavorable. The advantages of operating at elevated pressure are the lower syngas compression costs and a smaller reformer size. The down side is a lower methane conversion. To counterbalance the negative effect on the equilibrium, higher temperatures are applied and more excess steam is used. Figure 5.6 shows the effect of pressure (left) and excess steam on the methane slip (percentage of unreacted methane).

QUESTIONS:

> *Estimate the percentage of CH_4 in the 'dry syngas' (i.e. after removal of H_2O) at the conditions given in Table 5.3.*
>
> *Compare the costs of compression for the following two cases:(i) syngas is produced at atmospheric pressure (for practical reasons slightly above 1 atm) and the product mixture is subsequently compressed to, say, 50 bar, and (ii) the feed is compressed and synthesis is carried out at elevated pressure.*
>
> *Excess steam depresses carbon formation. Explain the underlying mechanisms.*

It is generally economically advantageous to operate at the highest possible pressure, and, as a consequence, at the highest possible temperature. The tube material, however, places constraints on the temperatures and pressures that can be applied: a maximum limit exists for the operating temperature at a given pressure because of the creep limit of the reformer tubes.

At typical reformer temperatures and pressures an appreciable amount of methane is still present in the produced syngas. Minimization of methane slip is crucial for the economics of the process. There is a trend towards higher operating temperatures

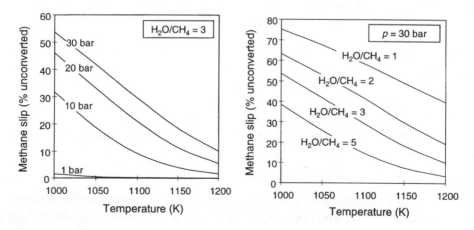

Figure 5.6 Effect of pressure (left) and steam-to-methane feed ratio (right) on methane slip as a function of temperature.

(exceeding 1200 K), which is made possible by the increasing strength of the tube materials [4].

5.2.3 Advances in Steam Reforming

Various modifications of the basic steam reforming process (Figure 5.4) can be implemented. For instance, if the natural gas feed is rich in higher hydrocarbons or if naphtha is used as the feedstock, a so-called pre-reformer may be installed upstream

---

Pre-reforming

When higher hydrocarbons (e.g. naphtha) are used as feedstock or if the natural gas contains appreciable amounts of higher hydrocarbons, it is advantageous to install a pre-reformer upstream the steam reformer. In the pre-reformer, which is a catalytic fixed-bed reactor operating adiabatically at about 770 K, the higher hydrocarbons are converted to methane and carbon dioxide (reaction (11)).

$$C_nH_{2n+2} + \tfrac{1}{2}(n-1)H_2O \rightarrow \tfrac{1}{4}(3n+1)CH_4 + \tfrac{1}{4}(n-1)CO_2$$

$$\Delta H^0_{298} < 0 \text{ kJ/mol}$$

(11)

In the actual reformer, which operates at higher temperature, reactions (1), (2), and (3) subsequently proceed to the desired equilibrium.

Figure B.5.1 shows a possible configuration of a steam reformer with pre-reformer. The most important advantage of a pre-reformer is that the steam reformer can be operated at lower steam/carbon ratio without danger of carbon deposition: higher hydrocarbons have a greater tendency to carbon formation, which is enhanced by low steam/carbon ratios and high temperature. By installing a pre-reformer operating at moderate temperature, carbon formation is minimized.

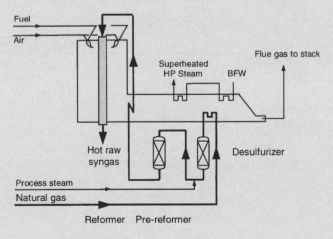

Figure B.5.1 Possible configuration of a steam reformer with pre-reformer

the reformer furnace (see box) to convert (part of) these compounds [4]. In this way the worst carbon precursors are removed in advance. This allows operation of the actual reformer at lower steam/carbon ratio.

Another option is the addition of CO_2 to the reformer feed. Steam reforming of natural gas results in a synthesis gas with a relatively high H_2/CO ratio (> 3 mol/mol). However, most applications call for a more CO-rich syngas. To increase the CO content of the produced gas, CO_2 can be added to the reformer inlet. Because CO_2 is an oxidizing agent, an additional advantage is that less steam is needed to prevent carbon deposition. Recently, several processes have been developed in which carbon dioxide reforming is used with little or no steam addition [5–7]. In this way H_2/CO ratios in the product gas below 0.5 can be achieved.

5.2.4 Autothermic Reforming

An alternative measure to achieve lower H_2/CO ratios is the addition of oxygen: steam/oxygen reforming or autothermic reforming [4,8,9]. Autothermic reforming is the reforming of light hydrocarbons in a mixture of steam and oxygen in the presence of a catalyst. Figure 5.7 shows a schematic of an autothermic reformer.

The reactor is a refractory lined vessel. Therefore, higher pressures and temperatures can be applied than in steam reforming. Part of the feed is oxidized in the combustion zone, reactions (6) and (7). In the lower part the remaining feed is catalytically reformed with the produced CO_2 and H_2O, reactions (1) and (3). The endothermic reforming duty is provided by the exothermic oxidation reaction.

Autothermic reforming of natural gas and light hydrocarbons is usually not applied on its own, because of the high investment and operating costs (oxygen). An auto-

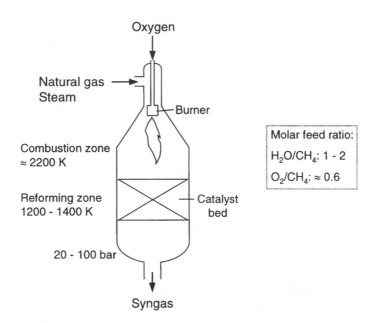

Figure 5.7   Schematic of autothermic reformer synthesis gas production.

Heat-Exchange Reformer

Since a large amount of heat is required in the steam reformer, while the autothermic reformer produces heat, an obvious thought would be to use the hot gas from the autothermic reformer to supply the heat input for the steam reformer in a so-called *heat-exchange reformer* or *gas-heated reformer* [4,9]. This would eliminate the expensive fired reformer, and thus reduce the investment costs. Of course, the consequence is that only medium pressure steam can then be recovered from the syngas plant and it would be necessary to import electric power for the large syngas compressor [4].

QUESTION:

> *Why can only medium pressure steam be recovered? (Hint: compare the flow schemes for syngas production in this chapter.)*

Figure B.5.2 shows two types of gas-heated reformer. The technology developed by ICI uses a gas-heated reformer, where about 75% of the methane is converted at about 970 K and 40 bar in combination with an autothermic reformer, which converts the remaining methane present in the exit stream of the gas-heated reformer.In the combined authothermic reforming (CAR) process developed by Exxon, steam reforming is combined with partial oxidation in a single fluidized-bed reactor. Heat is recovered from the syngas by the generation of steam (not shown).

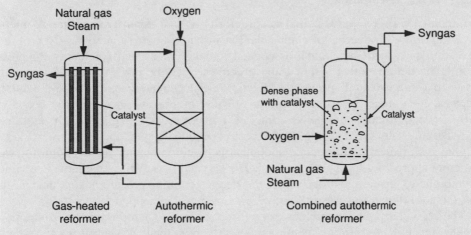

Figure B.5.2 ICI combined reforming process (left) and Exxon CAR process (right).

thermic reformer is frequently used, however, in the production of synthesis gas for ammonia manufacture. In that case it is installed downstream the steam reformer and air is used to supply the required oxygen. The advantage of this arrangement is that unreacted methane from the steam reformer is converted in the autothermic reformer, so that the steam reformer can be operated at lower temperature and higher pressure (ca. 1070 K and 30 bar). These conditions are thermodynamically unfavorable. However, a high outlet pressure is advantageous because it decreases the costs for compression of the synthesis gas to the pressure of the ammonia synthesis loop (> 100 bar). Since air supplies the required nitrogen, no expensive oxygen plant is needed in this special case.

Synthesis gas for methanol plants is also increasingly produced by combined steam and autothermic reforming, with the advantage of operation at high pressure (methanol synthesis takes place at 50–100 bar), but the disadvantage of pure oxygen requirement. Therefore, still at least 80% of the methanol production is based on steam reforming of natural gas.

Two novel reactor systems combining steam reforming and autothermic reforming are presented in the box 'Heat-Exchange Reformer'.

QUESTIONS:

> *Synthesis gas produced by autothermic reforming has a lower H_2/CO ratio than synthesis gas produced by steam reforming. What is the reason? Why shouldn't air be used in the production of syngas for methanol production? In the process of Figure 5.7 the combustion zone is not catalyzed. Presently, a lot of interest exists in catalyzing the oxidation reactions also (CPO, Catalytic Partial Oxidation). What would be the reason?*

5.2.5 Novel Developments

Producing syngas by conventional steam reforming is an expensive and energy-intensive step in the production of chemicals. For instance, approximately 60–70% of the investment in a methanol plant is associated with the syngas plant (including compression). The developments outlined above could reduce the investment and operating costs to some extent. Especially Catalytic Partial Oxidation has a large potential, because it is attractive from the point of view of *exergy efficiency* (see box and [10–13]) and materials usage. Nevertheless, a major breakthrough would be required to make any substantial savings in the cost of producing syngas.

One such radically different development is based on membrane technology. Two different approaches have been pursued. The first makes use of a palladium or ceramic membrane to separate hydrogen from the steam reforming process *in situ*, thus pushing the equilibrium towards full conversion even at low temperatures [4].

The second approach combines oxygen separation and autothermic reforming in a single step [14]. A ceramic membrane selectively extracts oxygen from air, which flows through the inside of the membrane. The oxygen is transported through the membrane and reacts with methane flowing over the outside.

Other approaches aim at the direct conversion of methane into valuable chemicals instead of first producing synthesis gas as an intermediate. Examples are the production of methanol by direct partial oxidation and the production of ethene by oxidative coupling [4] (see also Chapter 4).

QUESTION:

> *How would direct production of methanol from methane compare to the syngas route in terms of exergy efficiency?*

Another interesting thought is to carry out syngas production at the scale of the consumer: hybrid car engines including a syngas production unit on board are under development (see box 'Small-scale hydrogen production – example: HEV').

Exergy

Exergy is the theoretically maximum amount of work that can be obtained from a system. It is a measure for the *quality* of the energy. Exergy losses ('lost work') in a process occur because the quality of the energy streams entering the process is higher than the quality of the energy streams leaving the process (while the *quantity* of the energy remains the same: the first law of thermodynamics). Exergy losses can be divided in external and internal losses. External exergy losses are analogous to 'energy losses' or more precisely 'energy spills', and result from emissions to the environment (e.g. exergy contained in off-gas). Internal exergy losses are either physical (losses due to heat exchange, compression, etc.) or chemical. Chemical exergy losses depend on the process route and the conditions at which reactions occur. An *exergy analysis* of chemical processes can provide the starting point for process improvement.

Figure B.5.3 clearly shows the different results obtained from an energy and an exergy analysis for the production of methanol by steam reforming of natural

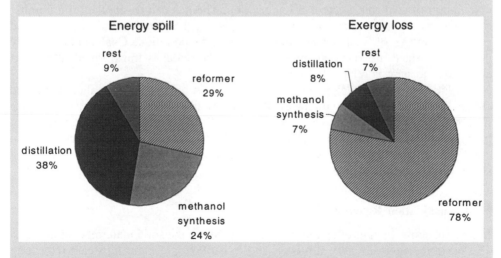

Figure B.5.3 Contribution of process sections in the ICI low-pressure methanol process to total energy spill and exergy loss. Based on data from [13].

gas followed by conversion of the synthesis gas using the using the low-pressure ICI process (see Section 6.2). Based on an energy analysis one would concentrate on improving the distillation section, while an exergy analysis points in a different direction, viz. the reformer. It has been demonstrated that replacing the steam reforming process by, in this case, the ICI combined reforming process (Figure B.5.2) results in a decrease in exergy losses of nearly 30% [13]. The main reason for this improvement in *exergy efficiency* is the fact that the irreversible combustion reaction, required for heating the reformer furnace has been avoided. A further improvement would result from eliminating the need for heat exchange, for instance by producing the synthesis gas in a single catalytic partial oxidation reactor.

5.3 COAL GASIFICATION

Gasification of coal to produce coal gas dates back to the end of the 18th century. During the mid-1800s, coal gas was widely used for heating and lighting in urban areas. The development of large-scale processes began in the late 1930s and the process was gradually improved. Following World War II, however, interest in coal gasification dwindled because of the increasing availability of inexpensive oil and natural gas. In 1973, when oil and gas prices increased sharply, interest was renewed and over the past 25 years much effort has been put in improving this process.

Coal gasification, like steam reforming of natural gas, produces synthesis gas. However, the incentive of the two processes is different: The goal of steam reforming is the *production* of CO and H_2 for chemical use, whereas coal gasification was primarily developed for the *conversion* of coal into a gas, which happens to predominantly contain CO and H_2.

In practice, transport of gas is much easier than transport of solids. Therefore, conversion of coal into a gas is a logical step. The gas can be applied in a number of processes, e.g. for heating or as a raw material for the chemical industry. A second major advantage of coal gasification is related to pollution. Coal contains many pollutants, which are released upon heating. When coal is transformed into a gas centrally, cleaning is relatively easy.

A relatively recent development is the application of coal in a so-called combined cycle. The name stems from the fact that gasification is combined with power generation by a gas turbine driven by the gases coming from the gasifier with the waste heat being used in a steam turbine (see Section 5.3.5). Compared to coal combustion combined with a steam turbine, coal gasification combined with a gas turbine and a steam turbine is more efficient.

5.3.1 Gasification Reactions

In a broad sense, coal gasification is the conversion of the solid material coal into gas. The basic reactions are very similar to those that take place during steam reforming (see Table 5.4). For simplicity, coal is represented by carbon, 'C'.

The first two reactions with carbon are endothermic, whereas the latter three are exothermic. It is common practice to carry out coal gasification in an 'autothermic'

Small-scale hydrogen production – example: HEV

Figure B.5.4 shows a schematic of a hybrid-electrical vehicle (HEV), in which hydrogen is produced by steam reforming of a fuel (e.g. natural gas, gasoline or methanol), and this hydrogen used in a PEM fuel cell (see also Section 3.7.4.2) for electricity generation.

The physical appearance of the on-board reformers will be different from large-scale reformers. One can think of a shell-and-tube heat exchanger with the tubes coated with catalyst, and combustion taking place at the shell, or a metallic monolith catalyst (see also Section 9.4.2). A membrane reactor can also be used, with the advantage of reaction and hydrogen separation taking place in one unit (see also Section 9.4.3).

Because PEM fuel cells require hydrogen of very high purity, CO has to be removed almost completely from the syngas produced in the reformer. Therefore, a gas cleaning unit is also present.

Most of the current development work is focused on methanol reforming, which is relatively easy; the temperature required is only about 530 K rather than the ca. 1000 K required for natural gas, gasoline, or other hydrocarbons. Prototypes of a methanol reforming fuel cell electric vehicle have been produced by, for example, Toyota, Mitsubishi, and DaimlerChrysler. These car companies all plan to reach commercialization around 2005.

QUESTIONS:

> *Which reactions occur during methanol reforming (reaction of CH_3OH with H_2O)? How could the CO be removed from the syngas to produce (nearly) pure hydrogen for use in the fuel cell (see also Section 5.4)? What are advantages and disadvantages of the fuels mentioned for use in a HEV? (Also think of logistics.)*

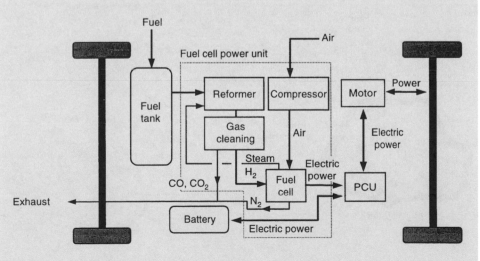

Figure B.5.4 HEV using steam reforming combined with a fuel cell; PCU, Power Control Unit. Adapted from [15].

Table 5.4 Reactions occurring during coal gasification

Reaction	ΔH^0_{800} (kJ/mol)	
Heterogeneous reactions		
$C + H_2O \rightleftarrows CO + H_2$	136	(12)
$C + CO_2 \rightleftarrows 2CO$	173	(-5)
$2C + O_2 \rightleftarrows 2CO$	-222	(13)
$C + O_2 \rightleftarrows CO_2$	-394	(14)
$C + 2H_2 \rightleftarrows CH_4$	-87	(-4)
Homogeneous reactions		
$2CO + O_2 \rightleftarrows 2CO_2$	-572	(8)
$CO + H_2O \rightleftarrows CO_2 + H_2$	-37	(2)

way. The reaction is carried out with a mixture of O_2 (or air) and H_2O, so that combustion of part of the coal produces the heat needed for heating up to reaction temperature and for the endothermic steam gasification reaction. Besides these heterogeneous reactions, also homogeneous reactions occur.

When coal is burned in air, strictly speaking also gasification occurs, but this process is usually not referred to as gasification. Coal is a complex mixture of organic and mineral compounds. When it is heated to reaction temperature (depending on the process, between 1000 and 2000 K), various thermal cracking reactions occur. Usually, the organic matter melts, while gases evolve (H_2, CH_4, aromatics, etc.). Upon further heating the organic mass is transformed into porous graphite-type mate-

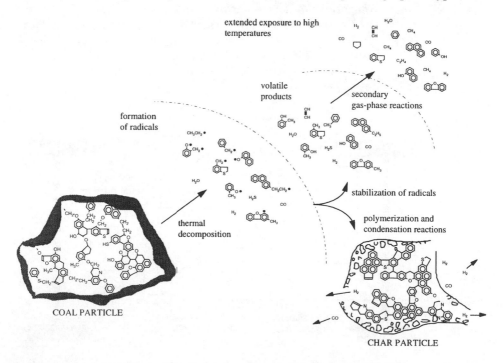

Figure 5.8 Pyrolysis network. Adapted from [16].

rial, called 'char'. This process is called pyrolysis. Figure 5.8 illustrates the complex pyrolysis network.

When in the gas phase reactive gases are present, several consecutive reactions occur. This is in particular the case in coal combustion. In a combustor, coal particles are introduced in the flame and they are heated very fast (>10000 K/s). Immediately, pyrolysis reactions occur, but the product mixture is not complex, because the products are essentially completely burned. When less than stoichiometric amounts of oxygen are present, as in coal gasification, the atmosphere is reducing, and the product contains a complex mixture of organic compounds.

Coal is not unique in this sense: many hydrocarbons and related compounds, for instance plastics, sugar, biomass, and heavy oil, exhibit similar behavior; heating of sugar leads to melting followed by complex sequences of reactions leading to volatiles and a solid material that might be referred to as caramel (mild conditions) or sugar char (severe conditions).

Depending on the temperature of the gasifier the mineral matter will be released as a liquid or a solid. When the mineral matter is released as a liquid, the gasifier is called a 'slagging' gasifier.

5.3.2 Thermodynamics

As has become clear, coal and its products are very complex. For convenience, in thermodynamic calculations we assume that coal solely consists of carbon. Furthermore, under practical gasification conditions, the following species are most abundant in the gas phase: H_2O, CO, CO_2, H_2, and CH_4. Of course, other hydrocarbons are also present, but in the thermodynamic calculations these will all be represented by CH_4. Figure 5.9 shows the equilibrium composition resulting from gasification of coal in an equimolar amount of steam as a function of temperature and pressure.

With increasing temperature, the CO and H_2 mole fractions increase due to the increasing importance of the endothermic H_2O-gasification reaction. Accordingly, H_2O decreases with temperature. Both CO_2 and CH_4 go through a maximum as a result of the exothermicity of their formation and endothermicity of their conversion. As can be seen from Figure 5.9, a low pressure is favorable for CO and H_2 forming

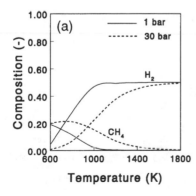
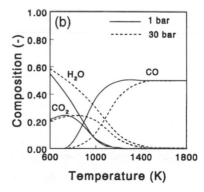

Figure 5.9 Effect of temperature and pressure on equilibrium gas composition in steam gasification of coal (represented as C); $H_2O/C = 1$ mol/mol; (a) CH_4 and H_2; (b) CO, CO_2, H_2O [17].

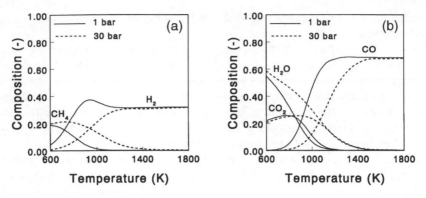

Figure 5.10 Effect of temperature and pressure on equilibrium gas composition in autothermic gasification of coal (C); (a) CH_4 and H_2; (b) CO, CO_2, H_2O; $H_2O/O_2/C$ ratio depends on temperature [17].

reactions due to the increase in number of molecules. The CO_2 and CH_4 maxima shift to higher temperature.

The heat required for coal gasification may either be supplied outside the reactor ('allothermic' gasification) or within the reactor by adding oxygen ('autothermic' gasification). Addition of oxygen not only supplies the necessary energy, but also changes the composition of the product gas. This can be seen from Figure 5.10, which shows the equilibrium composition resulting from autothermic gasification as a function of temperature and pressure.

The equilibrium composition in autothermic gasification shows the same trend concerning temperature and pressure as in allothermic steam gasification. However, the fractions of CO and CO_2 are much higher, while those of H_2 and CH_4 are lower. This obviously is the result of the higher oxygen content (lower H/O ratio) of the feed, particularly at high temperature.

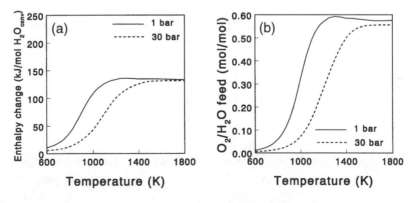

Figure 5.11 (a) Enthalpy change associated with allothermic steam gasification and (b) O_2/H_2O feed ratio (mol/mol) in autothermic gasification [17].

'Chemical Cooling'

Autothermic coal gasification supplies and uses heat in one reactor. One could also think of examples where the reactor itself does not operate in an autothermic mode, but the overall process does. For example, the riser/regenerator combination in the FCC process (Chapter 3) approaches autothermic operation, although this is not the case for the individual reactors.

Coupling of exo- and endothermic processes has often been considered as a means to upgrade thermal energy. Examples are the use of solar energy, waste heat from nuclear energy plants, and high-temperature exhaust gases from mineral processing plants for providing heat to chemical processes. In such a way, heat is transformed into chemical energy. This could be referred to as 'chemical cooling'. It is needless to say that one of the options mentioned constitutes an easy target for environmental pressure groups.

QUESTION:

> *The partial pressure of H_2 goes through a maximum (Figure 5.10a). What is the explanation?*

Figure 5.11 shows the enthalpy change (energy required for reactions) as a result of allothermic gasification (left) and the O_2/H_2O ratio required for autothermic operation.

The enthalpy change reflects the total change in reaction enthalpies of all reactions involved. Thus, at 1600 K, per mole H_2O converted, 135 kJ is required. This is the reaction enthalpy of the steam gasification reaction, the predominant reaction at high temperature. With higher temperature the endothermic reactions become increasingly important, resulting in larger enthalpy changes. Similarly, these enthalpy changes determine the required O_2/H_2O ratio for the system to become autothermic. At higher temperature more oxygen is required to compensate for the energy consumed in the endothermic steam gasification reactions.

5.3.3 Current Coal Gasification Processes

Coal gasification processes differ widely. Figures 5.12–5.14 show schematics of the three basically different reactor technologies that are applied, viz. the moving-bed, the fluidized-bed, and the entrained-flow gasifier.

The *moving-bed gasifier* (Lurgi) [18], in Figure 5.12, is operated counter-currently; the coal enters the gasifier at the top and is slowly heated and dried (partial pyrolysis) on its way down while cooling the product gas as it exits the reactor. The coal is further heated and devolatized as it descends. In the gasification zone part of the coal is gasified in steam and carbon dioxide, which is formed in the combustion zone upon burning of the remaining part of the coal.

The highest temperatures (ca. 1300 K) are reached in the combustion zone, near the bottom of the reactor. All that remains of the original coal is ash. For most coals 1300 K is below the 'slagging' temperature (the temperature at which the mineral matter becomes sticky or even melts) of the ash, so the ash leaving the reaction zone is dry.

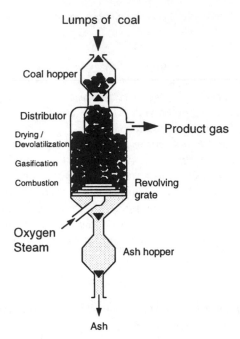

Figure 5.12 Moving-bed gasifier (Lurgi) [18].

The coal bed is supported on a rotating grate where the ash is cooled by releasing heat to the entering steam and oxygen. In this type of reactor the temperature has to be kept low in order to protect the internals of the reactor. The consequence is that a rather large excess of steam has to be fed to the gasifier. Of course, this reduces the efficiency.

A disadvantage of the moving-bed reactor is that a large amount of by-products are formed such as condensable hydrocarbons, phenols, ammonia tars, oils, naphtha, and dust. Therefore, gas cleaning for the Lurgi gasification process is much more elaborate than for other gasification processes and makes it uneconomic in many applications [19]. On the other hand, the counter-current operation makes the process highly efficient.

Not all types of coal can be processed in a moving-bed reactor. Some coals get sticky upon heating and form large agglomerates leading to pressure drop increases and even plugging of the reactor. Coals showing this behavior are called 'caking coals'.

A novel development is the so-called slagging Lurgi reactor [20] in which the temperature in the combustion zone is above 2000 K, resulting in slagging of the ash. To enable the application of such a high temperature, the construction of the reactor had to be changed. The conventional Lurgi reactor contains a grate but in the slagging Lurgi reactor the combustion zone is without internals.

QUESTIONS:

> *Conventional moving-bed reactors cannot withstand high temperatures in the combustion zone. Why? Guess the construction of the combustion zone in a slagging Lurgi reactor.*

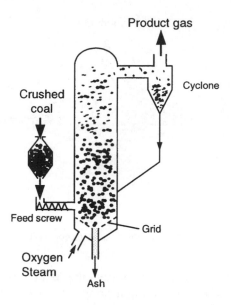

Figure 5.13   Fluidized-bed gasifier (Winkler) [21].

A *fluidized-bed gasifier* (Winkler) [21], Figure 5.13, is a back-mixed reactor in which coal particles in the feed are well mixed with coal particles already undergoing gasification. This gasifier operates at atmospheric pressure and moderate, uniform temperature. Char particles that leave the reactor with the product gas are to a large extent recovered in cyclones and recycled to the reactor. Dry ash leaves the reactor at the bottom.

As a result of the back-mixed character of the reactor, significant amounts of unreacted carbon are removed with the product gas and the ash, which lowers the conversion. The preferred feed for the Winkler gasifier is a highly reactive coal, as this disadvantage becomes greater when less reactive coal or coal with a low ash melting point is used [19]. A highly reactive coal is brown coal (in Anglo-Saxon countries usually referred to as lignite). This coal is geologically very young and, as a consequence, is highly functionalized, and therefore very reactive. In fact, the Winkler reactors were the first large-scale fluidized-bed reactors, applied in Germany for brown coal gasification in the 1930s. Due to the higher operating temperature the Winkler gasifier produces much less impurities than the Lurgi gasifier.

The *entrained-flow gasifier* (Koppers-Totzek) [19,22] (Figure 5.14) is a plug-flow system in which the coal particles react co-currently with steam and oxygen at atmospheric pressure. The residence time in the reactor is a few seconds. The temperature is high in order to maximize coal conversion. At this high temperature only CO and H_2 are formed. Ash is removed as molten slag.

QUESTION:

> *Draw a qualitative axial temperature profile for the entrained-flow gasifier. Is the statement that the reactor can be described as a plug-flow reactor correct for the whole reactor?*

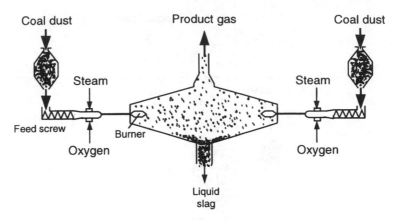

Figure 5.14 Entrained-flow gasifier (Koppers-Totzek) [19,22].

In fluidized-bed and entrained-flow reactors the coal particles are heated very fast, typically at rates over 1000 K/s. During heating extensive pyrolysis occurs and when mixing is poor considerable amounts of volatiles, coke, etc. will be formed on a microsecond scale. The process design should be such that this does not lead to lowered efficiency and problems downstream of the reactor.

Table 5.5 presents general characteristics of the three reactor types. The lowest exit gas temperature is obtained in the moving-bed reactor because of the counter-current mode of operation.

An important aspect is the state of the ash produced. When it melts (as in the slagging Lurgi and Koppers-Totzek process), it forms a vitrified material, that does not exhibit leaching upon disposal. In a low-temperature process, the ash does not go through a molten phase and, as a consequence, it is less inert causing extensive leaching upon disposal.

Not all processes can handle every type of coal. Only the entrained-flow reactor can handle all types.

QUESTION:

> *Explain why the conventional Lurgi reactor is not fit for all types of coal.*

The three processes result in quite different coal gas compositions. This is mainly due to their different operating temperature, which in turn depends on the reactor design, and is controlled by the amount of oxygen and steam fed to the gasifier. For

Table 5.5 Characteristics of moving bed, fluidized-bed and entrained-flow coal gasifiers

Reactor type	$T_{reactor}$ (K)	T_{exit} (K)	Through-put (t/d)	O_2-cons. (kg/kg coal)	d_p (mm)	Coal type
Moving bed	1250–1350	700	1500	0.5	20	Non-caking
Fluidized bed	1250–1400	1150	3000	0.7	2	Reactive
Entrained flow	1600–2200	1300	5000	0.9	<0.1	All

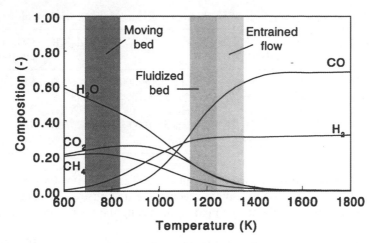

Figure 5.15 Equilibrium exit gas composition of autothermic coal gasification at 30 bar as a function of temperature; T_{exit} (gas) of industrial gasifiers indicated [17].

instance, the maintenance of a relatively low combustion temperature is essential for moving-bed gasifiers, since if the ash melts it cannot be handled on a supporting grate. Therefore, excess steam is added and very little oxygen. In an entrained-flow reactor with its small residence time, on the other hand, high temperatures are required to ensure high reaction rates. Hence, the oxygen-to-steam ratio is high.

Figure 5.15 indicates the composition of the gas phase, derived from thermodynamic equilibrium data, as a function of temperature at a pressure of 30 bar. The operating ranges of industrial gasifiers are also indicated.

The gas composition resulting from the low-temperature moving-bed process is the most complex. The hydrocarbon conversion is far from complete and CO_2 is present. In fact, the composition is more complex than shown because also many other hydrocarbons (in the figure only CH_4 is considered) are present. Furthermore, the product from moving-bed reactors also contains large amounts of tars, oils, and phenolic liquors due to the counter-current mode of operation. In most applications this is a disadvantage. In fact, moving-bed gasifiers are unsuitable for the production of synthesis gas for the production of chemicals. However, the large content of CH_4 in the gas makes it suitable for fuel gas applications and for the production of substitute natural gas (SNG). It has also been considered to use coal as the feedstock for a 'coal refinery' analogously to an oil refinery. In principle, this idea is feasible but at present coal cannot compete with oil in this respect.

An entrained-flow reactor is operated at higher O_2/coal ratio and lower amount of steam, and thus lower H/O ratio. This type of gasifier usually produces a gas with higher CO and H_2 content, while no hydrocarbons are present in the product. Such a gas is much more suitable for the production of chemicals.

5.3.4 Recent Developments

Coal gasification is currently under consideration for many applications. A large

Table 5.6 Typical performance data of gasification processes at the demonstration stage

	O_2-cons. (kg/kg coal)	C-conv (%)	Type	$T_{reactor}$ (K)	T_{exit} (K)	Cold gas eff. (%)	p (bar)
Slagging Lurgi	0.52	99.9	Moving	>2200	700	88	25
High-temp. Winkler	0.54	96	Fluidized	<1300	1200	85	10
Texaco	0.90	97	Entrained	1600–1800	<1100	74	41
DOW	0.86	99	Entrained	1600–1700	<1300	77	22
Shell	0.89	99	Entrained	2200	1200	81	30

number of processes are in the development stage. Table 5.6 presents typical performance data of processes at least having reached the demonstration stage.

Developments have mainly been aimed at increasing the operation temperature and pressure of the gasifiers.

The cold gas efficiency, a key measure of the efficiency of coal gasification, represents the chemical energy in the syngas relative to the chemical energy in the incoming coal. It is based on the heating values at standard temperature (at which H_2O is condensed, explaining the term 'cold' gas efficiency). The cold gas efficiencies of the moving-bed and fluidized-bed reactors are high compared to that of the entrained-flow reactor. This is explained by the fact that the product gas of the former two processes contains methane and higher hydrocarbons, which have a higher heating value than CO and H_2.

An important aspect of gasification processes is the way coal is fed to the reactor. In the Texaco and Dow processes a slurry of coal in water is added to the reactor. The reliability of this feeding system is very high. However, evaporation of the water requires energy resulting in efficiency loss. Therefore, in many processes dry feeding systems are applied.

Fluidized-bed reactors always show a limited carbon conversion, typically 95%. Therefore, the ash particles containing the unconverted carbon are fed to a boiler system, where the carbon is burned, providing heat for steam generation. This additional step is needed in all fluidized-bed reactor systems.

Currently, most processes are still under development and large-scale experience is limited. Therefore, it is not yet clear which process is best. Moreover, the choice of the optimal process will depend on the specific application of the gasifier gas and on local circumstances.

QUESTIONS:

Why has the moving-bed reactor the highest cold gas efficiency?
Why is the cold gas efficiency not a good parameter for comparing combined cycle power generation processes (see below)?
Discuss dry versus slurry feeding. What do you prefer? Explain.

5.3.5 Applications

Figure 5.16 shows a summary of the applications of coal gasification. When coal is gasified in air/steam mixtures low calorific gas is produced ('lean gas'), which is usually applied for heating or power generation. Coal gas is also applied in gas supply

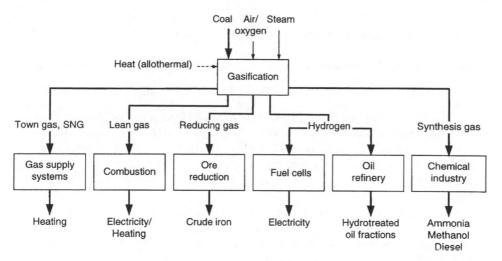

Figure 5.16 Coal gasification applications. Adapted from [23].

systems ('Substitute Natural Gas', SNG). Another application is as a raw material for the chemical industry, for instance in Fischer–Tropsch type reactions (see Chapter 6):

$$coal \rightarrow CO/H_2 \rightarrow hydrocarbons + alcohols \tag{15}$$

Coal gasification used to be the primary technology for the production of hydrogen:

$$coal \rightarrow CO/H_2 \rightarrow water\text{-}gas \; shift \rightarrow H_2 \tag{16}$$

Hydrogen can be used in a wide variety of processes applied in the oil refining and chemical industry. However, as a result of the much higher (three times! [24]) investment costs for a coal gasification plant compared to a methane steam reforming plant, coal gasification is not applied for this purpose, except maybe in situations where natural gas is not available.

Most applications of coal gasification are not unique for coal. The technology for coal gasification in principle is also applicable to other hydrocarbon type resources. Gasification is used for producing synthesis gas from gas oils and heavy residues. With these feedstocks usually the term partial oxidation is used. Probably in the future gasification will also be used to convert biomass, plastic waste, etc.

Coal is the primary fuel for electrical power generation. In a conventional power plant, steam is generated by heat exchange of boiler feed water with the flue gas from a coal combustor. This steam then drives a steam turbine to generate electricity.

A new development in power generation is the integrated coal gasification combined cycle (ICGCC) system, which consists of a coal gasifier and a combination of a gas turbine and a steam turbine for power generation. A notable example of the demonstration of combined cycle technology is a 250 MW power plant that is being run successfully in Buggenum, The Netherlands. Figure 5.17 shows a simplified flow scheme of this process.

The coal gasification process is based on an oxygen blown, dry feed, entrained-flow gasifier developed by Shell [27]. Oxygen for the gasifier is obtained in an air separa-

A thermoneutral process for methane production

In principle, coal can be converted directly into methane by reaction with steam:

$$2C + 2H_2O \rightleftharpoons CH_4 + CO_2 \qquad \Delta H_{800}^0 = 11.4 \text{ kJ/mol} \tag{17}$$

An attractive aspect of this reaction is that the process is nearly autothermic without adding oxygen. From the thermodynamic data (refer to Figure 5.9) it is clear that suitable process conditions are a relatively low temperature (<600–800 K) and a pressure depending on the temperature (higher temperature requires higher pressure for maximum CH_4 selectivity as apparent from Figure 5.9).

However, coal is not sufficiently reactive at these temperatures. Practical processes have been suggested (and developed) based on the use of a catalyst, high pressure and recycle of produced hydrogen and carbon monoxide. Figure B.5.5 shows a block diagram of a process developed by Exxon [25].

The process appeared to be technically feasible, but not economically. In the time that the process was being developed (1970s), it was concluded that the market was good. Later, large amounts of natural gas were discovered (see Chapter 2) and the process lost its economic appeal, before the intended large-scale plants had been built.

QUESTION:

When would conversion of coal into methane be useful?

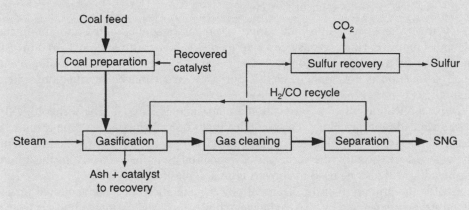

Figure B.5.5 Exxon catalytic gasification process; after [25].

tion unit. The gas from the gasifier flows into a gas cooler, after which particulates are removed and sulfur compounds are converted into sulfur.

The purified gas is sent to the gas turbine, where it is burned with compressed air to provide a stream of hot, high-pressure gas. The gas expands and conducts work on the turbine blades to turn the shaft that drives both the compressor and an electricity generator. The exhaust gases from the gas turbine are sent to a waste-heat boiler, where high-pressure steam is generated by heat exchange with boiler feed water. This steam is used in the steam turbine for the generation of additional electricity.

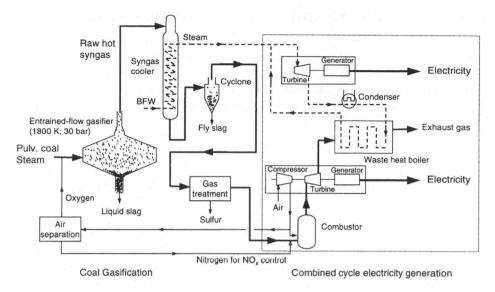

Figure 5.17   Flow scheme of integrated coal gasification combined cycle power generation

ICGCC power generation has great advantages over 'classical' pulverized coal combustors combined with a steam turbine, i.e. a high efficiency and minimal environmental impact. The efficiency of a coal gasification based combined cycle is 41–43%, much better than the efficiency of 34–35% achieved by conventional coal combustion steam cycle power plants [26].

Gasifier SO_2 and NO_x emissions are much lower than those for traditional coal combustion technologies. The sulfur compounds that are formed during gasification are converted to H_2S, whereas in coal combustion SO_2 is formed, which has to be removed from the stack gas. In the latter case the gas flow is much more diluted and, as a consequence, separation is less efficient.

During gasification N_2 and NH_3 are formed. N_2 is harmless and NH_3 is more easily removed than NO_x in the stack gases from a coal combustor. The lower amount of harmful emissions from gasification compared to combustion is the major incentive for considering this expensive technology.

QUESTIONS:

> *Why is NH_3 in gasifier gas more easily removed than NO_x in stack gases? How does nitrogen addition to the combustor in Figure 5.17 help control NO_x emissions from the process?*
>
> *Why is S more efficiently removed in gasification than in combustion? Compare oil and coal with respect to the preferred technologies for reducing S emissions.*
>
> *Extensive research and development studies are carried out with the aim to clean the gases of the gasifier at high temperature rather than cooling them to enable scrubbing at low temperature. What is the reason?*

5.4 PURIFICATION AND ADJUSTMENT OF SYNTHESIS GAS

It is self-evident that the gas produced in the steam reformer or in the gasifier in general does not have the desired composition. Nearly always extended treatment steps are required. As an example of the treatment of the raw synthesis gas, the steps to produce synthesis gas suitable for ammonia production are discussed next. Ammonia production requires a gas with hydrogen and nitrogen present in a ratio of 3 to 1:

$$3H_2 + N_2 \rightleftarrows 2NH_3 \qquad \Delta H^0_{298} = -91 \text{ kJ/mol} \tag{18}$$

The raw synthesis gas is assumed to be produced by steam reforming of natural gas followed by autothermic reforming (see Figure 5.18). The product gas from the reforming section, besides hydrogen and nitrogen, contains CO, CO_2, H_2O, and CH_4 in smaller or larger quantities. Especially CO and CO_2 have to be removed because they are poisonous to the catalyst in the ammonia production reactor (in most cases the catalyst contains metallic iron; it is not surprising that O-containing molecules are detrimental to this catalyst). Separation is possible in principle, but it is very costly. Furthermore, a large part of the feedstock would be lost as less valuable products. Therefore, adjustment and purification of the raw syngas takes place mainly by chemical conversion steps. Figure 5.18 shows the order of chemical steps involved, including temperatures, catalyst volumes, and gas compositions.

QUESTIONS:

> *Explain the temperatures of the various steps in Figure 5.18. In a plant something can go wrong during operation. When the HT shift catalyst would suddenly deactivate (because of malfunctioning somewhere upstream) what would happen? Answer the same question for when the CO_2 removal unit would break down.*

5.4.1 Conversion of Carbon Monoxide

The synthesis gas from the autothermic reformer contains appreciable amounts of

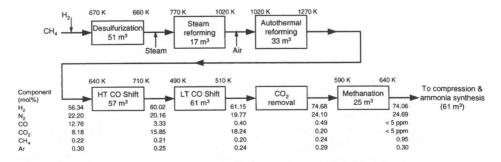

Figure 5.18 Synthesis gas production and treatment for ammonia production (1000 t/d); HT and LT indicate high temperature and low temperature, respectively; temperatures, catalyst volumes and compositions based on data from [3].

carbon monoxide and carbon dioxide, which have to be removed. In the water-gas shift reaction (2) CO is converted to CO_2:

$$CO + H_2O \rightleftharpoons CO_2 + H_2 \qquad \Delta H^0_{298} = -41 \text{ kJ/mol} \tag{2}$$

Although the reaction is only moderately exothermic, the temperature has a great impact on the equilibrium constant, as demonstrated in Figure 5.19.

Clearly, the temperature has to be low! In the past, catalysts were only active above temperatures of 590–620 K. Later, more active catalysts were developed which are active in the temperature range of 470–520 K. These catalysts, however, require extremely pure gases, because they are very sensitive to poisoning. Moreover, they are not stable at higher temperatures. In practice, the water-gas shift reaction is carried out using two adiabatic fixed-bed reactors with cooling between the two reactors. The first reactor ('High Temperature Shift') operates at high temperature and contains a classical catalyst. The second reactor contains a more active catalyst and operates at lower temperature ('Low Temperature Shift').

QUESTION:

> *Why are usually two reactors used instead of one large reactor with the superior LT catalyst?*

5.4.2 Gas Purification

The gases have to be purified to remove carbon dioxide and other impurities. It is also advantageous to remove water before the ammonia synthesis loop is entered.

QUESTION:

> *Give the reason for the latter statement.*

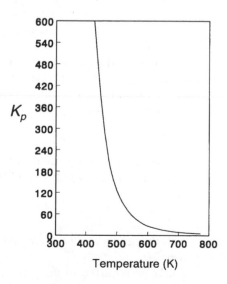

Figure 5.19 Water-gas shift equilibrium constant; $K_p = p_{H2}p_{CO2}/p_{H2O}p_{CO}$.

Table 5.7 Energy requirements for CO_2 removal systems [2]

Removal system	GJ/mol CO_2
MEA	210
MEA with inhibitors	93–140
K_2CO_3 with additives	62–107
MDEA[a] with additives	60–40

[a] Methyl diethanolamine.

CO_2 is present in large quantities after the shift reactors. It is usually removed by absorption in aqueous amines, e.g. MEA (monoethanolamine) or DEA (diethanolamine), see also Section 3.6. These amine solutions exhibit high mass-transfer rates, but regeneration is energy intensive. Therefore, at present solvents requiring lower heats of regeneration are being used. Refer to Table 5.7 for a comparison of the energy requirements of various CO_2 removal processes.

QUESTION:

Explain the trends in Table 5.7.

MEA forms corrosive compounds on decomposition, and therefore its concentration has to be relatively low (about 15–20 wt%). Adding a compound that prevents MEA from decomposing, called an inhibitor, permits higher amine concentrations (25–35 wt%), thus allowing lower circulation rates, and consequently lower regeneration duty.

The CO_2-removal processes based on potassium carbonate require even less regeneration energy because the bonding between CO_2 and the carbonate is relatively weak. Various additives are used to improve mass-transfer rates and to inhibit corrosion.

MDEA (methyl diethanolamine) has gained wide acceptance for CO_2 removal. MDEA requires very low regeneration energy because the CO_2-amine bonding is weaker than that for MEA. The additives mainly serve to improve mass transfer, as MDEA does not form corrosive compounds.

After the low temperature shift reactor the fraction of CO is small, approximately 0.4%. However, even such a small amount is poisonous to the ammonia synthesis catalyst. Therefore, the remaining CO (and residual CO_2) has to be removed. This can be done by adsorption but often methanation is applied:

$$CO + 3H_2 \rightleftarrows CH_4 + H_2O \qquad \Delta H^0_{298} = -206 \text{ kJ/mol} \qquad (-1)$$

The methanation reaction is the reverse of the steam reforming reaction and is highly exothermic. It takes place over a nickel catalyst at 620 K and about 30 bar.

QUESTIONS:

Why would a methanation step be preferred over additional water-gas shift conversion?

What steps would be required in the processing of raw syngas for methanol synthesis?

What will change when the synthesis gas is produced by coal gasification?
In HEV applications CO also often has to be removed in order to protect the catalyst. That is done by selective oxidation. Why would this not be attractive in ammonia production?

References

1 Rostrup-Nielsen JR (1984) 'Catalytic steam reforming' in: Anderson JR and Boudart M (eds.) *Catalysis. Science and Technology*, Springer-Verlag, Berlin, pp. 1–117.
2 Czuppon TA, Knez SA and Rovner JM (1992) 'Ammonia' in: Kroschwitz JI, Howe-Grant M (eds.) *Kirk-Othmer Encyclopedia of Chemical Technology* vol. 2, 4th ed., Wiley, New York, pp. 638–691.
3 Pearce BB, Twigg MV and Woodward C (1989) in: Twigg MV (ed.) *Catalyst Handbook*, 2nd ed., Wolfe Publishing Ltd.
4 Rostrup-Nielsen JR (1993) 'Production of synthesis gas' *Catal. Today* 18 305–324.
5 Dibbern C, Olesen P, Rostrup-Nielsen JR, Tottrup PB and Udengaard NR (1986) 'Make low H_2/CO syngas using sulfur passivated reforming' *Hydroc. Process.* 65(1) 71–74.
6 Udengaard NR, Bak Hansen JH, Hanson DC and Stal JA (1992) 'Sulfur passivated reforming process lowers syngas H_2/CO ratio' *Oil Gas J.* 90(10) 62–67.
7 Teuner S (1987) 'A new process to make oxo-feed' *Hydroc. Process.* 66(7) 52.
8 Ma L and Trimm DL (1996) 'Alternative catalyst bed configurations for the autothermic conversion of methane to hydrogen' *Appl. Catal.* 138 265–273.
9 Peña MA, Gomez JP and Fierro JLG (1996) 'New catalytic routes for syngas and hydrogen production' *Appl. Catal.* 144 7–57.
10 Smith JM amd Van Ness HC (1987) *Introduction to Chemical Engineering Thermodynamics*, McGraw-Hill, New York.
11 Kotas TJ (1985) *The Exergy Method of Thermal Plant Analysis*, Buttersworth, London.
12 De Swaan Arons J and Van der Kooi HJ (1993) 'Exergy analysis. Adding insight and precision to experience and intuition' in: Weijnen MPC and Drinkenburg AAH (eds.) *Precision Process Technology*, Kluwer, Dordrecht.
13 Stougy L, Dijkema GPJ, Van der Kooi HJ and Weijnen MPC (1994) 'Vergelijkende exergie-analyse van methanolprocessen' *NPT Procestechnologie* Aug, 19–23 (in Dutch).
14 Balachandran U, Dusek JT, Maiya PS, Ma B, Mieville RL, Kleefisch MS and Udovich, CA (1997) 'Ceramic membrane reactor for converting methane to syngas' *Catal. Today* 36 265–272.
15 *Nissan begins driving tests of a methanol fuel cell vehicle* (1999) Nissan Motor Co., Ltd, Corporate Communications Department,
http://global.nissan.co.jp/Japan/NEWS/19990513_0e.html
16 Tromp JJ and Moulijn JA (1988) 'Slow and rapid pyrolysis in coal' in: Yürüm Y (ed.) *New Trends in Coal Science*, Kluwer, Dordrecht, pp. 305–338.
17 Van Diepen AE and Moulijn JA (1998) 'Effect of process conditions on thermodynamics of gasification' in: Atimtay AT and Harrison DP (eds.) *Desulfurization of Hot Coal Gas*, NATO ASI Series, vol. G42, Springer-Verlag, Berlin Heidelberg, pp. 57–74.
18 Meyers RA (ed.) (1984) *Handbook of Synfuels Technology*, McGraw-Hill, New York, Chapter 3-7.
19 Cornils B (1987) 'Syngas via coal gasification' in: KR Payne (ed.) *Chemicals from Coal: New Processes*, Wiley, New York, pp. 1–31.
20 Meyers RA (ed.) (1984) *Handbook of Synfuels Technology*, McGraw-Hill, New York, Chapter 3-4.

21 Meyers RA (ed.) (1984) *Handbook of Synfuels Technology*, McGraw-Hill, New York, Chapter 3-6.

22 Speight JG (1994) ' Fuels, synthetic' in: Kroschwitz JI and Howe-Grant M (eds.) *Kirk-Othmer Encyclopedia of Chemical Technology*, vol. 12, 4th ed., Wiley, New York, pp. 126–155.

23 Van Heek KH (1986) 'General aspects and engineering principles for technical application of coal gasification' in: Figueiredo JL and Moulijn JA (eds.) *Carbon and Coal Gasification*, Martinus Nijhoff Publishers, pp. 383–401.

24 Czuppon TA, Knez SA and Newsome DS (1995) ' Hydrogen' in: Kroschwitz JI and Howe-Grant M (eds.) *Kirk-Othmer Encyclopedia of Chemical Technology*, vol. 12, 4th ed., Wiley, New York, pp. 838–894.

25 Teper M, Hemming DF and Ulrich WC (1983) 'The economics of gas from coal' EAS report E2/80.

26 Mahagaokar U and Krewinghaus AB (1993) 'Coal conversion processes (gasification)' in: Kroschwitz JI and Howe-Grant M (eds) *Kirk-Othmer Encyclopedia of Chemical Technology*, vol. 6, 4th ed., Wiley, New York, pp. 541–568.

27 Meyers RA (ed.) (1984) *Handbook of Synfuels Technology*, McGraw-Hill, New York, Chapter 2-3.

Chapter 6

Bulk Chemicals and Synthetic Fuels Derived from Synthesis Gas

The Chemical Process Industry (CPI) to a large extent aims at producing value-added chemicals from various raw materials. A convenient route is via synthesis gas, the production of which has been discussed in Chapter 5. Historically the most important processes have been ammonia and methanol production. As these processes have been developed along a path of instructive innovations they are treated here in detail. Some of their main derivatives, viz. urea from ammonia, and formaldehyde and MTBE both from methanol are also briefly discussed.

Synthesis gas is also a convenient intermediate in the conversion of natural gas and coal into transportation fuels (gasoline, diesel, and jet fuel) via the Fischer–Tropsch synthesis. Although this process is currently not applied much, in the future it will probably gain importance as the natural gas and coal reserves well exceed those of oil. In addition, large natural gas fields are often located far from possible markets, which makes conversion into a more readily transportable liquid fuel attractive.

6.1 AMMONIA

6.1.1 Background Information

Ammonia is a major product of the chemical industry. Already in the beginning of the 19th century its synthesis was subject of much research. The Industrial Revolution and the related growth of the population generated a large demand for nitrogen fertilizers.

Natural resources of N-containing fertilizers were saltpeter (KNO_3), Chile saltpeter ($NaNO_3$), and Guano (sea bird droppings). In the beginning of this century, ammonia was produced as by-product in coke ovens and gas works. In these industries, ammonia is formed during the distillation of coal. This source of ammonia is no longer important today, but it explains why industries that originally were coal based often are still ammonia producers.

It was recognized as early as the turn of the 20th century that supplies were not sufficient for agricultural needs. Moreover, the explosives industry was developing into a large volume industry due to the beginning of the First World War.

It is understandable that the direct synthesis of ammonia from N_2 in the air was attempted. However, the difficulty in the synthesis of ammonia is that nitrogen is very stable and inert. This can be understood from the data on the binding energy, the ionization potential, and the electron affinity (Table 6.1).

All values are very high. The bond dissociation energy is extremely high and, as a consequence, dissociation is expected to be very difficult. Activation by ionizing the

Table 6.1   Energy data for nitrogen

	Value (kJ/mol)	Compare with (kJ/mol)
Bond dissociation energy	945	C–H in CH_4: 439
Ionization enthalpy	1503	O_2: 1165
Electron affinity	34900	O_2: 43

molecule is also expected to be nearly impossible: the value for the electron affinity of nitrogen is very high, as illustrated by the value for oxygen, which is about 1000 times lower.

The history of the ammonia synthesis is very interesting because it was the first large-scale synthesis in the chemical industry at high pressure ($>$ 100 bar) and high temperature (670–870 K). It is remarkable that scientists and engineers succeeded in realizing a safe and reliable process in a short time, even to modern standards. The reaction to be carried out is:

$$N_2 + 3H_2 \rightleftarrows 2NH_3 \qquad \Delta H^0_{298} = -91.44 \text{ kJ/mol} \tag{1}$$

A crucial question concerned thermodynamics. Precise thermodynamic data were not known and, as a consequence, a strong controversy existed as to where the equilibrium is located at the conditions at which attempts to synthesize ammonia were made.

Around 1910 it was well documented that at atmospheric pressure in a mixture of nitrogen, hydrogen, and ammonia hardly any NH_3 is present. For instance, Haber found that at 1290 K the fraction of NH_3 in an equilibrium mixture of N_2, H_2, and NH_3 (N_2:H_2 = 1:3) was only 0.01%. These data convinced many experts, in the first instance including Haber, that industrial synthesis based on N_2 would not be economically feasible.

Meanwhile, more reliable thermodynamic data became available. Haber extrapolated these data to lower temperatures, and thus concluded that an industrial process was feasible, if suitable catalysts could be developed. It is interesting that Haber also pointed out that an industrial process is possible even if conversion is not complete. He proposed a recycle loop under pressure and also suggested the use of a feed-effluent heat exchanger. These concepts are still the basis of modern ammonia synthesis plants.

In the unbelievably short period of 5 years (1908–1913) Haber developed a commercial process in operation (30 t/d) in co-operation with BASF (Bosch and Mittach). The group at BASF tested over 6500 (!) catalysts and discovered that an iron based catalyst exhibits superior catalytic activity. The catalysts used today do not differ essentially from this catalyst, although recently a new ruthenium based catalyst has been commercialized by Kellogg [1]. Figure 6.1 shows that ammonia production has grown exponentially, although especially in Europe saturation has become visible.

Capacities of modern plants are high (see Figure 6.1). Two trends in plant scale are apparent. The first is a continuation of the present plant capacities. There is unlikely to be a trend to build larger capacity plants, except for special cases like China. Although single-stream plants of over 2000 t/d are feasible, such large projects are more difficult to finance and are likely to encounter constraints of feedstock availability and market. A second trend is to build medium-scale plants (400–

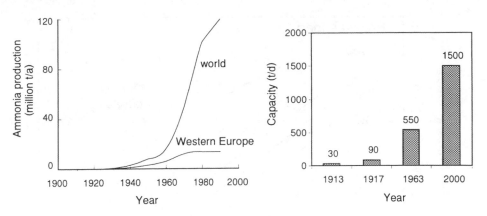

Figure 6.1   Growth of ammonia production (left) and plant capacity (right) [1,2].

600 t/d). Plants of this scale can be more easily matched to feedstock supplies and local markets.

The production of ammonia requires a mixture of H_2 and N_2 in a ratio of 3:1. The source of nitrogen is invariably air, but hydrogen can be produced from a variety of fossil fuels (see Chapter 5). Steam reforming of natural gas followed by autothermic reforming with air is most often employed in ammonia plants and accounts for approximately 80% of ammonia production [2]. Section 6.1.5 describes a complete ammonia plant based on natural gas, but first we will consider the ammonia synthesis reaction.

6.1.2 Thermodynamics

Because of the high practical relevance very reliable thermodynamic data are available on the ammonia synthesis reaction. Figure 6.2 shows the ammonia equilibrium content in equilibrium synthesis gas as a function of temperature and pressure.

Clearly, favorable conditions are low temperature and high pressure. However, kinetic limitations exist: at temperatures below about 670 K the rate of reaction is very low [3]. Therefore, a minimum temperature is required. Regarding pressure the optimal situation is a compromise between thermodynamically favorable conditions and minimum investment. Typical practical conditions are:

temperature: 675 K (inlet); 720–770 K (exit);
pressure: 100–250 bar.

QUESTION:

> *Estimate the maximum attainable (single-pass) conversion under industrial conditions (hint: do not forget that the reaction is associated with a large heat of reaction).*

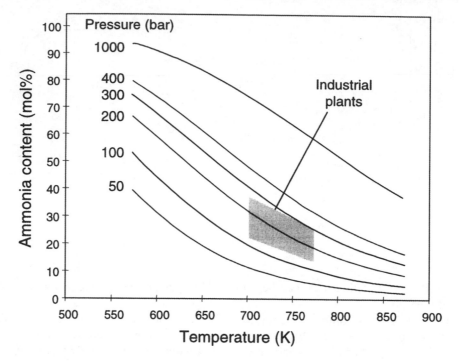

Figure 6.2 Ammonia content (mol%) in equilibrium synthesis gas; $H_2/N_2 = 3$ mol/mol; based on data from [2].

6.1.3 Commercial Ammonia Synthesis Reactors

Temperature control is crucial in ammonia synthesis: the reaction is exothermic and the heat produced needs to be removed. Two methods are applied:

1. in so-called quench reactors cold feed gas is added at different heights in the reactor;
2. the heat produced is removed between the catalyst beds by heat exchangers. Hence, heat is recovered at the highest possible temperature.

An example of a quench reactor is the ICI quench reactor [2,4], which is shown in Figure 6.3 together with the temperature-concentration profile.

Part of the cold feed is introduced at the top and passes downward between the catalyst-filled section and the converter shell, thus maintaining the shell at relatively low temperature. This is a feature used in most ammonia converter designs. It is then heated in the feed-effluent heat exchanger before entering the catalyst bed. The remainder of the feed is injected at the inlet of the second and third catalyst zones resulting in a reduction of the temperature. The single catalyst bed consists of three zones separated only by gas distributors, facilitating catalyst unloading. The effluent gas is heat-exchanged with the feed and leaves the reactor at approximately 500 K. An additional feature of this reactor is a separate inlet to which preheated feed gas can be fed to start up the reactor.

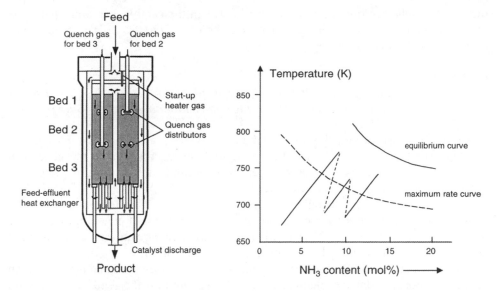

Figure 6.3   ICI quench reactor and temperature–concentration profile.

QUESTIONS:

> *Explain the temperature profile. Why are the dashed lines not parallel to the temperature axis? Are the solid lines parallel? Why or why not?*

Another quench reactor is the Kellogg vertical quench converter (see Figure 6.4) [2 4]. The Kellogg vertical reactor consists of four beds held on separate grids. Quench-gas distributors are placed in the spaces between the beds. The feed enters

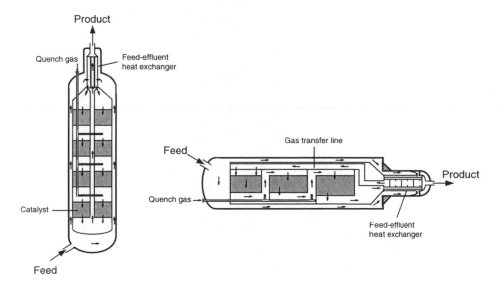

Figure 6.4   Kellogg vertical (left) and Kellogg horizontal quench reactor.

the reactor at the bottom and flows upward between the catalyst bed and shell to be heated in a feed-effluent exchanger located at the top of the vessel. The product gas outlet is also located at the top.

QUESTIONS:

> *Compare the ICI and Kellogg quench reactors. Which reactor is most convenient concerning catalyst loading and unloading?*

Kellogg has also designed a horizontal converter, for plant capacities in excess of 1700 t/d (see Figure 6.4) [2,4]. The flow pattern in this reactor remains the same as in the vertical converter, except that a larger cross-sectional area is possible, affording reduced pressure drop and catalyst particle size.

It is attractive to apply as small catalyst particles as possible to increase the efficiency of the catalyst. Of course, the pressure drop is a point of concern. In a reactor designed by Haldor Topsøe [2,3] (see Figure 6.5) two annular catalyst beds are applied and the gas flows radially. This results in a lower pressure drop.

After passing downward through the usual annulus between shell and internals, then up again through a heat exchanger located at the bottom, the feed gas passes up through a central pipe, from which it flows radially through the top catalyst bed. The gas then passes downward around the bottom bed to flow inward radially together with the quench gas. Radial flow reduces the pressure drop and permits smaller particles and more catalytic area per unit volume of reactor.

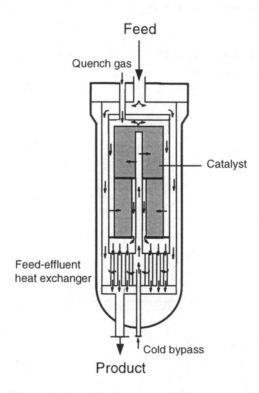

Figure 6.5   Haldor Topsøe radial-flow reactor.

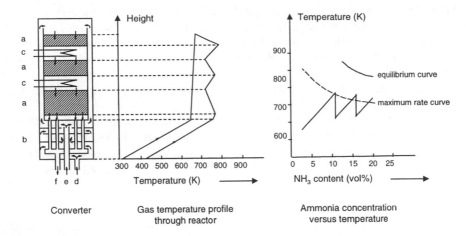

Figure 6.6 Multi-bed converter with indirect cooling; (a) catalyst, (b) feed-effluent heat exchanger, (c) cooling section, (d) feed, (e) cold bypass, (f) product [1,2].

All the reactors described so far are quench reactors. In the second type of reactor the catalyst bed is cooled by heat exchangers. The cooling medium can be the reactant gas or steam can be generated. The indirectly cooled reactor is preferred when heat is to be recovered at high temperature. Figure 6.6 shows an example of such a reactor. A typical temperature profile and a temperature-concentration diagram are also shown. The advantage of this type of reactor is that heat is recovered at the highest possible temperature. A disadvantage is that the investment cost is higher due to the cost of the interstage heat exchangers.

QUESTIONS:
> *Compare the reactor designs given. What are the similarities and what are the differences?*
> *Compare the ammonia concentration profile of Figure 6.6 with that of Figure 6.3.*

6.1.4 Ammonia Synthesis Loop

Because of the far from complete conversion (20–30% per pass), unconverted synthesis gas leaves the reactor together with NH_3. Ammonia is condensed and the unconverted synthesis gas is recycled to the reactor (see box: Synthesis loop designs).

6.1.5 Integrated Ammonia Plant

Modern ammonia plants are 'single-train' plants. All major equipment and machinery are single units. The disadvantage, of course, is that failure of a unit leads to plant shut down. The advantages are lower investments and simplicity. Fortunately, modern equipment is very reliable, and on-stream factors of over 90% are common. Figure 6.7 shows an example of an integrated ammonia plant. Accompanying flow rates and process conditions are given in the box. Ammonia production capacity is generally standardized to a production of 1360 t/d.

After desulfurization the natural gas is fed to the steam reformer (or primary reformer), where part of the feed is converted. In the autothermic reformer (also referred to as the secondary reformer), air, containing the required amount of nitrogen for ammonia synthesis is introduced. The secondary reformer operates at a temperature between 1100 and 1270 K and a pressure close to that of the primary reformer, about 30 bar. The catalyst is similar to that employed in the primary reformer (i.e. a supoported nickel catalyst).

Carbon oxides are highly poisonous to the ammonia synthesis catalyst. Therefore, the reformed gas is shifted and scrubbed for CO_2 removal. The remaining CO_2 and CO are removed by reaction with hydrogen to produce methane and water in a methanation step. The synthesis gas then is compressed and converted in the ammonia synthesis reactor.

Synthesis loop designs

Various synthesis loop arrangements are being used. Figure B.6.1 shows some typical arrangements. In all four cases the mixture leaving the reactor is separated in a condenser that removes the ammonia in the liquid state. They differ with respect to the points in the loop at which the make-up gas is added and the location of ammonia condensation. If the make-up gas is free of catalyst poisons, such as water, carbon dioxide, carbon monoxide, and perhaps some H_2S it can flow directly to the reactor (a). This represents the most favorable arrangement from a kinetic and an energy point of view: it results in the lowest ammonia content at the entrance of the reactor and the highest ammonia concentration for condensation.

When the make-up gas is not sufficiently pure, the condensation stage is partly or entirely located between the make-up gas supply and the reactor in order to remove the impurities before the gas enters the reactor. Figure B.6.1b shows the simplest form of such an arrangement. It has the disadvantage that the ammonia produced in the reactor has to be compressed together with the recycle gas. This disadvantage is avoided by using Figure B.6.1c in which the recycle gas is compressed after ammonia separation. In Figure B.6.1c, however, ammonia is condensed at lower pressure, resulting in a higher ammonia concentration in the gas stream, and thus at the entrance of the reactor. Another way to avoid compression of ammonia is to split the ammonia condensation as shown in Figure B.6.1d. The disadvantage of this scheme is the need for an additional condenser.

QUESTIONS:

Why is scheme (a) in Figure B.6.1 the most favorable from a kinetic point of view (at equal condensing temperature)? In this scheme draw the heat exchangers that you feel should be placed.

Analyze the four arrangements with respect to energy consumption.

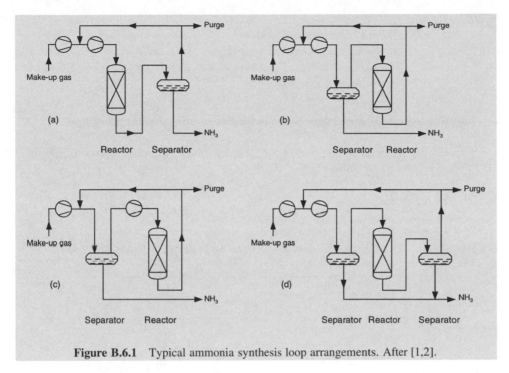

Figure B.6.1 Typical ammonia synthesis loop arrangements. After [1,2].

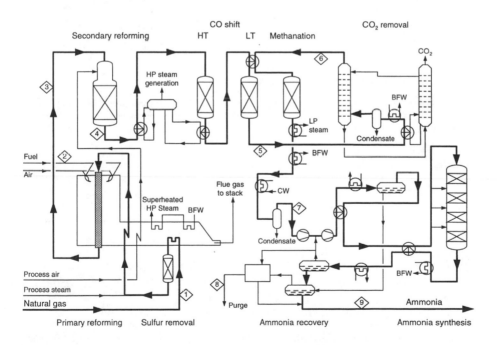

Figure 6.7 Flow scheme of an integrated ammonia plant. After [2]. Used with permission of Wiley-VCH

QUESTIONS:

*Which reactions take place in the reactors in Figure 6.7?
At many places heat is exchanged. Make an analysis with respect to
the production of HP, LP steam, etc. Referring to Figure B.6.1, what
arrangement is applied?*

Compositions and flow rates of an actual ammonia production plant

Data on flow rates and compositions of the streams in the plant given in Figure
6.7 are presented in Tables B.6.1 and B.6.2.

QUESTION:

Estimate the efficiency of the process.

Table B.6.1 Feed and product streams in Figure 6.9. After [2]. Used with permission of Wiley-VCH.

Stream no. Composition (vol%)	1 Feed: natural gas	2 Fuel[a]	9 Ammonia
CH_4	82.0	62.2	–
C_nH_m	3.4	2.4	–
N_2	14.3	15.1	–
H_2	–	18.7	–
NH_3	–	–	100
Others	0.3	1.6	–
Flow rate (kmol/h)	1770	1300	56.67 (t/h)
p (bar)	40	4	25
T (K)	303	288	284

[a] Fuel = natural gas + purge and flash gas.

Table B.6.2 Other streams in Figure 6.9. After [2]. Used with permission of Wiley-VCH.

Stream no.	3		4	5	6	7	8
Composition (vol%)	Wet	Dry					
H_2	40.8	67.9	56.7	61.4	74.8	74.2	54.0
CO	6.4	10.6	12.9	0.3	0.4	5 ppm	–
CO_2	5.8	9.6	7.5	17.8	0.01	–	–
N_2	2.6	4.4	22.4	19.9	24.3	24.8	18.8
CH_4	4.6	7.5	0.3	0.2	0.3	0.7	20.0
Ar	–	–	0.2	0.2	0.2	0.2	7.2
H_2O	39.8	$(66.1)^a$	$(56.5)^a$	$(39.3)^a$	$(1.06)^a$	–	–
Flow rate (kmol/h)	9535	5731	7691	8639	7069	6917	250
p (bar)	32	32	31.3	28.3	26.6	25	5
T (K)	1103	1103	1263	518	341	308	293

[a] vol% of H_2O in stream; the other amounts are on a dry basis, i.e. neglecting the water content; e.g., stream 4 actually contains only 32.0 vol% H_2.

Table 6.2 Specific energy requirements of various ammonia processes [1,2]

Process	GJ (LHV)/t ammonia[a]
Classical Haber-Bosch (coke)	80–90
Reformer pressure 5–10 bar (1953–1955)	47–53
Reformer pressure 30–35 bar (1965–1975)	33–42
Low-energy concepts (1975–1984)	27–33
State of the art (since 1991)	24–26

[a] LHV = lower heating value.

> *Do you expect that for the ammonia synthesis process further improvements are possible?*
> *List possible energy saving measures.*

Table 6.2 shows specific energy requirements for various ammonia processes. The classical Haber–Bosch ammonia process was based on coal as feedstock. This process is rather inefficient for production of hydrogen, because of the low H/C ratio in coal. Also, gas purification was much less efficient then. The introduction of steam reformers operating on a natural gas feed much improved the energy efficiency because of the high H/C ratio in the feed and the higher purity of the raw synthesis gas. During the 1960s high-pressure reformers came on stream, reducing the syngas compression power required (and investment costs). At that time also low-temperature shift and methanation catalysts were introduced, resulting in a significant improvement in gas quality, and thus higher ammonia yields per amount of natural gas processed.

The most important improvement of that time, however, was the use of centrifugal instead of reciprocating compressors. Centrifugal compressors have many advantages over reciprocating compressors, such as low investment and maintenance costs, high reliability and low space requirement. Centrifugal compressors are very efficient at high flow rates, which makes them well suited for use in the large capacity single-train ammonia plants.

Low-energy concepts resulted in better heat integration, e.g. recovering of most of the reaction heat by preheating boiler feed water and raising high-pressure steam. With state of the art technology, using different ammonia converter designs, even lower energy consumptions are possible.

An intrinsic inefficiency in the production of ammonia is the need for a purge in the recycle stream. This stream contains large amounts of hydrogen so recovery is attractive. Because the gas is available at high pressure, membrane technology (see also Section 3.6.2) can be a favorable option.

6.1.6 Applications of Ammonia

At present, about 85% of ammonia production is used for nitrogen fertilizers [6]. Direct application of ammonia represents the largest single consumption (about 30%). Of the solid fertilizers urea is the most important, accounting for about 40% of ammonia usage. Other solid nitrogen fertilizers are ammonium nitrate, ammonium sulfate, and ammonium phosphates.

Example of a membrane based H$_2$ recovery form the purge

Figure B.6.2 shows a flow diagram for a membrane based H$_2$ recovery process in an ammonia plant. The hydrogen permeates preferentially through the membrane. Usually, a two-stage process such as shown in Figure B.6.2 is applied. The two membrane units each consist of several modules. Table B.6.3 gives the stream compositions.

Table B.6.3 Composition of streams in ammonia purge separation [5]

Composition (mol%)	1	2	3	4	5
H$_2$	61.0	62.2	87.3	84.8	20.8
N$_2$	20.5	20.9	7.1	8.4	43.8
CH$_4$	10.5	10.7	3.6	4.3	22.5
Ar	6.0	6.1	2.0	2.5	12.9
NH$_3$	2.0	<0.02	<0.02	<0.02	0.02
Flow rate (m^3 h^{-1})	3868	3821	1591	820	1410
Pressure (bar)	140	–	70	25	–

QUESTIONS:

Why are two sets of membrane modules used?
What is the purpose of the scrubber?
Every process consumes energy. Analyze this scheme in terms of energy consumption.

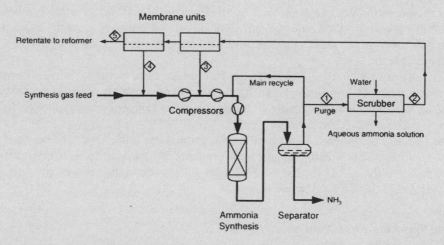

Figure B.6.2 Flow diagram of hydrogen recovery in ammonia synthesis. Adapted from [5].

Industrial applications of ammonia include the production of amines (e.g. H$_3$C—NH$_2$, (H$_3$C)$_2$—NH), nitriles (e.g. acrylonitrile: CH$_2$=CH-C≡N), and organic nitrogen compounds for use as intermediates in the fine chemicals industry.

Ammonia becomes increasingly important in environmental applications, such as removal of NO_x from flue gases of power plants.

6.1.6.1 Production of urea

Urea is produced by reacting ammonia and carbon dioxide. The latter is a by-product of the ammonia production process.

Reactions and thermodynamics The formation of urea occurs through two uncatalyzed equilibrium reactions. The first reaction, the formation of ammonium carbamate (liquid at reaction conditions, solid at standard temperature and pressure), is exothermic, and the second reaction is endothermic. Since more heat is produced in the first reaction than is consumed in the second reaction, the overall reaction is exothermic.

$$2\,NH_3 + CO_2 \overset{fast}{\rightleftharpoons} H_2NCONH_4 \qquad \Delta H^0_{298} = -117\,kJ/mol \qquad (2)$$

ammonium carbamate

$$H_2NCONH_4 \overset{slow}{\rightleftharpoons} H_2NCNH_2 + H_2O \qquad \Delta H^0_{298} = 16\,kJ/mol \qquad (3)$$

The most important side reaction is the formation of biuret:

$$2\,H_2NCNH_2 \overset{slow}{\rightleftharpoons} H_2NCNCNH_2 + NH_3 \;(\text{slightly endothermic}) \qquad (4)$$

Biuret is detrimental to crops already at very low concentrations, so its formation should be minimized [7].

QUESTIONS:

> *What is the influence of temperature and pressure on the three reactions?*
> *What would be the exact source of CO_2?*

Thermodynamically, both CO_2 conversion and the urea yield as a function of temperature go through a maximum, which, of course, is dependent on the initial composition, but usually lies between 450 and 480 K at practical conditions.

Increasing the NH_3/CO_2 ratio leads to a higher CO_2 conversion (but a lower NH_3 conversion). The yield on urea has a maximum at a NH_3/CO_2 ratio of somewhat above the stoichiometric ratio (2:1). A large excess of ammonia results in a reduced yield (which is defined as mole urea produced per mole feed).

Water has a negative effect on the conversion of ammonia and carbon dioxide and consequently on urea yield.

QUESTIONS:

Why do the CO_2 conversion and the urea concentration go through a maximum as a function of temperature?
Why does the presence of water have a negative effect on conversion and urea yield?
Why is the urea yield not maximum at the stoichiometric NH_3/CO_2 ratio but somewhat above it?

From a kinetic viewpoint, both the carbamate forming reaction and the urea forming reaction proceed faster with increasing temperature. The carbamate-to-urea conversion is much slower than carbamate formation; temperatures of over 420 K are required for a sufficiently high reaction rate. At these temperatures, pressures of over 130 bar are required to prevent ammonium carbamate from dissociation into ammonia and carbon dioxide, i.e. the reverse of reaction (2). No catalyst is used because of the corrosiveness of the reaction mixture, which would result in too many technical difficulties with respect to catalyst stability.

At economically and technically feasible conditions, the conversion to urea is only between 50 and 75%. Hence, the reactor effluent will always contain a considerable amount of carbamate, so a separation step is necessary.

In the separation of urea from the unconverted carbamate the high rate of reaction (2) presents an opportunity for process simplification. Upon reducing the pressure of the reactor effluent (by flashing through an expansion valve), the carbamate will decompose to yield the initial reactants. The remaining liquid phase is a urea solution containing up to 80 wt% urea [8]. Fortunately, urea is rather stable and if the residence time at a high temperature is sufficiently short, hardly any urea will hydrolyze to form carbamate.

Conventional processes   The key challenge in the design of a urea plant is efficient handling of the gases from the carbamate decomposition stage. The major problem is that in downstream processing low temperatures are encountered and this favors formation of by-products, such as biuret.

There are many different approaches, which is illustrated by the remarkably large number of urea processes. These may be divided in two groups, viz. once-through processes and total recycle processes.

The first commercial urea processes were *once-through* processes (Figure 6.8), in which the unconverted reactants were used in downstream processes, for instance in the production of ammonium nitrate and nitric acid. This is the simplest way of dealing with the NH_3/CO_2 gas mixture from the decomposition stage. However, a major disadvantage of this type of plant is that the economics of the plant are not only dependent on the market of the main product, i.e. urea, but also on that of the by-products, such as ammonium nitrate and nitric acid [9].

In *total recycle* processes no by-products are produced. However, recycling of the

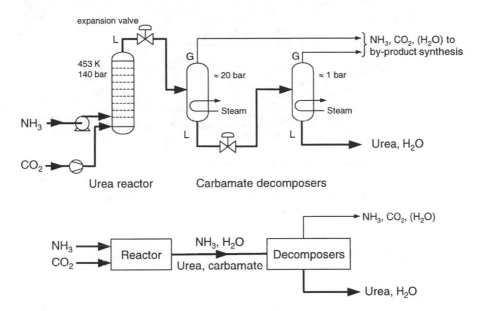

Figure 6.8 Once-through process for urea production; L, liquid; G, gas.

NH$_3$/CO$_2$ mixture is not as straightforward as it may seem. The reason is that recompression to reaction pressure is required, which would result in the recombination of NH$_3$ and CO$_2$ to form liquid carbamate (or solid carbamate at lower temperature). Normal compressors cannot handle gases containing droplets or crystals [10]. There are essentially two ways to overcome this problem.

The first solution is to separate the gases and recycle them separately. Separation of NH$_3$ and CO$_2$ generally takes place by scrubbing with a selective solvent. Acid solvents like NH$_4$NO$_3$ selectively absorb ammonia, while alkaline solvents such as aqueous amines (e.g. MEA and DEA, see Section 5.4.2) selectively absorb CO$_2$.

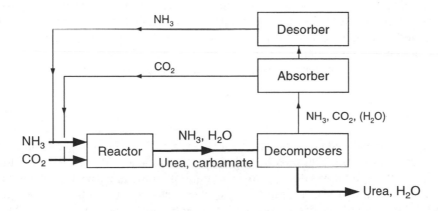

Figure 6.9 Total recycle process for urea production – recycle of separated NH$_3$ and CO$_2$.

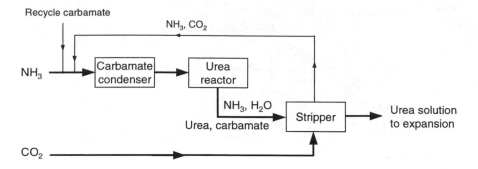

Figure 6.10 CO_2 stripping process for urea production.

Figure 6.9 shows a block diagram of a process in which ammonia is selectively absorbed.

The second solution is to condense the gases and recycle them in the liquid phase.

Stripping processes   In the processes described so far decomposition of carbamate is accomplished by a combination of pressure reduction and heating. However, pressure reduction is not the only method: lowering of the CO_2 or NH_3 concentration in the solution leaving the reactor also leads to decomposition of the carbamate. This can be accomplished by stripping of the product mixture from the urea reactor with either NH_3 (Snamprogretti) or CO_2 (Stamicarbon).

Figure 6.10 shows the main features of a CO_2 stripping process. CO_2 contacts the effluent from the urea reactor in a stripper, while heat is supplied at the same time (see box). Unconverted NH_3 is stripped from the solution, resulting in the decomposition of carbamate. The gases leaving the stripper flow to the carbamate condenser where the NH_3/CO_2 ratio is brought to a value of 3 by supplying fresh NH_3.

The major advantage of stripping processes is that the unconverted reactants are mainly recycled via the gas phase, which eliminates the large water recycle to the reactor section; about 85% of the carbamate is removed in the stripper. The remainder is removed by expansion and further processing (see [7–11] for more details).

The urea product is a concentrated solution of ca. 72 wt% urea in water. It is usually processed further in a number of steps to produce prills (particles).

Stripper

The stripper in the Stamicarbon urea process is a falling-film type shell and tube heat exchanger (see Figure B.6.3). The reactor effluent is distributed over the stripper tubes and flows downward as a thin film on the inner walls of the tubes counter-currently to the stripping CO_2, which enters the tubes at the bottom. The tubes are externally heated by steam flowing at the shell side of the heat exchanger.

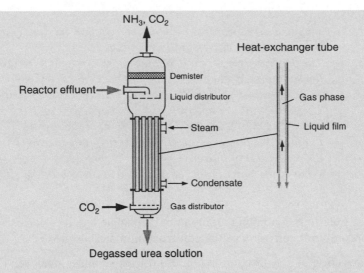

Figure B.6.3 Schematic of stripper used in CO_2 stripping process.

Pool reactor

A recent development by Stamicarbon, the so-called 'pool reactor' has greatly simplified the reactor section, and thereby reduced the investment costs. The pool reactor combines the condensation of carbamate from ammonia and carbon dioxide with the dehydration of carbamate into urea in one vessel (see Figure B.6.4) [11].

The off-gasses from the stripper, the carbamate recycle stream, and the ammonia feed are introduced in the pool reactor. The liquid phase is agitated by the gases from the stripper entering through a gas divider. The heat of condensation is partly used to supply the heat for the endothermic urea forming reaction and partly for raising steam in the tube bundle.

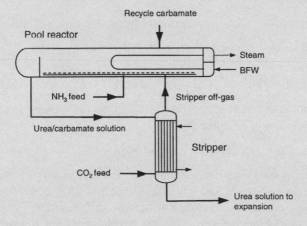

Figure B.6.4 Pool reactor synthesis section for the Stamicarbon CO_2 stripping process.

QUESTIONS:

> *Why is a large water recycle to the reaction section, typical of solution recycle processes, a disadvantage?*
> *The process using recycle of the separate reactants (Figure 6.9) and the solution recycle process are not very energy-efficient. What is the reason for this? (Hint: What are the heat effects of the various steps?)*
> *The carbamate condenser, the urea reactor, and the stripper all operate at a pressure of 140 bar. What is the reason?*
> *Evaluate the flow scheme of the CO_2 stripping process with respect to the formation of biuret.*
> *What is the function of the demister at the top of the stripper?*

Corrosion   Solutions containing ammonium carbamate are very corrosive. Some of the measures to limit corrosion of the construction materials are:

- proper selection of construction materials, at least stainless steel, see [7];
- addition of a small amount of air with the CO_2 feed. By reaction of oxygen with the stainless steel, the protective metal oxide layer is kept intact.
- minimal exposure of materials to the corrosive carbamate solution. Stripping processes are preferred over solution recycle processes from this point of view.

6.2 METHANOL

6.2.1 Background Information

Methanol synthesis is the second large-scale process involving catalysis at high pressure and temperature (BASF, 1923). The same team that developed the ammonia synthesis process also developed a commercial process for the production of methanol based on synthesis gas ($H_2/CO/CO_2$).

6.2.2 Thermodynamics

The main reactions for the formation methanol from synthesis gas are:

$$CO + 2H_2 \rightleftharpoons CH_3OH \qquad \Delta H^0_{298} = -90.8 \text{ kJ/mol} \tag{5}$$

$$CO_2 + 3H_2 \rightleftharpoons CH_3OH + H_2O \qquad \Delta H^0_{298} = -49.6 \text{ kJ/mol} \tag{6}$$

The two methanol forming reactions are coupled by the water-gas-shift reaction, which has been discussed in Chapter 5:

$$CO + H_2O \rightleftharpoons CO_2 + H_2 \qquad \Delta H^0_{298} = -41 \text{ kJ/mol} \tag{7}$$

Table 6.3 shows typical equilibrium data for the methanol forming reactions. Clearly, the temperature should be low. Note that the CO_2 conversion increases with temperature as a result of the reverse water-gas shift reaction.

Compared with ammonia synthesis, catalyst development for methanol synthesis was more difficult, because besides activity, *selectivity* is crucial. It is not surprising

Table 6.3 CO and CO$_2$ equilibrium conversion data [12]

Temp. (K)	CO conversion			CO$_2$ conversion		
	Pressure (bar)			Pressure (bar)		
	50	100	300	10	100	300
525	0.524	0.769	0.951	0.035	0.052	0.189
575	0.174	0.440	0.825	0.064	0.081	0.187
625	0.027	0.145	0.600	0.100	0.127	0.223
675	0.015	0.017	0.310	0.168	0.186	0.260

that in CO hydrogenation other products can be formed such as higher alcohols and hydrocarbons. Thermodynamics show that this is certainly possible. Figure 6.11 shows thermodynamic data for the formation of methanol and some possible by-products resulting from the reaction of CO with H$_2$, with water as by-product.

Clearly, methanol is thermodynamically less stable and in that sense less likely to be formed from CO and H$_2$ than other possible products, such as methane, which can be formed by the methanation reaction (see Chapter 5). Therefore, the catalyst needs to be very selective. The selectivity of modern catalysts (Cu/ZnO/Al$_2$O$_3$) is over 99%, which is remarkable considering the large number of possible by-products [13].

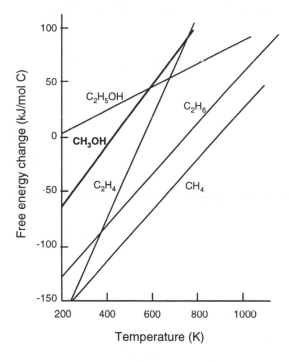

Figure 6.11 Products from CO and H$_2$; change of standard free energy of formation (ΔG^0) as a function of temperature.

QUESTION:

> *Why can in methanol synthesis in theory a poor catalyst (having a low selectivity) lead to a 'runaway'?*

The original catalysts ($ZnO–Cr_2O_3$) were only active at high temperature. There-fore, the pressure had to be high (250–350 bar) to reach acceptable conversions. Until the end of the 1960s the original catalyst was used. More active catalysts were known, but they were not resistant to impurities such as sulfur. In modern plants the synthesis gas is very pure, which allows the use of very active catalysts. This has led to a new generation of plants called 'Low-Pressure Plants'.

Figure 6.12 shows the equilibrium conversion of CO to methanol as a function of temperature and pressure. The typical temperature ranges for the classical and modern methanol processes are indicated in the figure. The figure illustrates that the develop-ment of catalysts that are active at lower temperature made it possible to operate at lower pressure (50–100 bar), while maintaining the same conversion as in the classical process. Temperature is critical. A low temperature is favorable from a thermody-namic point of view. Moreover, the high activity catalysts are sensitive to 'sintering' which increases progressively with temperature.

Catalyst deactivation will always occur. The usual practice for restoring activity, increasing temperature, is normally applied to maintain the rate of reaction, but pressure increase is also applied. The temperature must not exceed 570 K because then unacceptable sintering of the catalyst will occur [13].

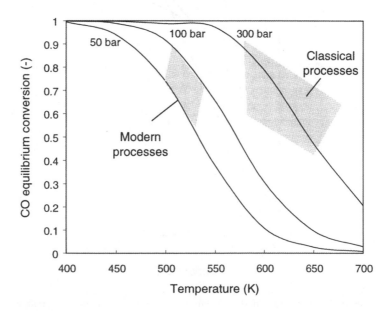

Figure 6.12   Equilibrium CO conversion to methanol; feed $H_2/CO = 2$.

Sintering

Metal catalysts often consist of metal crystallites deposited on a porous carrier. Sintering of these catalysts occurs by the growth of the metal crystallites, resulting in the loss of active surface area (see Figure B.6.5). Sintering is usually irreversible. Sintering phenomena in most cases are a result of too high temperatures, i.e. they lead to thermal deactivation. However, the chemical environment can also play a role. For instance, the presence of steam or traces of chlorine often accelerates thermal deactivation.

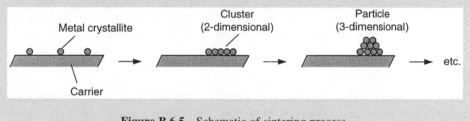

Figure B.6.5 Schematic of sintering process.

6.2.3 Synthesis Gas for Methanol Production

Nowadays, most synthesis gas for methanol production is produced by steam reforming of natural gas. The production of synthesis gas has been dealt with in Chapter 5. The ideal synthesis gas for methanol production has a H_2/CO ratio of about 2. A small amount of CO_2 (about 5%) increases the catalyst activity. A H_2/CO ratio lower than 2 leads to increased by-product formation (higher alcohols, etc.), a higher ratio results in a less efficient plant due to the excess hydrogen present in the syngas, which has to be purged.

The composition of synthesis gas depends on the feedstock used. When naphtha is the raw material, the stoichiometry is approximately right. When CH_4 is used, however, H_2 is in excess.

QUESTIONS:

> *Why, despite the better stoichiometry of naphtha reforming, would natural gas still be the most applied feedstock for methanol production?*
> *In the purification of synthesis gas for methanol, would a methanation reactor be included? And a shift reactor?*

In practice, either the excess H_2 is burned as fuel, or the carbon oxides content in the syngas is increased. This can be done by one of two methods:

1. When available, CO_2 addition to the process is a simple and effective way to balance the hydrogen and carbon oxides content of the syngas. CO_2 addition can be implemented by injecting CO_2 either in the reformer feed stream or in the raw syngas. In both cases the stoichiometric ratio for methanol synthesis is achieved, although the compositions will be somewhat different.

QUESTION:

> *Why will the synthesis gas compositions be different?*

2. Installing an oxygen-fired autothermic reformer downstream the reformer, as discussed in Chapter 5 or using only autothermic reforming. In the latter case, the synthesis gas is too rich in carbon oxides, so adjustment by CO_2 removal is required.

QUESTIONS:

> *In syngas production recently a large explosion has taken place. Discuss safety aspects of this process. Where are the most probable process steps where explosions in principle might occur?*

6.2.4 Methanol Synthesis

The first industrial plants were based on a catalyst that was fairly resistant to impurities but not very active and selective. Therefore, to achieve a reasonable conversion to methanol these plants were operated at high pressure. Figure 6.13 shows a flow scheme of such a high-pressure plant.

The heart of the process is the reactor with a recirculation loop according to the same principle as in ammonia synthesis. All methanol processes still follow this principle. The crude methanol is distilled to separate the methanol from water and impurities. The conversion per pass is low, which facilitates good temperature control. Nevertheless special reactor designs are required. Quench reactors and cooled multi-tubular reactors are most commonly applied.

The high operating pressure (ca. 300 bar) of this classical methanol process results in high investment and high syngas compression costs. In addition, large amounts of

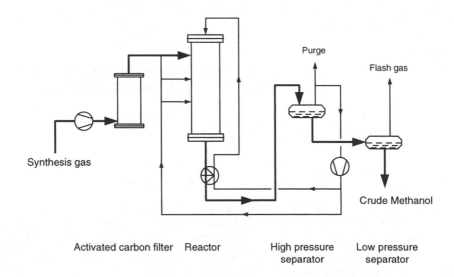

Figure 6.13   Simplified flow scheme of a classical methanol process.

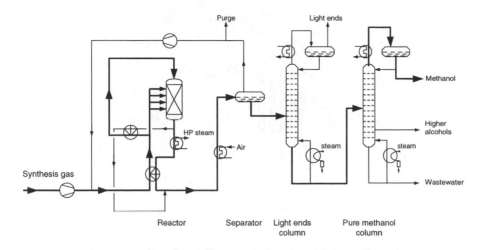

Figure 6.14   Flow scheme of the ICI low-pressure methanol process.

by-products are formed due to the low selectivity of the catalyst. These by-products include ethers, hydrocarbons, and higher alcohols.

All modern processes are low-pressure processes with plant capacities ranging from 150 to 6000 t/d, although plants using remote natural gas may have a capacity as large as 10000 t/d. The plants differ mainly in reactor design, and, interrelated with this, in the way the heat of reaction is removed.

In the ICI process [14] an adiabatic reactor is used with a single catalyst bed. The reaction is quenched by adding cold reactant gas at different heights in the catalyst bed (see Figure 6.3). The temperature in the bed has a saw-tooth profile (compare Figure 6.3). Figure 6.14 shows a flow scheme of the ICI process.

Fresh synthesis gas is compressed and mixed with recycle gas. The mixture is heated by heat exchange with reactor effluent. Then about 40% of the stream is sent to the reactor after undergoing supplementary preheating also by the reactor effluent. The rest is used as quench gas. The reactor effluent is cooled by heat-exchange with the feed and with water required for high-pressure steam generation, and then by passage through an air-cooled exchanger in which the methanol and water are condensed. Gas/liquid separation is carried out in a vessel under pressure. The gas is recycled after purging a small part to keep the level of inerts in the loop within limits. The crude methanol is purified in two columns. The first column removes gases and other light impurities, the second separates methanol from heavy alcohols (side stream) and water.

QUESTIONS:

> *The crude methanol is purified with two distillation columns. Are there more configurations possible than the one given in Figure 6.14? Which one do you prefer and why?*

The Lurgi process is very similar to the ICI process. The most important difference is the reactor. In the Lurgi process a cooled multi-tubular reactor is applied (see Figure 6.15) [15,16].

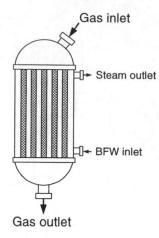

Figure 6.15   Schematic of Lurgi reactor for methanol production.

The catalyst particles are located in the tubes and cooling takes place by boiling water (BFW). The Lurgi reactor is nearly isothermal. In the Lurgi process the heat of reaction is directly used for generation of high-pressure steam (ca. 40 bar), which is used to drive the compressors and subsequently as distillation steam.

In the Haldor Topsøe process (see Figure 6.16) several adiabatic reactors are used, arranged in series [17]. Intermediate coolers remove the heat of reaction. The synthesis gas flows radially through the catalyst beds, which results in reduced pressure drop compared to axial flow. Crude methanol purification is the same as in the other processes.

The choice of methanol process design depends on various factors, such as type of feedstock, required capacity, energy situation, local situation, etc.

QUESTIONS:

> *Compare the three reactor designs. Give advantages and disadvantages. Which reactor system has the smallest total catalyst volume for the same methanol production and which the largest? Explain.*

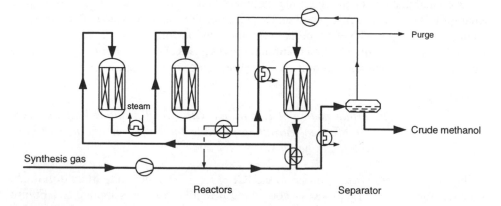

Figure 6.16   Flow scheme of the reaction section of the Haldor Topsøe methanol process.

6.2.5 Applications of Methanol

The major application of methanol is in the chemical industry, where it is used as a solvent or as an intermediate. It is also increasingly used in the energy sector. Figure 6.17 shows a survey of the most important reactions involving methanol.

6.2.5.1 Formaldehyde

One of the largest applications, accounting for about 25% of methanol consumption, is the synthesis of formaldehyde, which is based on the following reactions:

$$CH_3OH \rightleftharpoons CH_2O + H_2 \qquad \Delta H_{298}^0 = 85 \text{ kJ/mol} \tag{9}$$

$$CH_3OH + \tfrac{1}{2}O_2 \rightarrow CH_2O + H_2O \qquad \Delta H_{298}^0 = -158 \text{ kJ/mol} \tag{10}$$

Methanol process based on slurry reactor

A relatively new development in methanol synthesis is a *three-phase* process developed by Air Products and Chem Systems [18,19] (see Figure B.6.6). In this process a fluidized-bed reactor is applied, in which the solid catalyst is suspended in an inert hydrocarbon liquid, while the syngas passes through the bed in the form of bubbles. The catalyst remains in the reactor, and the hydrocarbon liquid, after separation from the gas phase, is recycled via a heat exchanger. The main advantage of this process is that the presence of the inert hydrocarbon limits the temperature rise as it absorbs the heat liberated by the reaction, while it also keeps the temperature profile in the reactor uniform. Therefore, a higher single-pass conversion can be achieved than in conventional processes, reducing the syngas compression costs.

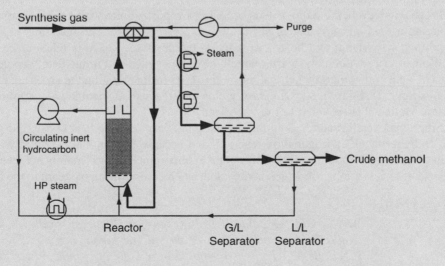

Figure B.6.6 Chem Systems three-phase methanol process [18].

The direct conversion of methane

The direct conversion of methane to methanol has been the subject of many academic research efforts during the past decade:

$$CH_4 + 0.5O_2 \rightleftarrows CH_3OH \qquad\qquad \Delta H^0_{298} = -126 \text{ kJ/mol} \qquad\qquad (8)$$

Such a process would be a spectacular event. In principle, the efficiency would be increased enormously, compared to the conventional process of steam reforming followed by methanol synthesis. In addition a real contribution to the reduction of the Greenhouse effect would be realized. This is linked to the formation of CO_2, a major greenhouse gas, from methane in the reformer feed and fuel [20]: during steam reforming of hydrocarbons (see Chapter 5) CO_2 is formed by the water-gas shift reaction in the reformer. Moreover, combustion of fuel for heating the reformer furnace yields a large amount of CO_2. Both sources of CO_2 would be eliminated in a direct conversion process. Unfortunately, to date the low yields achieved are a major obstacle to the commercialization of this route [21].

QUESTION:

Evaluate the difference between the direct oxidation (Eqn. (8)) and the route via syngas with respect to CO_2 emissions. Assume the most optimal case (perfect catalysts, etc).

So, formaldehyde can be produced by dehydrogenation (9) or by partial oxidation (10). The former is endothermic, while the latter is exothermic. As should be expected, several side reactions take place.

In industrial practice often dehydrogenation and partial oxidation are carried out in a single reactor in which the heat produced by the exothermic partial oxidation reaction supplies the heat for the endothermic dehydrogenation reaction. Figure 6.18 shows a flow scheme of such a process.

Fresh and recycle methanol is evaporated in a vaporizer into which air is sparged to generate a feed mixture that is outside the explosive limits. The resulting vapor is combined with steam and heated to reaction temperature. Reaction takes place in a shallow bed of silver catalyst, after which the product gases are immediately cooled in a waste heat boiler, thereby generating steam. After further cooling the gases are fed to an absorber. Distillation subsequently yields a 40–55 wt% formaldehyde solution in water. Methanol is recycled to the vaporizer.

Since the overall reaction is highly exothermic, despite some heat consumption by the endothermic dehydrogenation reaction, and because it occurs at essentially adiabatic conditions, temperature control is of great importance. This is mainly achieved by the use of excess methanol, which is recycled, and by the addition of steam to the feed.

QUESTIONS:

Why would the catalyst bed (Figure 6.18) be so shallow? What would be the composition of the off-gas (What are possible side reactions)? Which percentage of methanol would have to be converted by the partial oxidation reaction for autothermic opera-

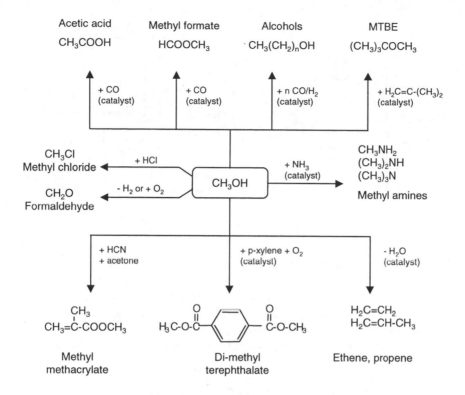

Figure 6.17   Methanol as base chemical.

tion? Would a fully balanced process (in terms of heat consumption) be possible in practice? (Hint: consider differences in the expected rates of the dehydrogenation and the oxidation reactions.)
What type of reactor would you choose for the partial oxidation process and why? What would be advantages and disadvantages of this process compared to the process described above?

A process of equal industrial importance is based on the partial oxidation reaction only (metal oxide catalyst) [22–24]. In contrast to the silver-catalyzed process, this process operates with excess air. Therefore, it has the advantage that the distillation column for methanol recovery can be omitted, because in this case methanol conversion is over 99%.

6.2.5.2 MTBE

Increasing amounts of methanol (currently about 30%) are used in combination with isobutene in the production of methyl *tert*-butyl ether (MTBE). MTBE has a high octane number (116) and is used as an octane booster in gasoline. It has been the fastest growing chemical in the bulk chemical industry of the last decade, see Figure

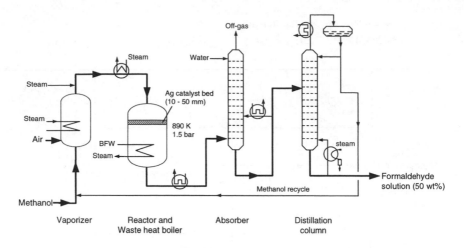

Figure 6.18 Flow scheme of typical formaldehyde process using a silver catalyst [22–24].

6.19. Recently, however, the environmental friendliness has been questioned. It appears that in case of leakage and spills MTBE enters the groundwater because it is well miscible with water. In California it might well be banned.

The etherification reaction is an equilibrium-limited exothermic reaction:

$$\Delta H^0_{298,\text{liq}} = -37.5 \text{ kJ/mol} \quad (11)$$

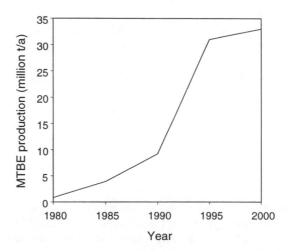

Figure 6.19 Growth in world production MTBE [25].

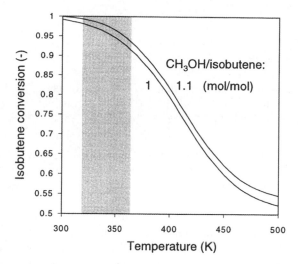

Figure 6.20 Thermodynamic equilibrium conversion of isobutene to MTBE in the liquid phase as a function of temperature and methanol/isobutene mole ratio at a pressure of 20 bar; shaded area represents typical operation temperatures.

The isobutene feed usually is a mixed C_4 hydrocarbon stream containing about 30% isobutene. The other constituents are mainly *n*-butane, isobutane, and *n*-butene, which are inert under typical reaction conditions (320–360 K and 20 bar). The reaction takes place in the liquid phase.

With conventional processes employing fixed-bed reactors and slight excess methanol it is possible to obtain isobutene conversions of 90–97%, which is slightly less than the thermodynamic attainable conversion (see Figure 6.20).

Separation of MTBE from the unconverted C_4-fraction and methanol is achieved by distillation. Methanol is recovered for recycle (see Figure 6.21). Because separation of isobutene from the other hydrocarbons is difficult it usually leaves the process with the inerts. Therefore, an isobutene loss of 3–10% must be accepted. This is not the case with catalytic distillation, see box.

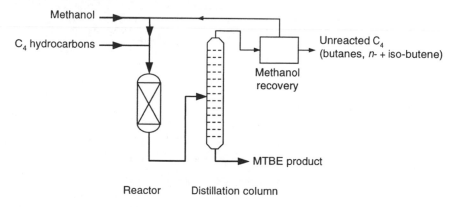

Figure 6.21 Production of MTBE by the conventional process. Adapted from [26].

Catalytic Distillation

The isobutene conversion can be enhanced by continuous removal of the product MTBE. Catalytic distillation offers this possibility, thus increasing the isobutene conversion to up to 99%. Figure B.6.7 shows a schematic of this process.

The catalyst is placed inside a distillation column. Feed methanol and isobutene come in contact with the catalyst. MTBE is formed and is continuously removed from the reaction section. Methanol, isobutene and the inert hydrocarbons flow upward through the catalyst bed and then to a rectification section, which returns liquid MTBE.

An additional advantage of catalytic distillation is that the exothermic heat of reaction is directly used for the distillative separation of reactants and products. Hence, perfect heat integration is achieved and the heat of reaction will not increase the temperature, and operation in the reaction zone is nearly isothermal.

In practice, the catalytic distillation column is preceded by a conventional fixed-bed reactor (which is smaller than the one in the conventional process). In this reactor the largest part of the conversion takes place at relatively high temperature (high rate, lower equilibrium conversion), while the remainder takes place in the catalytic distillation column at lower temperature (lower rate, high equilibrium conversion).

A catalytic distillation column is a nice example of a so-called multifunctional reactor. This technology is very promising [27]. Other examples are discussed in Chapter 9.

QUESTIONS:

> *Where do you think the C_4 feed comes from? Why is a mixture processed instead of pure isobutene? Why is a conventional reactor placed upstream the catalytic distillation reactor?*

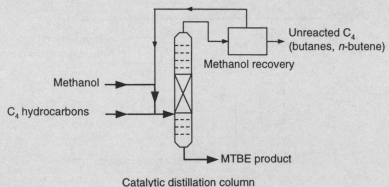

Figure B.6.7 Production of MTBE by combined reaction and distillation. Adapted from [26].

6.2.5.3 Other applications

The reaction of methanol with synthesis gas to produce a mixture of higher alcohols has recently regained interest. Like MTBE and other ethers, these higher alcohols have a positive effect on the gasoline octane number.

A success story is the homogeneously catalyzed carbonylation of methanol to acetic acid, which is treated in detail in Chapter 8. Since its introduction, this process has very rapidly replaced the then available technologies.

Methanol is also used in the production of monomers such as dimethyl terephthalate (DMT, see Chapter 8) and methyl methacrylate.

The recent development of processes for the dehydration of methanol to produce alkenes (mostly ethene and propene) presents an opportunity to base the petrochemical industry on natural gas derived methanol instead of oil in the future. The natural gas reserves are much greater than the oil reserves. The overall reaction to produce ethene proceeds as follows:

$$2CH_3OH \rightleftarrows C_2H_4 + 2H_2O \qquad \Delta H^0_{298} = -29.1 \text{ kJ/mol} \tag{12}$$

This so-called Methanol-to-Olefins (MTO) process is described in Chapter 4 and elsewhere [28–31].

6.3 FISCHER–TROPSCH PROCESS

The Fischer–Tropsch (F-T) process was named after F. Fischer and H. Tropsch, the German coal researchers who discovered in 1923 that synthesis gas can be converted catalytically into a wide range of hydrocarbons and/or alcohols. Until recently, however, in hydrocarbon synthesis it has never been able to compete economically with conventional petroleum-based fuels and chemicals. In fact, its use was limited to special cases. For instance, during World War II, Germany used F-T synthesis to make fuels from coal-derived gas. Coal-rich South Africa has used coal-based F-T plants for the production of fuels (mainly gasoline) and base chemicals since the 1950s to reduce the dependence on imported oil (Sasol). Still, the now three Sasol plants (commissioned in 1955, 1980, and 1983, respectively) are the only indirect coal liquefaction plants in the world currently producing liquid fuels by the F-T process.

In recent years, F-T has come into the picture again, especially as a means to convert natural gas from remote gas fields into liquid fuels. The transportation of the gas to possible consumer markets, either by pipeline or as liquefied natural gas (LNG) in special tankers, is either too costly or logistically difficult. An interesting option then may be to convert this gas into more readily transportable liquids, such as the bulk chemicals ammonia and methanol, or liquid fuels. The latter have a much larger market, and are therefore more attractive from the viewpoint of economy of scale. Shell has been the first to build an F-T plant for the production of middle distillates based on remote natural gas. The Shell Middle Distillate Synthesis (SMDS) plant using natural gas from offshore fields has been in operation in Malaysia since 1994.

Table 6.4   Major overall reactions in Fischer–Tropsch synthesis

Main reactions		
Alkanes	$nCO + (2n + 1)H_2 \rightarrow C_nH_{2n+2} + nH_2O$	(13)
Alkenes	$nCO + 2nH_2 \rightarrow C_2H_{2n} + nH_2O$	(14)
Water-gas shift	$CO + H_2O \rightleftarrows CO_2 + H_2$	(7)
Side reactions		
Alcohols	$nCO + 2nH_2 \rightarrow H(-CH_2-)_nOH + (n - 1)H_2O$	(15)
Boudouard reaction	$2CO \rightarrow C + CO_2$	(16)

6.3.1 Reactions and Thermodynamics

Although the chemistry of F-T synthesis is rather complex, the fundamental aspects can be represented by the generalized stoichiometric relationships in Table 6.4.

A characteristic of the F-T reactions is their high exothermicity. For example, the formation of 1 mole of $-CH_2-$ is accompanied by a heat release of 165 kJ/mol.

From Figure 6.13 in Section 6.2 it is clear that thermodynamically both hydrocarbons and alcohols can be formed. Therefore, the choice of catalyst and process conditions is very important (see [32]). The product molecular weight distribution depends on the catalyst, the temperature and pressure, and the H_2/CO ratio.

The hydrocarbons are formed by a chain-growth process (see box), with the length of the chain being determined by the catalyst selectivity and the reaction conditions.

The selective production of hydrocarbons other than methane is not (yet?) possible; a mixture of hydrocarbons with various chain lengths is always formed (see, e.g. [32–34] and box). However, an appropriate choice of catalyst and reaction conditions

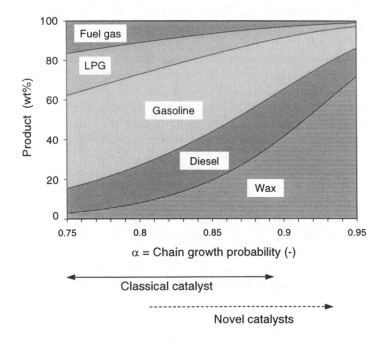

Figure 6.22   Product distribution as function of chain-growth probability. After [33].

enables the value of the chain-growth probability (α) to be shifted. So, different product ranges can be produced (see Figure 6.22).

QUESTION:

> *Complete Figure 6.22 for lower values of α.*

Chain-growth probability in F-T synthesis

Various mechanisms have been proposed for the F-T synthesis reactions (see [34–38]). However, for an understanding of the product distribution it is sufficient to only consider the basics of the reaction mechanism. No matter what the exact mechanism is, growth of a hydrocarbon chain occurs by stepwise addition of a one-carbon segment derived from CO at the end of an existing chain. For a certain minimum chain length, say $n \geq 4$, it is plausible that the relative probability of chain growth and chain termination (α and $1 - \alpha$, respectively) is independent on the chain length and hence constant [34]. Often it is assumed that this applies to all values of n. Then, the carbon-number distribution of F-T products can be represented by a simple statistical model, the Anderson–Flory–Schulz (AFS) distribution [32–34,39] (Scheme B.6.1).

QUESTION:

> *Based on the model above, what would be the effect of an increase in the H_2-to-CO ratio in the feed on the average chain length of the produced hydrocarbons?*

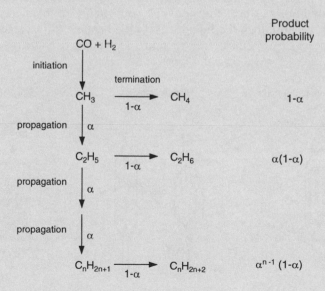

Scheme B.6.1 Chain-growth mechanism for F-T synthesis; Anderson–Flory–Schulz kinetics. Adapted from [33].

6.3.2 Reactors used in Fischer-Tropsch Synthesis

Efficient removal of the heat of reaction is a major consideration in the design of suitable F-T reactors, because the reactions taking place are strongly exothermic. Many reactors have been proposed and developed for proper heat management, see [34]. The currently used reactors are multi-tubular fixed-bed, riser and slurry reactors (see Figure 6.23 and Table 6.5).

In the multi-tubular fixed-bed reactor (FBR) small-diameter tubes containing the catalyst are surrounded by circulating boiling water, which removes the heat of reaction. A high linear gas velocity is applied and unconverted synthesis gas is recycled to enhance heat removal.

The FBR is suitable for operation at low temperature. There is an upper temperature limit of about 530 K, above which carbon deposition would become excessive, leading to blockage of the reactor. This reactor can be considered a trickle-flow reactor, since a large part of the products formed at the relatively low operating temperatures are liquid (waxes: C_{19+}).

In riser (entrained-flow) reactors for F-T synthesis two banks of heat exchangers remove a significant part of the heat of reaction, while the remainder is removed by the recycle and product gases, which are then separated from the catalyst by cyclones. The formation of heavy wax must be prevented, because this condenses on the catalyst particles causing agglomeration of the particles and thus defluidization. Therefore, the riser reactor is preferably used at temperatures of over 570 K.

In the slurry reactor, a finely divided catalyst is suspended in a liquid medium, usually the F-T wax product, through which the synthesis gas is bubbled. This also provides agitation of the reactor contents. As the gas flows upwards through the slurry, it is converted into more wax by the F-T reaction. The fine particle size of the catalyst reduces diffusional mass- and heat-transfer limitations. The heat generated by reaction is removed by internal cooling coils. The liquid medium surrounding the catalyst particles greatly improves heat transfer. The temperature in the slurry reactor must not be too low, because then the liquid wax would become too viscous, while above about 570 K the wax hydrocracks (see Section 3.4), leading to a less favorable product

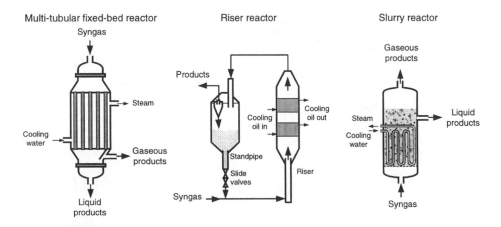

Figure 6.23   Reactors used for Fischer–Tropsch synthesis.

spectrum. Separation of the product wax from the suspended catalyst has been a problem; a heavy product is produced so distillation and flashing are no options, while separation methods such as filtration and centrifugation are expensive. The riser reactors at Sasol are gradually being replaced by slurry reactors.

QUESTIONS:

> *Give advantages and disadvantages of the three reactor types. (Consider operability, economics, ease of catalyst replacement, product flexibility, possibility of runaway, ease of scale-up, etc.)*
> *H_2S is a strong poison for the F-T catalysts. Therefore, syngas purification is necessary. In the case of an upset in the syngas purification plant, which reactor would suffer least from activity loss and why? (Hint: think of flow characteristics, mixed or plug flow?)*

The three reactors yield rather different product distributions (see Table 6.5). As a result of the relatively low temperature in the fixed-bed and in the slurry reactor, in these reactors the selectivity towards heavy products (waxes) is high. The low H_2/CO ratio in the slurry reactor results in a relatively high selectivity towards liquid products, but the higher temperature compared to the fixed-bed reactor has the opposite effect.

In the riser reactor, which has to operate at much higher temperature, gasoline is the major product. In addition, the riser reactor produces a larger quantity of light products, such as methane.

QUESTIONS:

> *Would the products be mainly linear or branched? What is desired for diesel fuel and why?*
> *Why is the conversion in the FBR relatively low?*
> *Estimate the chain-growth probability α for each reactor (see Figure 6.22).*
> *Why does a low H_2/CO ratio lead to high liquid product selectivity?*
> *Regarding the H_2/CO feed ratios what would be the best raw materials and processes for the production of the syngas for the three reactors?*

6.3.3 Carbon Removal

As in any process involving hydrocarbon reactions, deposition of carbon on the catalyst is inevitable. Riser and slurry reactors allow catalyst to be replaced during operation, but this is impossible in multi-tubular fixed-bed reactors. Usually, carbon removal is carried out periodically (e.g. after a year of operation) once signs of catalyst deactivation, such as loss in conversion, begin to manifest. Two methods are applied, i.e. replacement of deactivated catalyst with fresh catalyst, or regeneration with hydrogen:

$$C + 2H_2 \rightarrow CH_4 \qquad \Delta H_{298}^0 = -75 \text{ kJ/mol} \qquad (18)$$

Table 6.5 Typical dimensions and operation and selectivity data of the three reactor types (Fe catalyst) [38,40,41]

	Multi-tubular fixed-bed reactor	Riser reactor	Slurry reactor
Dimensions			
Reactor length (m)	12	46	22
Reactor diam. (m)	–	2.3	5
Tube diam. (m)	0.05	–	–
Number of tubes	>2000	–	–
Catalyst	Extrudates (1–3 mm)	40–150 μm	≈50 μm
Conditions			
Inlet T (K)	496	593	533
Outlet T (K)	509	598	538
Pressure (bar)	25	23	15
H_2/CO feed ratio	1.7	2.54	0.68[a]
Conversion (%)	60–66	85	87
Products (wt%)			
CH_4	2.0	10.0	6.8
C_2H_4	0.1	4.0	1.6
C_2H_6	1.8	4.0	2.8
C_3H_6	2.7	12.0	7.5
C_3H_8	1.7	2.0	1.8
C_4H_8	2.8	9.0	6.2
C_4H_{10}	1.7	2.0	1.8
C_5–C_{11} (gasoline)	18.0	40.0	18.6
C_{12}–C_{18} (diesel)	14.0	7.0	14.3
C_{19}^+ (waxes)	52.0	4.0	37.6
Oxygenates	3.2	6.0	1.0

[a] Higher ratio is also possible.

6.3.4 Processes

Various companies are active in F-T technology (see [39]) but only Sasol and Shell have commercial scale processes in operation (1999).

The Sasol I plant, which came on stream in 1955, utilizes both fixed-bed and riser reactors. Figure B.6.8. illustrates that the Sasol I plant can be seen as a 'coal refinery' analogous to an oil refinery.

The Sasol II and III plants, commissioned in 1980 and 1983, respectively, are virtually identical but use only slurry-reactor technology. The main objective is the production of gasoline, so the product distribution of these plants is much narrower in comparison with Sasol I. Extensive use of downstream catalytic processing of the F-T products, such as catalytic reforming, alkylation, hydrotreating (see Chapter 3), oligo-merization, and isomerization maximizes the production of transportation fuels (see Figure 6.24).

Hydrodewaxing involves the removal of long-chain hydrocarbons by cracking and isomerization in the presence of hydrogen. The goal of this process is to improve the properties of the diesel fuel.

QUESTIONS:

> *What is the purpose of catalytic reforming? Why are pentanes and hexanes subjected to another process? What is an alternative process for the production of gasoline from the C_3/C_4 fraction? How can CO_2 be removed?*

Sasol I plant

Figure B.6.8 shows a simplified block diagram of the Sasol I plant.The synthesis gas feed for the F-T reactors is produced by gasification of coal with steam and oxygen in 13 Lurgi gasifiers (see Section 5.3). The required oxygen is produced by cryogenic air separation. The raw synthesis gas is cooled to remove water and tars. The aqueous phase is treated to recover phenols. The tar is distilled to produce a range of products.

The synthesis gas is then purified to remove H_2S, CO_2, and hydrocarbons in the naphtha boiling range. The latter, together with the naphtha from the tar work-up, are hydrotreated and the products are either blended into the gasoline pool or sold as aromatic solvents (benzene, toluene, and xylenes: BTX).

The purified syngas is sent to the F-T reactor sections. The effluent from the F-T reactors is cooled and water and oil are condensed. The water is sent to the oxygenate recovery and refining plant, where various alcohols and ketones are produced. The hydrocarbon product from the fixed-bed reactors is distilled to produce gasoline, diesel, and waxes. The latter are distilled in a vacuum column to produce medium (590–770 K) and hard wax (>770 K), which are then hydrogenated to remove the remaining oxygenated compounds and alkenes.

The oil fraction from the riser reactors is treated over an acid catalyst to convert the oxygenates to alkenes, which are then isomerized to improve the octane number of the gasoline. The C_3 and C_4 products are oligomerized to produce gasoline components.

The off-gas from the reactors contains methane, ethane, and unconverted synthesis gas. It is consumed in several different ways. Part is treated in a cryogenic separation unit to recover hydrogen, which together with nitrogen from the oxygen plant is used for the production of ammonia (see Section 6.1). The CH_4-rich stream from the cryogenic separation is used as town gas. Another part is recycled to the fixed-bed reactors. The remaining off-gas is catalytically reformed with steam and oxygen (see Section 5.2) to produce more syngas, which is recycled to the riser reactors.

QUESTIONS:

> *Discuss the logic of the material streams in the power plant. Why is the off-gas from the fixed-bed reactors recycled directly, while the off-gas from the riser reactors is reformed first?*

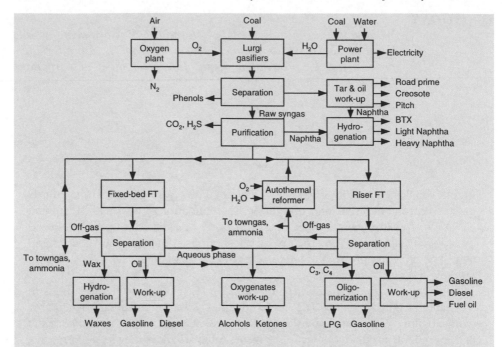

Figure B.6.8   Simplified block diagram of Sasol I Fischer–Tropsch process. After [32,35].

SMDS process

The Shell Middle Distillate Synthesis (SMDS) [33,34] uses remote natural gas (from offshore fields) as the feedstock. A plant at Bintulu, Malaysia was commissioned in early 1993. The process consists of several steps shown schematically in Figure B.6.9.

The overall process starts with the conversion of natural gas into synthesis gas, for which there are several processes available (see Chapter 5). Within the SMDS process Shell's own technology based on the non-catalytic partial oxidation of natural gas with oxygen, at about 1600 K and 30–50 bar (see also Section 5.2) is being used.

Synthesis gas manufacture from natural gas is followed by hydrocarbon synthesis, which is a modernized version of the classical F-T synthesis. Using Shell's proprietary catalyst and a multi-tubular fixed-bed F-T reactor and operating at 570–620 K and 30–50 bar [33], the synthesis of long-chain waxy alkanes is favored, so that a high overall selectivity towards distillates is achieved. The heavy alkanes are subsequently converted in a combined hydro-isomerization and hydrocracking step to produce the desired products, predominantly transportation fuels. In the final stage of the process, the products, mainly gas oil, kerosene, and naphtha, are separated by distillation. By varying the operating conditions in the hydrocracking and the subsequent distillation step, the product slate can be shifted towards a maximum kerosene mode or towards a maximum gas oil mode, depending on market requirements. Shell also sells several wax grades directly [39].

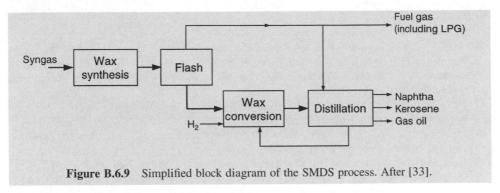

Figure B.6.9 Simplified block diagram of the SMDS process. After [33].

6.3.5 Future Developments

There currently is a lot of attention for the F-T process, especially as a means to convert natural gas located far from consumer markets into liquid fuels. The F-T process produces one of the most desirable diesel fuels: F-T has a high cetane number and a negligible sulfur content (see Section 3.7 and references therein). In the future the F-T process will probably gain increasing importance as a clean liquid fuel supplier. Perhaps in the further future the process will also be applied in other countries than South Africa for the conversion of coal.

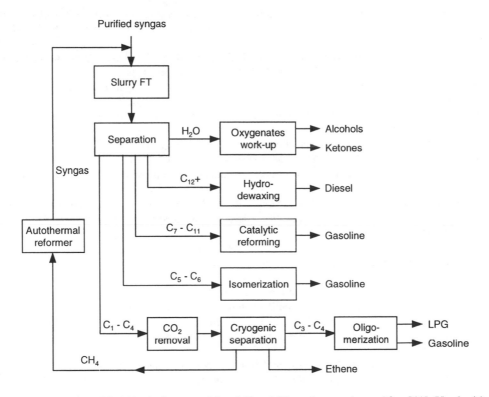

Figure 6.24 Simplified block diagram of Sasol II and III product work-up. After [41]. Used with permission of Wiley-VCH.

References

1   Appl M (1999) *Ammonia. Principles and Industrial Practice*, Wiley-VCH, Weinheim.
2   Bakemeier H, Huberich T, Krabetz R, Liebe W, Schunck M and Mayer D (1985) 'Ammonia' in: Gerhartz W, et al. (eds.) *Ullmann's Encyclopedia of Industrial Chemistry*, vol. A2, 5th ed., VCH, Weinheim, pp. 143–242.
3   Chauvel A and Lefebvre G (1989) *Petrochemical Processes 1. Synthesis Gas Derivatives and Major Hydrocarbons*, Technip, Paris, Chapter 1.
4   Jennings RJ and Ward SA (1996) 'Ammonia synthesis' in: Twigg MV (ed.) *Catalyst Handbook*, 2nd ed., Manson Publishing Ltd., London, Chapter 8.
5   Rautenbach R and Albrecht R (1989) *Membrane Processes*, Wiley, New York, pp. 422–454.
6   Czuppon TA, Knez SA, Rovner JM (1992) 'Ammonia' in: Kroschwitz JI and Howe-Grant M (eds) *Kirk-Othmer Encyclopedia of Chemical Technology*, vol. 2, 4th ed., Wiley, New York, pp. 638–691.
7   Meessen JH and Petersen H (1996) 'Urea' in: Gerhartz W, et al. (eds.) *Ullmann's Encyclopedia of Industrial Chemistry*, vol. A27, 5th ed., VCH, Weinheim, pp. 333–365.
8   Powell R (1968) 'Urea Process Technology' *Chemical Process Review*, NDC, NJ.
9   Van den Berg PJ and de Jong WA (1980) *Introduction to Chemical Process Technology*, Delft University Press, Delft.
10 (1969) 'Process Survey: Urea' *Eur. Chem. News*, Jan 17.
11 Mavrovic I, Shirley AR and Coleman GR (1998) in: Kroschwitz JI and Howe-Grant M (eds.), *Kirk-Othmer Encyclopedia of Chemical Technology*, Supplement, 4th ed., Wiley, New York, pp. 597–621.
12 Chang T, Rousseau RW and Kilpatrick PK (1986) 'Methanol synthesis reactions: calculations of equilibrium conversions using equations of state' *Ind. Eng. Chem. Process Des. Dev.* 25 477–481.
13 Hansen JB (1997) 'Methanol synthesis' in: Ertl G, Knözinger H and Weitkamp J (eds.) *Handbook of Heterogeneous Catalysis*, VCH, Wcinheim, pp. 1856–1876.
14 Royal MJ and Nimmo NM (1969) 'Why LP methanol costs less' *Hydroc. Process.* 48(3) 147–153.
15 Hiller H and Marchner F (1970) 'Lurgi makes low pressure methanol less' *Hydroc. Process.* 49(9) 281–285.
16 Supp E (1981) 'Improved methanol process' *Hydroc. Process.* 80(3) 71–75.
17 'Methanol – Haldor Topsøe A/S' (1983) *Hydroc. Process.* 62(11) 111–175.
18 Sherwin MB and Frank ME (1976) 'Make methanol by three phase reaction' *Hydroc. Process.* 55(11) 122–124.
19 Brown DM, Leonard JJ, Rao P and Weimer RF (1990) US Patent 4,910,227, High Volumetric Production of Methanol in a Liquid Phase Reactor (assigned to Air Products and Chemicals, Inc., Allentown, PA).
20 Rostrup-Nielsen JR (1993) 'Production of synthesis gas' *Catal. Today* 18 305–324.
21 Cheng WH and Kung HH (1994) *Methanol, Production and Use*, Marcel Dekker, New York, Chapter 1.
22 Davies P, Donald RT and Harbord NH (1996) 'Catalytic oxidations' in: Twigg MV (ed.) *Catalyst Handbook*, 2nd ed., Manson Publishing Ltd., London, Chapter 10.
23 Gerberich HR and Seaman GC (1994) 'Formaldehyde' in: Kroschwitz JI and Howe-Grant M (eds.) *Kirk-Othmer Encyclopedia of Chemical Technology*, vol. 11, 4th ed., Wiley, New York, pp. 929–951.
24 Reuss G, Disteldorf W, Grundler O and Hilt A (1988) 'Formaldehyde' in: Gerhartz W, et al. (eds.) *Ullmann's Encyclopedia of Industrial Chemistry*, vol. A11, 5th ed., VCH, Weinheim, pp. 619–651.

25 Maxwell IE, Naber JE and de Jong KP (1994) 'The pivotal role of catalysis in energy related environmental technology' *Appl. Catal. A: General* 113 153–173.

26 DeGarmo JL, Parulekar VN and Pinjala V (1992) 'Consider reactive distillation' *Chem. Eng. Prog.* March 43–50.

27 Moulijn JA and Stankiewicz AI (eds.) (1999) 'Special Issue, 1st International Symposium on Multifunctional Reactors, Amsterdam, 25–28 April 1999' *Chem. Eng. Sci.* 54 1297–1594.

28 Chang CD (1984) 'Methanol conversion to light olefins' *Catal. Rev.-Sci. Eng.* 26 323–345.

29 Inui T and Takegami T (1982) 'Olefins from methanol by modified zeolites' *Hydroc. Process.* 61(11) 117–120.

30 Kaiser SW (1985) 'Methanol conversion to light olefins over silico-aluminophosphate molecular sieves' *Arab. J. Sci. Eng.* 10(4) 361–366.

31 Inui T, Phatanasri S and Matsuda H (1990) 'Highly selective synthesis of ethene from methanol on a novel nickel-silicoaluminophosphate catalyst' *J. Chem. Soc. Chem. Commun.* 205–206.

32 Baldwin RM (1993) 'Liquifaction' in: Kroschwitz JI and Howe-Grant M (eds.) *Kirk-Othmer Encyclopedia of Chemical Technology*, vol. 6, 4th ed., Wiley, New York, pp. 568–594.

33 Sie ST, Senden MMG and Van Wechem HMH (1991) 'Conversion of natural gas to transportation fuels via the Shell middle distillate synthesis process (SMDS)' *Catal. Today* 8 371–394.

34 Sie ST (1998) 'Process development and scale up: IV. Case history of the development of a Fischer-Tropsch synthesis process' *Rev. Chem. Eng.* 14(2) 109–154.

35 Dry ME (1981) 'The Fischer-Tropsch synthesis' in: Anderson JR and Boudart M (eds.) *Catalysis Science and Technology*, vol. 1, Springer-Verlag, Berlin, pp. 159–255.

36 Biloen P and Sachtler WMH (1981) 'Mechanism of hydrocarbon synthesis over F-T catalysts', in: *Advances in Catalysis*, vol. 30, Academic Press, New York, pp. 165–216.

37 Ponec V (1978) 'Some aspects of the mechanism of methanation and Fischer-Tropsch synthesis' *Catal. Rev. Sci. Eng.* 18(1) 151–171.

38 Dry ME (1990) 'The Fischer-Tropsch process – commercial aspects' *Catal. Today* 6(3) 183.

39 Van der Laan GP (1999) *Kinetics, Selectivity and Scale up of the Fischer-Tropsch Synthesis*, PhD Thesis, University of Groningen.

40 Jager B and Espinoza R (1995) 'Advances in low temperature Fischer-Tropsch synthesis' *Catal. Today* 23 17–28.

41 Derbyshire F and Gray D (1986) 'Coal Liquefaction' in: Gerhartz W, et al. (eds.) *Ullmann's Encyclopedia of Industrial Chemistry*, vol. A7, 5th ed., VCH, Weinheim, pp. 197–243.

Chapter 7

Inorganic Bulk Chemicals

This chapter discusses the production of two important inorganic bulk chemicals, viz. sulfuric acid and nitric acid.

Sulfuric acid (H_2SO_4) is by far the largest-volume chemical commodity produced (>130 million t/a worldwide) [1]. It is used in the production of all kinds of chemicals, of which fertilizers are the most important. Other major uses are in alkylation in petroleum refining (see Chapter 3), copper leaching, and the pulp and paper industries.

Nitric acid (HNO_3) is mainly used for the production of ammonium nitrate for fertilizer use. Ammonium nitrate is also used for chemicals, explosives, and various other uses. Other relatively large uses of nitric acid are in the production of cyclohexanone, dinitrotoluene, and nitrobenzene. The current worldwide production of nitric acid is about 65 million t/a [2].

7.1 SULFURIC ACID

Sulfuric acid production involves the catalytic oxidation of sulfur dioxide (SO_2) into sulfur trioxide (SO_3). The process has developed via the 'Lead Chamber' process, which was used in the 18th and 19th centuries and used a homogeneous catalyst (nitrogen oxides), to the modern 'contact process'. The contact process originally used a supported platinum catalyst, but from the 1920s on this has been gradually replaced by a supported liquid-phase vanadium catalyst. Another significant change occurred in the early1960s with the introduction of the so-called double-absorption process, in which part of the formed SO_3 is removed between two conversion stages.

Sulfur dioxide can be obtained from a wide variety of sources, including elemental sulfur, spent sulfuric acid (contaminated and diluted), and hydrogen sulfide. In the past, iron pyrites were also often used, while large quantities of sulfuric acid are also produced as a by-product of metal smelting. Nowadays, elemental sulfur is by far the most widely used.

Sulfuric acid is commercially produced in various acid strengths, ranging from 33.33 to 114.6 wt%. Sulfuric acid with a strength of over 100 wt% is referred to as oleum, which consists of sulfuric acid with dissolved sulfur trioxide. The concentration of oleum is expressed as wt% dissolved SO_3 ('free SO_3') in 100 wt% sulfuric acid.

QUESTIONS:

> *The lead chamber process is one of the few examples of homogeneous catalysis in the gas phase. Do you know another example? Why is the process called the 'Lead Chamber Process'? At present, what is the source of elemental sulfur?*

7.1.1 Reactions and Thermodynamics

The reactions in the production of sulfuric acid from elemental sulfur involve generation of SO_2 from elemental sulfur (1), followed by catalytic oxidation of SO_2 (2) and absorption of the formed SO_3 in water (3):

$$S \text{ (l)} + O_2 \text{ (g)} \rightarrow SO_2 \text{ (g)} \qquad \Delta H^0_{298} = -298.3 \text{ kJ/mol} \tag{1}$$

$$SO_2 \text{ (g)} + \tfrac{1}{2}O_2 \text{ (g)} \rightarrow SO_3 \text{ (g)} \qquad \Delta H^0_{298} = -98.5 \text{ kJ/mol} \tag{2}$$

$$SO_3 \text{ (g)} + H_2O \text{ (l)} \rightarrow H_2SO_4 \text{ (l)} \qquad \Delta H^0_{298} = -130.4 \text{ kJ/mol} \tag{3}$$

Since both the oxidation of sulfur and of sulfur dioxide require oxygen, excess air is used to burn S, which has the advantage of complete oxidation of sulfur.

QUESTIONS:
> *What is the (theoretically) maximum attainable SO_2 concentration (mol%) in the outlet stream of a sulfur burner, given that air is used as the oxidizing medium? The actual SO_2 content varies from about 7 to 12 mol%. How much excess air is used? (Assume 100% sulfur conversion).*

Figure 7.1 shows the influence of temperature on the equilibrium conversion of SO_2. The oxidation of SO_2 is thermodynamically favored by low temperature.

The conversion is (nearly) complete up to temperatures of about 700 K. As with all exothermic equilibrium reactions, however, the ideal temperature must be a compromise between achievable conversion (thermodynamics) and the rate at which this

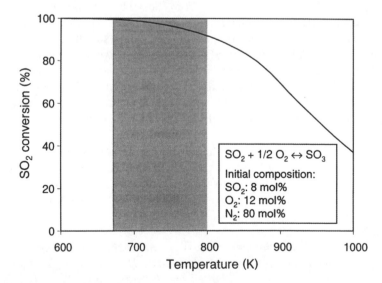

Figure 7.1 Equilibrium conversion of SO_2 to SO_3; $p = 1$ bar; shaded area indicates range of practical operating temperatures; based on data from [2].

conversion can be attained (kinetics). With the current SO_2 oxidation catalysts this means a minimum temperature of 680–715 K.

Elevated pressure is thermodynamically favorable, but the effect of pressure is small.

7.1.2 SO_2 Conversion Reactor

It is important that SO_2 is converted to SO_3 nearly quantitatively, not only for plant economic reasons but, more importantly, for environmental reasons. For instance, in the United States SO_2 emissions from newer sulfuric acid plants must be limited to 2 kg SO_2 per metric ton of sulfuric acid (as 100% acid) produced. This is equivalent to a sulfur dioxide conversion of 99.7% [1].

The catalytic oxidation of SO_2 is carried out in adiabatic fixed-bed reactors. Initially, platinum was used as the catalyst because of its high activity. However, platinum is expensive and very sensitive to poisons such as arsenic (often present in pyrite). Therefore, nowadays commercial catalysts consist of vanadium and potassium salts supported on silica.

As discussed in the previous section, the oxidation of SO_2 is an exothermic equilibrium reaction. Consequently, with increasing conversion the temperature rises, leading to lower attainable equilibrium conversions. For example, the equilibrium conversion at 710 K is about 98%, but due to the adiabatic temperature rise conversions of only 60–70% are obtainable in a single catalyst bed. In practice, this is overcome by using multiple catalyst beds (usually four) with intermediate cooling, as shown in Figure 7.2. Cooling can be achieved by heat-exchange or by quenching with air.

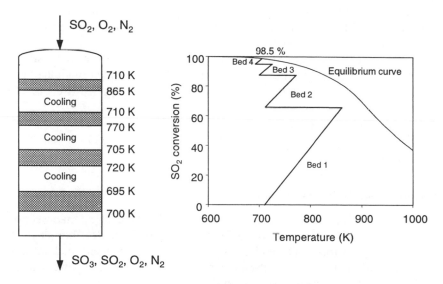

Figure 7.2 SO_2 oxidation in reactor with four catalyst beds with intermediate cooling; initial composition same as in Figure 7.1; after [2]. Used with permission of M.V. Twigg.

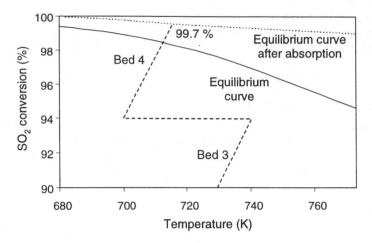

Figure 7.3   Effect of interstage absorption of SO_3 on SO_2 oxidation; initial composition: 11.5 mol% SO_2; 9.5 mol% O_2; after [2]. Used with permission of M.V. Twigg.

QUESTIONS:

> *Give examples of other processes using this type of reactor configuration.*
> *What do the diagonal lines and horizontal lines represent? Are the diagonal lines parallel to each other? Explain.*

As can be seen from Figure 7.2, the major part of the conversion is obtained in the first bed. The inlet temperature of the first bed is 695–710 K and the exit temperature is 865 K or more, depending on the concentration of SO_2 in the gas. The successive lowering of the temperature between the beds ensures an overall conversion of 98–99%. Still, this is not enough to meet current environmental standards. Therefore, modern sulfuric acid plants use intermediate SO_3 absorption after the second or, more commonly, the third catalyst bed.

The intermediate removal of SO_3 from the gas stream enables the conversion of SO_2 'beyond thermodynamic equilibrium'. Other examples where the principle of carrying out a reaction beyond its thermodynamic equilibrium is applied are discussed in Chapter 9.

Figure 7.3 shows the equilibrium curves for SO_2 conversion with and without intermediate removal of SO_3. It illustrates that an overall conversion exceeding 99% is obtained.

7.1.3 Modern Sulfuric Acid Production Process

Figure 7.4 shows a schematic of a modern sulfuric acid plant. Three main sections can be distinguished, i.e. combustion of sulfur to produce SO_2 (sulfur burner), catalytic oxidation of SO_2, to form SO_3, and absorption of SO_3 in concentrated sulfuric acid (absorption towers). Because of the presence of two absorption towers it is often referred to as a 'double absorption process'.

Dried air and atomized molten sulfur are introduced at one end of the sulfur burner,

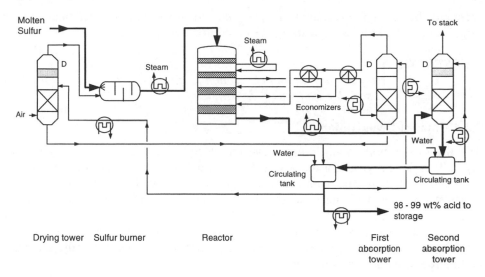

Figure 7.4 Flow scheme of sulfuric acid plant with two absorption towers; D = demister.

which is a horizontal, brick-lined combustion chamber. The degree of atomization and good mixing are key factors in producing efficient combustion. Atomization typically is accomplished by pressure spray nozzles or by mechanically driven spinning cups. Some sulfur burner designs contain baffles or secondary air inlets to promote mixing. Handling of molten sulfur requires that the temperature be kept between 410 and 425 K, where its viscosity is lowest. Therefore, it is transported through heated lines.

Most of the heat of combustion is used to generate high pressure steam in a waste heat boiler, which reduces the gas temperature to the desired inlet temperature of the first catalyst bed (typically 690 K). The heat of reaction of SO_2 oxidation raises the temperature in the first bed significantly. The hot exit gases are cooled before entering the second bed, where further conversion takes place. The gas leaving this bed is cooled by heat-exchange with cold gas leaving the first absorption tower. It subsequently undergoes further conversion in the third bed, after which it is cooled and fed to the first absorption tower. Here, most of the SO_3 produced is absorbed in a circulating stream of sulfuric acid. The gas leaving the absorption tower is reheated by heat-exchange with gas from the third and second converter bed and then enters the fourth catalyst bed, where most of the remaining SO_2 is converted to SO_3 (overall conversion > 99.7%). The final absorption tower removes this SO_3 from the gas stream before release to the atmosphere.

Small amounts of sulfuric acid mists or aerosols are always formed in sulfuric acid plants when gas streams are cooled or SO_3 reacts with water below the sulfuric acid dew point. Formation of sulfuric acid mists is highly undesirable because of corrosion and process stack emissions. Therefore, the absorbers in sulfuric acid plants are equipped with demisters, consisting of beds of small-diameter glass beads or Teflon fibers.

Sulfuric acid plants can be considered co-generation plants. Much of the heat produced in the combustion of sulfur is recovered as high-pressure steam in waste heat boilers, while some of the heat produced in the catalytic SO_2 oxidation is also recovered by steam production in so-called economizers. Steam production in modern large sulfuric acid plants exceeds 1.3 t/t of sulfuric acid produced [3].

QUESTIONS:
> *How are drying and absorption performed? What liquid is used for absorption and why?*

7.1.4 Catalyst Deactivation

The life of the modern vanadium/potassium catalysts may be as long as 20 years, typically at least 5 years for the first and second bed and at least 10–15 years for the third and fourth bed. The main reasons for loss of activity of vanadium catalysts are physical breakdown giving dust, which could plug the catalyst bed, and chemical changes within the catalyst itself [2]. The former is overcome by regular screening of the catalyst (first bed annually, others less frequently). The latter results from migration of the molten vanadium salt from catalyst particles into adjacent dust, and increases with increasing temperature.

7.2 NITRIC ACID

The history of nitric acid production goes back to the eighth century. From the Middle Ages onward, nitric acid was primarily produced from saltpeter (potassium nitrate) and sulfuric acid. In the 19th century, Chile saltpeter (sodium nitrate) largely replaced saltpeter. However, at the beginning of the 20th century new production technologies were introduced. The rapid development of a process for the production of ammonia (first from coal, later from natural gas or naphtha (see Chapter 6)) made possible a commercial route for the production of nitric acid based on the catalytic oxidation of ammonia with air. This route, although modernized, is still used for all commercial nitric acid production.

7.2.1 Reactions and Thermodynamics

Nowadays, all nitric acid production processes are based on the oxidation of ammonia, for which the overall reaction reads:

$$NH_3 + 2O_2 \rightarrow HNO_3 + H_2O \qquad \Delta H^0_{298} = -330 \text{ kJ/mol} \qquad (4)$$

Many reactions are involved in the oxidation, a simplified representation of which is the following sequence of reactions: burning of ammonia to nitric oxide (5), oxidation of the nitric oxide (6), and absorption of nitrogen dioxide in water (7):

$$4NH_3 + 5O_2 \rightarrow 4NO + 6H_2O \qquad \Delta H^0_{298} = -907 \text{ kJ/mol} \qquad (5)$$

$$2NO + O_2 \rightarrow 2NO_2 \qquad \Delta H^0_{298} = -113 \text{ kJ/mol} \qquad (6)$$

$$3NO_2 + H_2O \rightarrow 2HNO_3 + NO \qquad \Delta H^0_{298} = -138 \text{ kJ/mol} \qquad (7)$$

7.2.1.1 Ammonia oxidation

Ammonia oxidation (5) is highly exothermic and very rapid. However, the reaction system has a strong tendency towards the formation of nitrogen:

$$4NH_3 + 3O_2 \rightarrow 2N_2 + 6H_2O \qquad \Delta H^0_{298} = -1261 \text{ kJ/mol} \tag{8}$$

Figure 7.5 illustrates this by presenting the Gibbs energy of formation of reactions (5) and (8). Clearly, the formation of N_2 is thermodynamically much more favorable. This can be overcome by using a selective catalyst (see below) and short residence time (10^{-4}–10^{-3} s) at high temperature.

QUESTION:

> *The NO production is catalyzed by a solid catalyst (a Pt-Rh alloy).*
> *What type of reactor would you choose?*

7.2.1.2 Oxidation of nitric oxide and absorption in water

The oxidation of NO to nitrogen dioxide (NO_2), Eqn. (6) is a non-catalyzed reaction. The thermodynamic data in Figure 7.6 show that the conversion of NO is favored by low temperature and high pressure. NO_2 is in equilibrium with its dimer N_2O_4.

The NO oxidation is a famous reaction because it seems to be one of the few third-order reactions known:

$$r_{NO} = -k \cdot p^2_{NO} \cdot p_{O_2} \tag{9}$$

Moreover, a peculiarity of NO oxidation is the fact that its reaction rate constant increases with decreasing temperature. Therefore, this reaction is favored by low temperature not only thermodynamically but also kinetically. The data suggest a stepwise reaction mechanism.

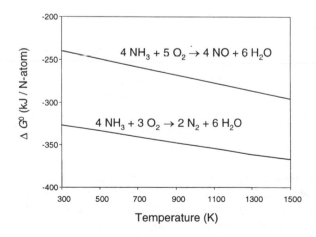

Figure 7.5 Gibbs free energy of formation, ΔG^0, in the oxidation of ammonia at 1 bar. After [4].

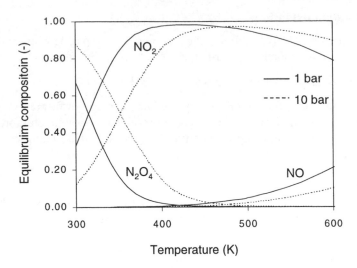

Figure 7.6 Influence of temperature and pressure on equilibrium composition (mol/mol total N compounds) of mixture of $NO/NO_2/N_2O_4$ (starting from stoichiometric mixture NO/O_2).

QUESTION:

> *Suggest a reaction mechanism that can explain the observed negative temperature coefficient of k.*

The absorption of NO_2 in water (7) is quite complex, because several reactions can occur, both in the liquid phase and in the gas phase. For practical purposes the simplified Eqn. (7) is satisfactory.

A peculiar point is that in the reaction the desired product (HNO_3) is formed, but also part of the reactant (NO) is formed back. As reaction (7) takes place in the liquid phase, while NO oxidation (6) occurs in the gas phase, mass transfer plays a major role.

7.2.2 Processes

The optimal conditions for ammonia oxidation are high temperature, short residence time (in order to minimize side and consecutive reactions) and low pressure. In practical processes this is realized by applying a reactor consisting of layers of a woven catalyst net, consisting of platinum strengthened with 10% rhodium. Due to crystallization and erosion at the high gas velocity applied, some Pt is lost with the product gas. The lost Pt is often recovered by filters. The catalyst life is about 2 years.

Because the absorption of NO_2 is a process that operates best at high pressure the oxidation is often also carried out under pressure. However, this lowers the oxidation efficiency (% ammonia converted to NO), which is reduced from 97% at 1 bar to 92% at 12 bar. Therefore, a dilemma exists: should we apply a low or a high pressure? In practice, several options are used, viz. both the oxidation and the absorption are carried out at atmospheric pressure, or a combination of low pressure for the oxidation reactor and high pressure for the absorption process is used.

Table 7.1 Typical data for nitric acid plants [2]. Used with permission of M.V. Twigg

Parameter	Atmospheric pressure	Intermediate pressure		High pressure	
		Single	Dual	Single	Dual
Converter pressure (bar)	1–1.4	3–6	0.8–1	7–12	4–5
Absorption pressure (bar)	1–1.3	3–6	3–5	7–12	10–15
Acid strength (wt%)	49–52	53–60	55–69	52–65	60–62
NH_3 to HNO_3 efficiency (%)	93	88–95	92–96	90–94	94–96
NH_3 to NO converter efficiency (%)	97–98	96	97–98	95	96
Pt loss (g/t HNO_3)	0.05	0.10	0.05	0.30	0.1
Gauze temperature, average (K)	1100	1150	1100	1200	1150
Typical gauze life (months)	8–12	4–6	8–12	1.5–3	4–6

In recent years atmospheric pressure plants have become obsolete. The choice of process depends on site requirements, economics of heat recovery and gas compression, acid strength, etc.

Table 7.1 shows some typical data on the various plant types.

QUESTIONS:

> *How do process conditions influence the acid strength? Can you explain the differences in the gauze temperature?*

Figure 7.7 shows a simplified flow scheme of a single-pressure nitric acid process. Most single-pressure processes operate at high pressure (7–12 bar), because of

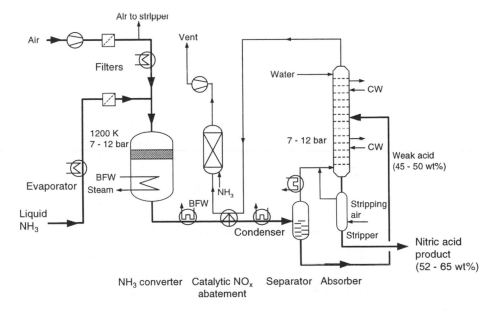

Figure 7.7 Flow scheme of single-pressure nitric acid plant (high pressure) using catalytic NO_x abatement; BFW = boiler feed water, CW = cooling water.

reduced equipment size and capital cost. The higher gauze temperature (see Table 7.1) and operating pressure enable more efficient energy recovery. Both ammonia and air are filtered to remove impurities such as rust and other particulates present in the feedstocks; rust particles promote the decomposition of ammonia.

The compressed air and gaseous ammonia are mixed such that there is excess oxygen (ca. 10 vol% ammonia in mixture), and are passed over a shallow bed of Pt catalyst to produce nitric oxide (NO) and water. This reaction is very exothermic, so much heat evolves. The resulting gases are rapidly cooled, thus generating steam that can be exported or used to power the steam turbine driving the compressors. Upon cooling, nitric oxide is further oxidized to form nitrogen dioxide (NO_2) in equilibrium with its dimer, dinitrogen tetroxide (N_2O_4). In the condenser, which must be made of materials that are resistant to corrosion by the hot acid, condensation of (weak) nitric acid takes place. This is subsequently separated from the gas mixture and fed to the absorption column. The gases leaving the separator are mixed with secondary air (from stripper) to enhance oxidation of NO, and then fed to the bottom of the absorber, where the equilibrium mix of NO_2 and N_2O_4 is absorbed into water, producing nitric acid. The nitric acid leaving the bottoms of the absorber is contacted with air in a stripper to strip dissolved NO_x (NO and NO_2) from the product acid.

The gas leaving the top of the absorber column contains residual nitrogen oxides, NO_x, which have to be removed before venting to the atmosphere for environmental reasons. Figure 7.7 shows NO_x removal by means of so-called selective catalytic reduction (SCR) (see Section 7.2.3).

Figure 7.8 shows a simplified flow scheme of a dual-pressure nitric acid plant (high pressure). The main change in the flow scheme is the addition of a compressor between the NH_3 conversion stage and the absorption stage. Another difference in this scheme is the way NO_x abatement is achieved, i.e. by extended absorption (see

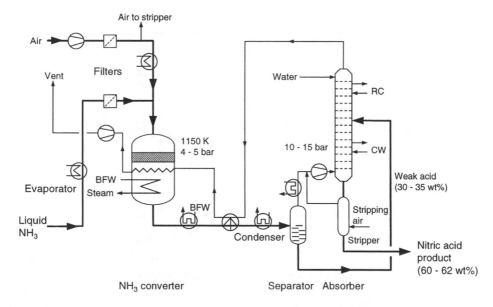

Figure 7.8 Flow scheme of dual-pressure nitric acid plant using extended absorption for NO_x abatement; BFW = boiler feed water, CW = cooling water, RC = refrigerated cooling.

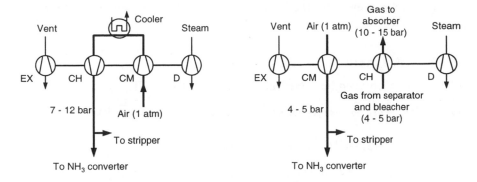

Figure 7.9 Compression/expansion section of nitric acid plants; left: single-pressure plant, right: dual-pressure plant; EX = expander; CM = medium level compression; CH = high level compression; D = makeup driver (steam turbine).

Section 7.2.3). Note, however, that the NO_x abatement methods are interchangeable between the single-pressure and dual-pressure processes.

QUESTIONS:

> *Explain the term 'extended' absorption. How is this done in Figure 7.8? Explain why in Figure 7.8 the vent gas passes through the heat-recovery section of the NH_3 converter, while this is not the case in Figure 7.7.*

For many reasons the compression and expansion operations may be considered the heart of the nitric acid process. Although the compressors and expander in Figures 7.7 and 7.8 are depicted as independent units, in fact they are intimately linked. Figure 7.9 shows the compressor-expander combinations of the single-pressure and dual-pressure plants.

As can be seen from Figure 7.9, the energy required for compression is partly delivered by expansion of the tail gas from the absorber. The makeup driver, which usually is a steam turbine, but can also be an electric motor, supplies the remainder of the required energy. Compressors, expander, and makeup driver all share a common shaft.

7.2.3 NO_x Abatement

Environmental regulations necessitate the removal of NO_x from gases to be vented to the atmosphere. The concentration of nitric oxides in the gas leaving a conventional absorber is in the range of 2000–3000 ppmv, while regulations require maximum concentrations of 200–230 ppmv. The most commonly employed methods for NO_x abatement in existing plants are SCR, non-selective catalytic reduction (NSCR), and extended absorption. For new nitric acid plants, NO_x emissions can be well controlled by using advanced processes, such as high inlet pressure absorption columns or strong acid processes. Recently, it has been realized that the emission of N_2O is also a point of concern. Novel processes for the reduction of N_2O emissions are under development [5].

7.2.3.1 Selective catalytic reduction (SCR)

In SCR NO is converted into nitrogen by reacting it with NH_3:

$$4NO + 4NH_3 + O_2 \rightarrow 4N_2 + 6H_2O \qquad \Delta H_{298}^0 = -408 \text{ kJ/mol NO}$$

$$2NO_2 + 4NH_3 + O_2 \rightarrow 3N_2 + 6H_2O \qquad \Delta H_{298}^0 = -669 \text{ kJ/mol NO}_2$$

The SCR process was developed in Japan in the 1970s. This process requires operation at temperatures of 520–670 K, although some processes can operate at lower temperatures (down to about 420 K) [6].

The reaction of ammonia and NO_x is favored by the presence of excess oxygen. The primary variable affecting NO_x reduction is temperature. A given catalyst exhibits optimum performance within a temperature range of plus or minus 30 K for when the oxygen concentration is greater than one percent. Below this optimum range, the catalyst activity is greatly reduced, allowing unreacted ammonia to slip through. Above the range, ammonia begins to be oxidized to form additional NO_x. Furthermore, excessive temperatures may damage the catalyst.

The original SCR catalysts were pellets or spheres and were used in fixed catalyst beds, but current catalysts are usually shaped as honeycombs (monolithic catalyst) (see Figure 7.10) or parallel plates. The main advantages of these shaped catalysts for SCR are their low pressure drop, large surface area and good resistance against dust (see also Chapter 9).

Selective catalytic reduction is used in many nitric acid plants in Europe and Japan, and also in some US plants. Several advantages of SCR make it an attractive control technique. Since the SCR process can operate at any pressure, it is a viable retrofit control alternative for existing low-pressure acid plants as well as for new plants. Another technical advantage is that because the temperature rise through the SCR reactor bed is small, energy recovery equipment (e.g. waste heat boilers and high-temperature turboexpanders) is not required, as is the case with the NSCR system, discussed below.

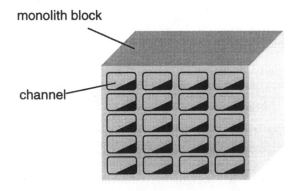

Figure 7.10   Honeycomb monolith catalyst.

The main disadvantages of this system are the high costs involved; the cost of the catalyst can amount to between 40 and 60% of operating costs, and it is necessary to replace the catalyst every 2–3 years.

QUESTIONS:

SCR has the fundamental disadvantage that a pollutant is removed by a stoichiometric reaction with a valuable chemical. Could you think of a process that is fundamentally much more elegant? Why is that process not applied in practice?

7.2.3.2 Non-selective catalytic reduction (NSCR)

In this process, absorber tailgas from nitric acid production is heated to ignition temperature using ammonia converter effluent gas in a heat exchanger, and fuel (usually natural gas) is added. The gas/fuel mixture then passes through the catalytic reduction unit where the fuel reacts in the presence of a catalyst with NO_x and oxygen to form elemental nitrogen, water, and carbon dioxide when hydrocarbon fuels are used. The process is called non-selective because the fuel first depletes all the oxygen present in the tailgas and then removes the NO_x. Despite the associated fuel costs, NSCR offers advantages that continue to make it a viable option for new and retrofit applications. Flexibility is one advantage, especially for retrofit considerations. Additionally, heat generated by operating an NSCR unit can be recovered in a waste heat boiler and a tailgas expander to supply the energy for process compression with additional steam for export. Test data for nitric acid plants using NSCR shows the NOx control efficiencies range from 95 to 99%.

7.2.3.3 Extended absorption

The final step in the production of weak nitric acid involves the absorption of NO_2 to form nitric acid. When NO_2 is being absorbed it releases gaseous NO. Extended absorption reduces NO_x emissions by increasing absorption efficiency (i.e. acid yield). This option can be implemented by installing a single large absorption tower, extending the height of an existing tower, or by adding a second tower in series with the existing tower. The increase in the volume and the number of trays in the absorber results in more NO_x recovered as nitric acid.

Extended absorption is particularly suited for high-pressure absorption in which abatement to less than 200 ppmv NO_x can be achieved in a single column. In general, medium-pressure plants use two absorption columns to achieve concentrations of about 500 ppmv NO_x. Extended absorption is typically used in retrofit applications by adding a second absorption tower in series with the existing tower. Compliance tests for seven new (post-1979) nitric acid plants using extended absorption showed NO_x control efficiencies to range from 93.5 to 97%.

References

1 Muller TL (1997) 'Sulfuric acid and sulfur trioxide' in: Kroschwitz JI and Howe-Grant M (eds.) *Kirk-Othmer Encyclopedia of Chemical Technology*, vol. 23, 4th ed., Wiley, New York, pp. 363–408.

2 Davies P, Donald RT and Harbord NH (1996) 'Catalytic oxidations' in: Twigg MV (ed.) *Catalyst Handbook*, 2nd ed., Manson Publishing Ltd., London, Chapter 10.

3 Austin GT (1984) *Shreve's Chemical Process Industries*, 5th ed., McGraw-Hill, New York, Chapter 19.

4 Van den Berg PJ and de Jong WA (1980) *Introduction to Chemical Process Technology*, Delft University Press, Delft.

5 Kapteijn F, Rodriguez-Mirasol J and Moulijn JA (1996) 'Heterogeneous catalytic decomposition of nitrous oxide' *Appl. Catal. B: Environmental* 9 25–64.

6 Beretta A, Tronconi E, Groppi G and Forzatti P (1997) 'Monolithic catalysts for the selective reduction of NO_x with NH_3 from stationary sources' in: Cybulski A and Moulijn JA (eds.) *Structured Catalysts and Reactors*, Marcel Dekker, New York, pp. 121–148.

Chapter 8

Homogeneous Catalysis

8.1 INTRODUCTION

In homogeneous catalysis soluble catalysts are applied, in contrast with hetero-geneous catalysis, where solid catalysts are used. One example of homogeneous catalysis has been discussed in Chapter 3, viz. alkylation of isobutane with alkenes, with acids such as HF and H_2SO_4 as catalyst. In this chapter, we will confine ourselves to homogeneous catalysts based on transition metals. These catalysts are becoming increasingly important, and already have found a large number of industrial applications. Homogeneous catalysis is applied in essentially all sectors of the chemical process industry, in particular in polymerization, in the synthesis of bulk chemicals (solvents, detergents, plasticizers), and in fine chem-istry.

It is useful to compare homogeneous and heterogeneous transition metal based catalysis. In homogeneous catalysis the reaction mixture contains the catalyst complex in solution. This means that all of the metal is exposed to the reaction mixture. In heterogeneous catalysis on the other hand, the metal is typically applied on a carrier material or as a porous metal sponge type material, and only the surface atoms are active (see Figure 8.1).

Thus, in terms of activity per metal center, homogeneous catalysts are often more active. The high dispersion of homogeneous catalysts also minimizes the effect of catalyst poisons; in homogeneous catalysis one poison molecule only deactivates one metal complex, whereas in heterogeneous catalysis a poison mole-cule can block a pore containing many active sites (pore plugging). Homogeneous

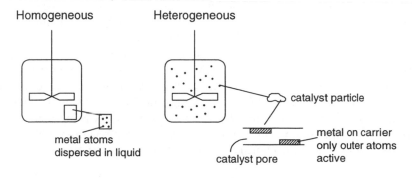

Figure 8.1 Catalyst dispersion in homogeneous and heterogeneous catalysis.

catalysts are also capable of being much more selective because with homogeneous catalysts there is usually only one type of active site, whereas heterogeneous catalysts often contain many different types of active sites, some of which could even catalyze undesired reactions. The discrete metal complexes in homogeneous systems provide a well-defined catalyst system, which is easy to study by conventional techniques (infrared spectroscopy, NMR). From this another advantage of homogeneous catalysis arises, namely the possibility to follow the effect of changes in ligands and/or reaction conditions accurately. Hence, the selectivity for a specific product can be controlled by modification of the ligands. The behavior of heterogeneous catalysts, with their complex surfaces, is much more difficult to understand, and therefore, effective modification is fairly difficult. Because homogeneous catalysis is usually carried out in the liquid phase, temperature control is relatively easy.

Why then, with all these advantages, is homogeneous transition-metal catalysis not applied in every known process? First of all, its high specificity is not always called for. In oil refinery processes especially, the complex feeds and products do not justify the use of a highly specific catalyst. Secondly, homogeneous catalysis of course also has disadvantages. The main problem is the separation of catalyst and product. This is often only feasible for low molecular weight products. Furthermore, the use of solvents adds an additional separation step. Thirdly, the temperature window that can be applied is limited, in particular because catalyst complexes are often not resistant to high temperature. In homogeneous catalysis often precious metals are used. Furthermore, the ligands are also expensive. Therefore, the catalyst productivity indeed should be high, and catalyst losses should be minimized, requiring nearly complete recovery of metal and ligands. Fouling of the reactor and other equipment is another problem specific for homogeneous catalysis.

Homogeneous catalysis is a highly innovative area. It also typically is an area where progress is accelerated when chemical engineers and chemists co-operate closely from the start. Novel developments in homogeneous catalysis are for a large part related to new insights in organometallic chemistry. Table 8.1 shows examples of industrially important reactions that are catalyzed by transition metals, which will be dealt with in this chapter. It is often thought that in bulk chemicals production only heterogeneous catalysis is applied. All examples in Table 8.1, however, belong to the class of bulk chemistry.

As stated previously, the mild reaction conditions are an advantage of homogeneously catalyzed processes. Table 8.2 shows these conditions for the reactions in Table 8.1. As can be seen from Table 8.2, all processes take place at moderate temperatures. For comparison Table 8.3 shows data of some important heterogeneously catalyzed processes. It is clear that the conditions are more severe.

Homogeneous catalysts are not suitable for strongly endothermic reactions such as cracking of $C-C$ bonds in Fluid Catalytic Cracking (FCC), because of their limited stability at higher temperatures. In most cases pressures are low but there are exceptions, for example the hydroformylation and carbonylation reactions with cobalt. It is fair to state that also heterogeneously catalyzed processes exist at moderate temperatures.

Table 8.1   Reactions of industrial importance catalyzed by transition metals

Reaction	ΔH^0_{298} (kJ/mol)	

Wacker synthesis of acetaldehyde

$$H_2C{=}CH_2 \ + \ \tfrac{1}{2}\,O_2 \ \longrightarrow \ H_3C{-}\overset{\overset{\displaystyle O}{\|}}{C}H$$

ethene   acetaldehyde

-244 (1)

Acetic acid by methanol carbonylation

$$CH_3OH \ + \ CO \ \longrightarrow \ CH_3\overset{\overset{\displaystyle O}{\|}}{C}{-}OH$$

methanol   acetic acid

-138 (2)

Hydroformylation (OXO reaction)

$$CH_3CH{=}CH_2 \ + \ CO \ + \ H_2 \ \longrightarrow \ CH_3\underset{\underset{\displaystyle CH_3}{|}}{\overset{\overset{\displaystyle O}{\|}}{C}H}CH \ + \ CH_3CH_2CH_2\overset{\overset{\displaystyle O}{\|}}{C}H$$

propene   iso aldehyde   normal aldehyde

-125 (3)

Witten synthesis of dimethyl terephthalate

$$H_3C{-}\langle\text{ring}\rangle{-}CH_3 \ + \ 3\,O_2 \ + \ 2\,CH_3OH \ \longrightarrow \ H_3C{-}O{-}\overset{\overset{\displaystyle O}{\|}}{C}{-}\langle\text{ring}\rangle{-}\overset{\overset{\displaystyle O}{\|}}{C}{-}O{-}CH_3 \ + \ 4\,H_2O$$

p-xylene   dimethyl terephthalate -1305 (4)

Synthesis of terephthalic acid

$$H_3C{-}\langle\text{ring}\rangle{-}CH_3 \ + \ 3\,O_2 \ \longrightarrow \ HO{-}\overset{\overset{\displaystyle O}{\|}}{C}{-}\langle\text{ring}\rangle{-}\overset{\overset{\displaystyle O}{\|}}{C}{-}OH \ + \ 2\,H_2O$$

-1360 (5)

terephthalic acid

8.2 WACKER OXIDATION

Selective catalytic oxidation plays a major role in the chemical process industry. Examples are the production of ethene oxide and acetaldehyde from ethene by oxidation with O_2. Originally, both examples involved more complex routes. For instance, ethyne used to be the feedstock for acetaldehyde production:

$$HC{\equiv}CH + H_2O \rightarrow CH_3CHO \qquad \Delta H^0_{298} = -138 \text{ kJ/mol} \qquad (6)$$

Table 8.2 Production and process conditions of homogeneously catalyzed processes [1–4][a]

Process	World capacity (million t/a)	Catalyst	Temperature (K)	Pressure (bar)
Acetaldehyde	2.5	Pd/Cu	375–405	3–8
Acetic acid	4.0	Co or Rh	425–475	≈45/500
Oxo-alcohols[b]	7.0	Co or Rh	335–470	200/30
Dimethyl terephthalate	3.3	Co	415–445	4–8
Terephthalic acid	9.4	Co	450–505	15–30

[a] 1992 capacities.
[b] By hydroformylation.

Before the Second World War, in Germany in particular, the use of ethyne as chemical feedstock was the subject of many research activities. When the oil era started, it became much cheaper (and safer) to use ethene as feedstock. This became feasible by the discovery of the 'Wacker' process based on 'direct oxidation' of ethene:

$$H_2C{=}CH_2 + \tfrac{1}{2}O_2 \rightarrow CH_3CHO \qquad \Delta H^0_{298} = -244 \text{ kJ/mol} \qquad (1)$$

The first industrial plant came on stream in 1960, and now this process has largely replaced the ethyne-based process.

Acetaldehyde is a highly reactive molecule. It is an important intermediate in the production of many chemicals, the most important being acetic acid (although its importance has decreased because nowadays methanol carbonylation is the preferred route (see Section 8.3)).

8.2.1 Background Information

The observation that ethene forms acetaldehyde in an aqueous palladium chloride solution dates back to the end of the 19th century. In this reaction Pd^{2+} reduces to Pd:

$$C_2H_4 + PdCl_2 + H_2O \rightarrow CH_3CHO + 2HCl + Pd \qquad (7)$$

Acetaldehyde is formed by reaction (7), but catalysis is not taking place because oxygen is not able to re-oxidize palladium. In the 1950s, Wacker Chemie and Hoechst developed a process in which re-oxidation of palladium takes place successfully. This

Table 8.3 Production and process conditions of heterogeneously catalyzed processes [5–10][a]

Process	World capacity (million t/a)	Catalyst	Temperature (K)	Pressure (bar)
Methanol	19.6	Cu	530	50–100
Ammonia	125	Fe	675	100–200
FCC	>500	Zeolite-Y	775	1
Ethylbenzene	17.0	Zeolite (ZSM-5)	700	15–25
		AlCl₃ (homogeneous)	440	10
Ethene oxide	9.6	Ag	550	20

[a] 1992 capacities.

progress became possible by the discovery that Cu^{2+} (as $CuCl_2$) is able to re-oxidize Pd:

$$Pd + 2CuCl_2 \rightarrow PdCl_2 + 2CuCl \tag{8}$$

Of course, the two reactions above still do not result in a catalytic process, but it is a well-known fact that Cu^+ readily oxidizes to Cu^{2+} in oxygen or air:

$$2Cu^+ + \tfrac{1}{2}O_2 + 2H^+ \rightarrow 2Cu^{2+} + H_2O \qquad \Delta H^0_{298} = -233 \text{ kJ/mol} \tag{9}$$

Thus, a series of oxidation-reduction reactions effects the continuous oxidation of ethene to acetaldehyde, for which the overall reaction is:

$$H_2C{=}CH_2 + \tfrac{1}{2}O_2 \rightarrow CH_3CHO \qquad \Delta H^0_{298} = -244 \text{ kJ/mol} \tag{1}$$

Although the reaction mechanism has not been elucidated completely, the following rate equation is generally accepted at low Pd^{2+} concentration [11,12] (see also box: catalytic cycle):

$$r_{CH_3CHO} = k \cdot \frac{\left[PdCl_4^{2-}\right] \cdot [C_2H_4]}{[Cl^-] \cdot [H^+]} \tag{10}$$

Catalytic cycle

Scheme B.8.1 below shows a possible catalytic cycle for the Pd/Cu catalyzed oxidation of ethene to acetaldehyde.

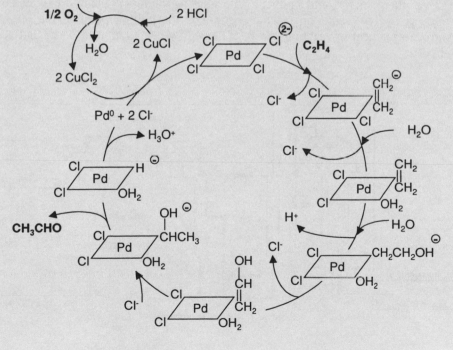

Scheme B.8.1 Catalytic cycle in the Wacker process.

QUESTIONS:

> *Why is this process called 'Direct Oxidation'? Why did its discovery*
> *evoke much enthusiasm? In practice, although not needed in view of*
> *the rate equation, small amounts of HCl are added to the reaction*
> *mixture. Explain this.*

8.2.2 Processes

Two processes are commercially available for the Wacker oxidation of ethene, both
jointly developed by Wacker Chemie and Hoechst:

- single-stage process using oxygen;
- two-stage process using air.

In the single-stage process the 'reaction' (formation of acetaldehyde) and the oxidation of Pd take place simultaneously with the 'oxidation' (re-oxidation of Cu^+) in one
reactor. The two-stage process employs two separate reactors; The reaction of ethene
with $PdCl_2$ (Pd^{2+}) and the oxidation of Pd with Cu^{2+} occur in the first reactor, while re-
oxidation of Cu^+ with air takes place in the second reactor.

8.2.2.1 Single-stage process

Figure 8.2 shows a flow scheme of the single-stage Wacker oxidation process. Ethene
and oxygen are fed into the lower part of the reactor in which circulation occurs by the
upward movement of the gas bubbles. The reactor content is a two-phase system, viz.
the gases ethene and O_2, and a rather acidic aqueous solution of catalyst in which part
of the gas is dissolved. The reactor operates at the boiling point of the reaction
mixture, so the heat of reaction can be removed by evaporation of water and acetaldehyde.

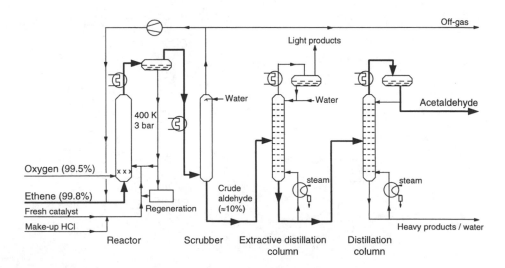

Figure 8.2   Flow scheme of the single-stage process for acetaldehyde production.

The gas stream leaving the top of the reactor is cooled and sent to a gas-liquid separator, where the catalyst solution is separated for recycle to the reactor. The gas phase, rich in unconverted ethene, is cooled and scrubbed with water to recover acetaldehyde, which is completely miscible with water. The remaining gases are recycled to the reactor. A small part is purged to prevent accumulation of inerts that are introduced with the feed (especially nitrogen) or formed during the reaction (e.g. CO_2).

The dilute acetaldehyde is purified and concentrated in two stages. In the first stage, light by-products are removed by extractive distillation (see box) with water (acetaldehyde has a much higher solubility but only slightly higher boiling point than the light by-products, such as chloromethane and chloroethane). In the second stage, heavy products (acetic acid and other organic compounds, including chlorinated by-products) and water are removed. A small amount of make-up HCl solution is added to the reactor to compensate for the chloride losses.

The catalyst solution deteriorates to some degree during oxidation. Therefore, part of the liquid phase from the separator is sent to the regenerator.

Ethene and oxygen in certain proportions form explosive mixtures. Therefore, ethene is chosen to be in such excess that the mixture is outside the higher explosion limit, resulting in an ethene conversion per pass of only about 30%.

Extractive distillation

Extractive distillation (Figure B.8.1) is applied when a binary mixture is difficult or impossible to separate by ordinary distillation because the relative volatility of the components is low. A third component, called solvent, is added, which alters the relative volatility of the original components. The solvent itself is of low volatility and is not appreciably vaporized in the column.

Extractive distillation is often applied for the separation of aromatics from alkanes that have similar volatility. For example, separation of toluene (b.p. 383.8 K) and iso-octane (b.p. 372.4 K) is achieved by adding phenol (b.p. 455 K), which strips toluene from the mixture so that iso-octane is obtained at the top. The separation of toluene and phenol is relatively easy.

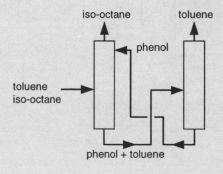

Figure B.8.1 Principle of extractive distillation.

QUESTIONS:

> *In many examples in this book first thermodynamics is treated. Why is this not done in this case? What is the preferred reactor, a plug-flow reactor or a CSTR?*
>
> *Why does the reactor temperature equal the boiling point of the reaction mixture? Guess the composition of the off-gas.*
>
> *Guess the composition of the light products from the first distillation column.*
>
> *Show that 'classical' distillation is not useful for removing the light products.*

8.2.2.2 Two-stage process

In the two-stage process ethene and oxygen react in different reactors. In the first reactor, ethene is converted. The conversion is almost complete and no recycle of ethene takes place. In the second reactor, re-oxidation of the catalyst ($Cu^+ \rightarrow Cu^{2+}$) takes place. Air can be used instead of pure oxygen, since the gas phase from this reactor only contains inert components (nitrogen), which can be vented. In the single-stage process these inerts would accumulate in the recycle stream.

Figure 8.3 shows a flow scheme of the two-stage process. In this scheme 'reaction' and 'oxidation' take place in two separate bubble-column reactors. Ethene and catalyst are introduced in the ethene oxidation reactor. The effluent is flashed to 1 bar. The flasher overhead, containing water and acetaldehyde in the vapor phase, is concentrated in the crude aldehyde column, after which the light and heavy compounds are removed by distillation.

Water discharged at the bottom of the crude aldehyde column is recycled to the flash vessel in order to maintain a constant catalyst concentration. A portion of the water is used for scrubbing acetaldehyde from the off-gases of the catalyst oxidation reactor and the ethene oxidation reactor.

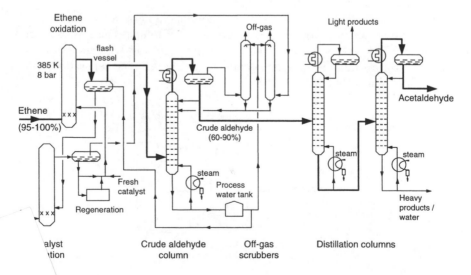

re 8.3   Flow scheme of the two-stage process for acetaldehyde production.

The catalyst solution leaves the flash vessel as the bottom product and is fed into the oxidation reactor in which Cu^+ is reoxidized in air. The conversion of oxygen is nearly complete. The bulk of the catalyst solution leaving the separator at the bottom is recycled to the first reactor. A small fraction is sent to a regenerator.

QUESTIONS:

> *High ethene conversion is desired in the two-stage process. Why?*
> *Why would a CSTR not be suitable for this process (kinetics)?*
> *Compare the distillation trains for the purification of the crude aldehyde solution for the single and two-stage process (hints: also compare the water flows, and the concentrations of ethene and acetaldehyde in the flows entering the distillation train).*

8.2.2.3 Process comparison

In the single-stage process oxygen is used, whereas in the two-stage process usually air is used. In both the one- and two-stage process the acetaldehyde yield is about 95% and the production costs are virtually the same. Expensive construction materials have to be used in both processes because the catalyst solution is very corrosive. In that respect the single-stage process is advantageous because here only one reactor is in contact with the corrosive solution, whereas in the two-stage process two reactors, piping and pumps have to be corrosion resistant. Furthermore, the use of air in the two-stage process requires larger separation equipment, etc., resulting in a much greater (about 1.5 times) investment. Main advantages of the two-stage process are the possibility to use cheap air instead of oxygen, and the absence of the need to recycle unconverted ethene (compression), reducing operation and investment costs. Another advantage of the two-stage process is that it is safer, because ethene and oxygen react in separate reactors and, thus, never form explosive mixtures. Many aspects govern the choice of process. These include the raw material and energy situation, rate considerations, safety requirements, as well as the availability of oxygen at a reasonable price.

QUESTIONS:

> *Why is the purity of the ethene feed for the single-stage process rather critical, whereas this is not the case for the two-stage process?*
> *How is the heat of reaction removed in the two-stage process? How does this differ from the single-stage process?*

8.2.2.4 Disadvantages of the Wacker process

Although the Wacker oxidation process is an elegant way of introducing oxygen selectively into hydrocarbons, disadvantages of the process limit its practical importance:

- By-products are formed. Both the single-stage and the two-stage process yield chlorinated hydrocarbons, chlorinated acetaldehydes, and acetic acid as by-products.

- Construction materials are expensive. The acidic catalyst solutions are extremely corrosive. Suitable construction materials are among others titanium, acid-proof ceramic, and tantalum.
- Wastewater is a problem. It contains chlorinated hydrocarbons, which are not biodegradable due to their toxicity. Chlorine-free catalysts have been studied but have not (yet) been commercialized.

8.3 ACETIC ACID PRODUCTION

Apart from food applications acetic acid is used for the production of vinyl acetate, cellulose acetate (thermoplastic and fiber constituent) and as a solvent. It can be produced from a number of raw materials. Commercially nearly all non-food acetic acid is produced by one of the three processes shown in Table 8.4.

All three reactions are based on homogeneous catalysis. The first two processes proceed through radical reactions (see Scheme 8.1). The chemistry of methanol carbonylation is fundamentally different (see Section 8.3.2).

8.3.1 Background Information

Until late in the 19th century, acetic acid was manufactured by the age-old process of sugar fermentation to ethanol and subsequent oxidation to acetic acid. Table vinegar occasionally is still produced by fermentation. In 1916 the first process for production of acetic acid on an industrial scale was commercialized. This process was based on the liquid-phase oxidation of acetaldehyde. The acetaldehyde process is still widely

Table 8.4   Processes for the production of acetic acid

Reaction	ΔH^0_{298} (kJ/mol)
Oxidation of acetaldehyde	

$$H_3C\text{-}\overset{\displaystyle O}{\overset{\|}{C}}H \;+\; {}^1\!/_2\,O_2 \;\longrightarrow\; CH_3\overset{\displaystyle O}{\overset{\|}{C}}\text{-}OH \qquad -292 \qquad (11)$$

Direct oxidation of hydrocarbons (mainly naphtha and n-butane)

$$CH_3CH_2CH_2CH_3 \;+\; O_2 \;\longrightarrow\; CH_3\overset{\displaystyle O}{\overset{\|}{C}}\text{-}OH \;+\; \text{by-products} \ll 0 \qquad (12)$$

Carbonylation of methanol

$$CH_3OH \;+\; CO \;\longrightarrow\; CH_3\overset{\displaystyle O}{\overset{\|}{C}}\text{-}OH \qquad -138 \qquad (2)$$

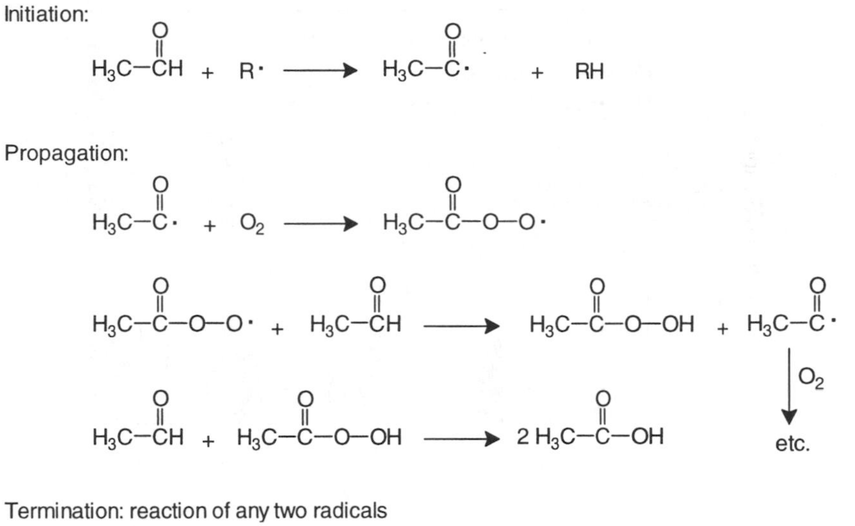

Initiation:

$$H_3C-\overset{\overset{\displaystyle O}{\|}}{C}H + R\cdot \longrightarrow H_3C-\overset{\overset{\displaystyle O}{\|}}{C}\cdot + RH$$

Propagation:

$$H_3C-\overset{\overset{\displaystyle O}{\|}}{C}\cdot + O_2 \longrightarrow H_3C-\overset{\overset{\displaystyle O}{\|}}{C}-O-O\cdot$$

$$H_3C-\overset{\overset{\displaystyle O}{\|}}{C}-O-O\cdot + H_3C-\overset{\overset{\displaystyle O}{\|}}{C}H \longrightarrow H_3C-\overset{\overset{\displaystyle O}{\|}}{C}-O-OH + H_3C-\overset{\overset{\displaystyle O}{\|}}{C}\cdot \ \Big\downarrow O_2$$

$$H_3C-\overset{\overset{\displaystyle O}{\|}}{C}H + H_3C-\overset{\overset{\displaystyle O}{\|}}{C}-O-OH \longrightarrow 2\,H_3C-\overset{\overset{\displaystyle O}{\|}}{C}-OH \qquad \text{etc.}$$

Termination: reaction of any two radicals

Scheme 8.1 Radical mechanism for acetaldehyde oxidation.

applied, but since the development of the Monsanto methanol carbonylation process it is progressively losing ground.

Direct liquid-phase oxidation of butane and/or naphtha was once the preferred route to acetic acid, because of the low cost of these hydrocarbons. The reaction mechanism is based on radicals. Scheme 8.1 lists the initiation, propagation and teminations steps. From the scheme it is not surprising that a major drawback of this process is that up to 50% of the feed goes to by-products (e.g. formic acid, higher acids, and aldehydes), many of which have very limited markets. Furthermore, the process requires a very complex purification train, adding to the investment and operating costs. Nowadays only a small portion of acetic acid is manufactured by this process.

The manufacture of acetic acid by carbonylation of methanol dates back to 1925, but the first commercial methanol carbonylation plant did not come on stream until 1963 when a new cobalt/iodide catalyst system was developed (BASF technology). In 1968 the Monsanto Chemical Co. introduced a new carbonylation process using a novel highly selective catalyst (rhodium iodide), which was commercialized success-fully only 2 years later. This process has now replaced all other methods for acetic acid

Table 8.5 Comparison of processes for acetic acid production

Process	Yield (mol%)	T (K)	p (bar)	Catalyst
Acetaldehyde oxidation	93—96	335–355	3–10	Mn/Co[a]
n-Butane/naphtha oxidation	70—80	395–475[b]	45–55	Mn/Co[a]
BASF	90 (CH_3OH & CO)	455–515	500	Co/HI
Monsanto	99 (CH_3OH), 90 (CO)	425–475	30–60	Rh/HI

[a] Salts (usually acetate salts) of these metals are applied.
[b] The lower temperatures are applied for naphtha, the higher for *n*-butane.

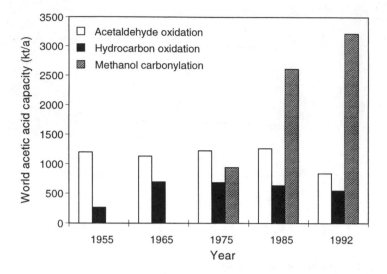

Figure 8.4 Breakdown of various technologies for acetic acid manufacture [1,12,13].

manufacture. Figure 8.4 shows that the Monsanto methanol carbonylation technology indeed has emerged surprisingly quickly.

The process conditions and yields of the various processes are very different. Table 8.5 gives a comparison. The BASF and Monsanto technologies are closely related. In the Monsanto process, however, the pressure is much lower, resulting in less severe, lower cost, operating conditions and higher acetic acid yields.

8.3.2 Methanol Carbonylation Process

Methanol carbonylation is an exothermic reaction and thus is favored by low temperature. Figure 8.5 shows the effect of temperature and pressure on the equilibrium conversion of methanol to acetic acid. The feed contains equal molar amounts of

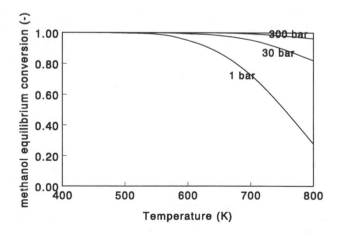

Figure 8.5 Methanol equilibrium conversion to acetic acid; feed $CH_3OH/CO = 1$ mol/mol.

methanol and carbon monoxide, and it is assumed that acetic acid is the only product formed.

The methanol equilibrium conversion is essentially complete at the reaction conditions employed in both the BASF and Monsanto processes. Indeed, at the temperatures employed the conversion is even nearly complete at atmospheric pressure. The reaction is carried out at elevated pressure, however, for the following reasons:

- to keep the reaction mixture in the liquid phase, thus facilitating temperature control.
- to generate and maintain the catalyst in its active form; the catalyst complex is not stable at low CO pressure.

In the development of the carbonylation process, corrosion has always been a point of concern. In the beginning only gold and graphite were found to be stable under the severe conditions tested (up to 600 K, up to 200 bar, highly corrosive catalyst solutions). BASF workers found that copper and cobalt catalyst systems in the presence of iodide could be handled by employing Hastelloy C[1].

The key to the success of the Monsanto process is the very active catalyst system, which allows operation at much less severe conditions than the BASF process. Still, the need for corrosion resistant materials like Hastelloy C, titanium, or even more 'exotic' materials remains. Titanium is most often used.

An essential part of the catalyst system is iodide, usually supplied as hydrogen iodide, which is very corrosive. Fortunately, hydrogen iodide rapidly reacts with methanol to form methyl iodide:

$$CH_3OH + HI \rightleftharpoons CH_3I + H_2O \qquad \Delta H^0_{298} = -53.1 \text{ kJ/mol} \qquad (13)$$

The formation of methyl iodide is the start of the catalytic cycle (see box).

Since addition of methyl iodide to the rhodium complex has been found to be the rate-determining step, the reaction rate is independent on the two reactants methanol and carbon monoxide. The rate equation is the following:

$$r_{CH_3COOH} = k \cdot [Rh] \cdot [CH_3I] \qquad (14)$$

The consequence of the zero order dependence of the reaction rate on methanol and carbon monoxide concentrations is that at any conversion level, the production rate is the same. Therefore, high conversions can be obtained even in a CSTR of relatively small volume.

Methanol is usually diluted with water to suppress the formation of methylacetate and dimethyl ether:

$$CH_3OH + CH_3COOH \rightleftharpoons CH_3COOCH_3 + H_2O \qquad \Delta H^0_{298} = -15.3 \text{ kJ/mol} \qquad (15)$$

$$2CH_3O \rightleftharpoons H_3COCH_3 + H_2O \qquad \Delta H^0_{298} = -23.6 \text{ kJ/mol} \qquad (16)$$

H_2O addition increases the methanol conversion to acetic acid but inevitably leads to loss of CO due to interference of the water-gas shift reaction (CO_2 is formed, which is a worthless by-product).

---

[1] Hastelloy C: one of a range of alloys, containing nickel (54%), molybdenum (17%), chromium (15%), and iron (5%), that were developed for corrosion resistance to strong mineral acids like HCl and HI.

Catalytic cycle

Following the formation of methyl iodide, an oxidative addition reaction occurs between methyl iodide and $[Rh(CO)_2I_2]^-$. This has been found to be the rate-determining step. Methyl migration results in the penta-co-ordinated acyl intermediate (acyl: H_3C-CO-). After addition of CO, acetyl iodide splits to yield the original rhodium complex. Acetyl iodide subsequently reacts with water to form acetic acid and hydrogen iodide, which closes the catalytic cycle (Scheme B.8.2).

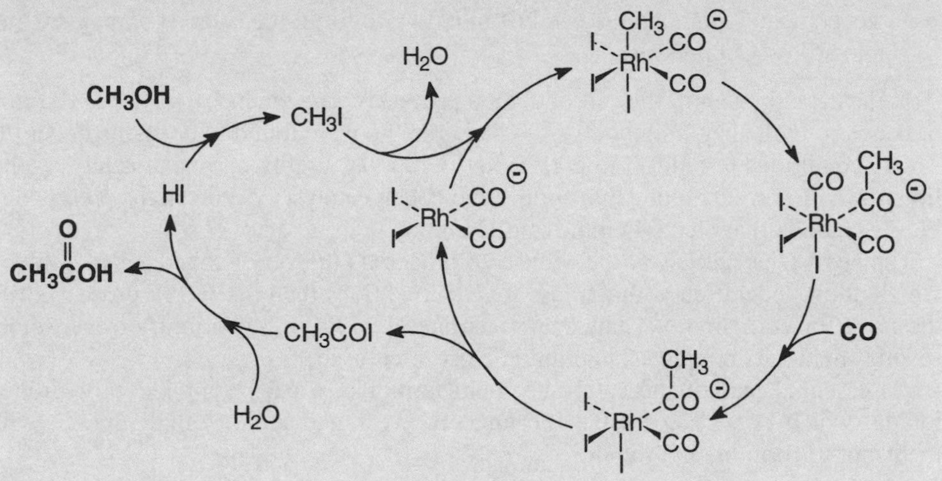

Scheme B.8.2 Catalytic cycle in methanol carbonylation to acetic acid.

QUESTIONS:

Why would HI be so corrosive?

Recall the design equation for a CSTR and show that only in the case of 0th order reactions complete conversion can be attained in a reactor of finite volume.

Figure 8.6 shows a flow scheme of the Monsanto acetic acid process. The reactor can either be a sparged stirred-tank reactor (as shown) or a bubble column.

Carbon monoxide and methanol (via scrubber) are fed to a CSTR containing the catalyst. Reaction takes place in the liquid phase under relatively mild conditions. The liquid reactor effluent is depressurized in a flash vessel resulting in a gas and a liquid phase. The liquid phase is sent to a light ends column. The overhead is combined with the gases from the flash vessel and scrubbed with methanol to recover methyl iodide. Off-gases from the scrubber are flared. The bottom stream of the light ends column, containing the catalyst complex, water, and acetic acid is recycled to the reactor. Wet ic acid is taken as a side stream from the light ends column and fed to the drying nn. Dry acetic acid is removed as bottoms product. The overheads, containing a re of acetic acid ($\approx 35\%$) and water, are recycled to the reactor. Thus, a fixed

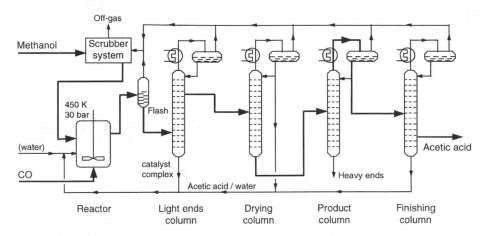

Figure 8.6 Flow scheme of Monsanto process for acetic acid production.

amount of acetic acid and water is continuously circulating through the plant. The dry acetic acid is fed to the product column from which heavy by-products are removed as bottoms. The overhead of the product column is sent to the finishing column from which ultra pure acetic acid is obtained as a side stream. Both overhead and bottoms from this column are recycled to the process.

QUESTIONS:

Why is such a large amount of acetic acid present in the overhead stream from the drying column?
Why is the acetic acid/water mixture not separated further so that only water would have to be recycled to the reactor?
Is the catalyst recycled? How is the heat of reaction dissipated?
Guess the composition of the off-gas.
What are the advantages and disadvantages of the Monsanto process?

An important requirement of the acetic acid is that it contains only a minimal amount of water. The complete separation of water and acetic acid is very difficult and thus energy intensive (5 kg steam/kg dry acetic acid [13]) owing to their small relative volatility. The separation also requires a large number of trays. The BASF process employs an azeotropic agent (esters, which are formed as by-products anyway) to enhance the relative volatility of water and acetic acid, thus facilitating the separation (see box). In this way, both a pure acetic acid and a pure water stream can be obtained. In the Monsanto process, which produces few by-products, adding an azeotropic agent would imply the use of an extra distillation column to recover this agent. Therefore, in this process it is not attempted to produce a water stream free of acetic acid. By choosing to recycle an acetic acid/water mixture instead of pure water and a clever combination of distillation columns a satisfactory product is produced by normal distillation.

QUESTIONS:

> *A major concern is the production of 'dry acetic acid', which may only contain a very small amount of water. Evaluate the two solutions presented. Can you think of alternatives?*

Azeotropic distillation

In azeotropic distillation (Figure B.8.2) of two components that form an azeotrope[2] or are difficult to separate otherwise, a third component (an azeotropic agent) is added to increase the relative volatility of the two components.

An example of industrially applied azeotropic distillation is the dehydration of ethanol. Ethanol and water form a minimum boiling azeotrope at a temperature of 351 K, where the mixture contains 89 mol% ethanol. Starting with a mixture containing a lower proportion of ethanol, it is not possible to obtain a product richer in ethanol than 89%. If benzene is added to the azeotrope, a ternary azeotrope is formed with a boiling point of 338 K. The relatively non-polar benzene serves to volatilize water (a highly polar molecule) more than ethanol (a moderately polar molecule).

A virtually pure ethanol product is obtained. The first tower gives nearly pure ethanol as the bottoms product and the ternary azeotrope at the top. Cooling results in the formation of two phases: a benzene-rich organic phase, which is used as reflux, and a second, aqueous phase, which also contains benzene and ethanol. The second column separates the ternary azeotrope, which is recycled to the first column, from water.

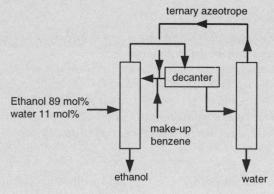

Figure B.8.2   Principle of azeotropic distillation.

The catalyst complex must be recovered with an extreme efficiency and returned to the reactor, because rhodium is a harmful metal and it is very expensive. Nearly complete recovery of rhodium is effected in the light ends column. Furthermore, the volatile and toxic methyl iodide has to be retained in the system. This is achieved in the methanol scrubber. Overhead gases from all columns are sent to this scrubber.

---

[2] Azeotrope: a liquid mixture that exhibits a maximum or minimum boiling point relative to the boiling points of the components and that distils without change in composition.

Table 8.6 Economic comparison of acetaldehyde and methanol process for acetic acid production. Adapted from [13]

	Acetaldehyde oxidation ($c/kg)	Methanol carbonylation ($c/kg)
Raw materials	63.7	21.6
Utilities	5.7	3.3
By-products	− 2.0	0.0
Labor	0.7	1.2
Other costs (capital, catalyst)	9.5	10.1
Total production costs	77.6	36.2

8.3.3 Process Economics

It is interesting to compare the economics of the acetaldehyde oxidation and the Monsanto carbonylation process for the production of acetic acid (300 kt/a). From the comparison in Table 8.6 it is obvious that carbonylation of methanol is the preferred route to acetic acid (current market price 79 $c/kg). It is not surprising that the share of the process based on acetaldehyde is shrinking.

8.4 HYDROFORMYLATION

Hydroformylation, also referred to as OXO-synthesis, was discovered in Germany before the Second World War in one of the many programs aimed at application of synthesis gas from coal (CO/H_2 mixtures). It was discovered that alkenes react with syngas, provided an appropriate catalyst is present. Formally, formaldehyde (CH_2O) is added to the double bond. Therefore, the reaction was called 'hydroformylation' (analogous to 'hydrogenation' being the addition of hydrogen). The most important application is the hydroformylation of propene yielding two isomers:

$$CH_3CH{=}CH_2 + CO + H_2 \longrightarrow \underset{\substack{|\\CH_3}}{CH_3CHCH} \overset{\displaystyle O}{\overset{\|}{}} + CH_3CH_2CH_2CH \overset{\displaystyle O}{\overset{\|}{}}$$

propene   iso-butyraldehyde   *n*-butyraldehyde

$$\Delta H^0_{298} = -125 \ \text{kJ/mol} \qquad (3)$$

In most cases the linear aldehyde is the preferred product because it enjoys a much larger market than the branched aldehyde. The major reason is that the aldehydes are for a large part used for the production of alcohols and linear alcohols show a relatively high biodegradability.

Hydroformylation is not limited to propene. Linear and branched C_3–C_{17} alkenes are common feedstocks.

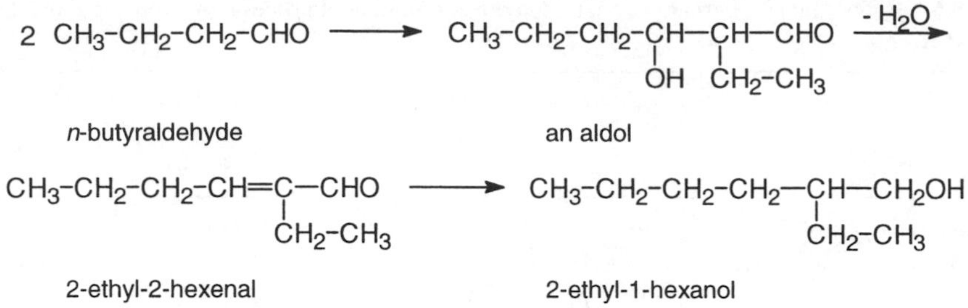

$$2 \ CH_3-CH_2-CH_2-CHO \longrightarrow CH_3-CH_2-CH_2-\underset{\underset{OH}{|}}{CH}-\underset{\underset{CH_2-CH_3}{|}}{CH}-CHO \xrightarrow{-H_2O}$$

     n-butyraldehyde   an aldol

$$CH_3-CH_2-CH_2-CH{=}\underset{\underset{CH_2-CH_3}{|}}{C}-CHO \longrightarrow CH_3-CH_2-CH_2-CH_2-\underset{\underset{CH_2-CH_3}{|}}{CH}-CH_2OH$$

    2-ethyl-2-hexenal   2-ethyl-1-hexanol

Scheme 8.2 Reaction sequence for the production of 2-ethyl-1-hexanol.

8.4.1 Background Information

The products of hydroformylation usually are intermediates for the production of several types of alcohols. For instance, an important practical process is the production of 2-ethyl-1-hexanol from *n*-butyraldehyde (formed by hydroformylation of propene) as shown in Scheme 8.2.

The alcohols produced by the OXO-synthesis are commonly known as OXO-alcohols. *n*-Butanol and 2-ethyl-1-hexanol are the most important products of hydroformylation. *n*-Butanol has two main uses, as a solvent in surface coatings and for the manufacture of butyl esters.

2-Ethyl-1-hexanol is mainly used in plasticizers. Other OXO-alcohols in the C_6–C_{11} range are also applied in the synthesis of plasticizers (in particular for PVC). A trend in this field of application is the use of longer chain alcohols in order to reduce the volatility of the plasticizers. The OXO-alcohols in the C_{12}–C_{18} range are applied in detergents (see Figure 8.7).

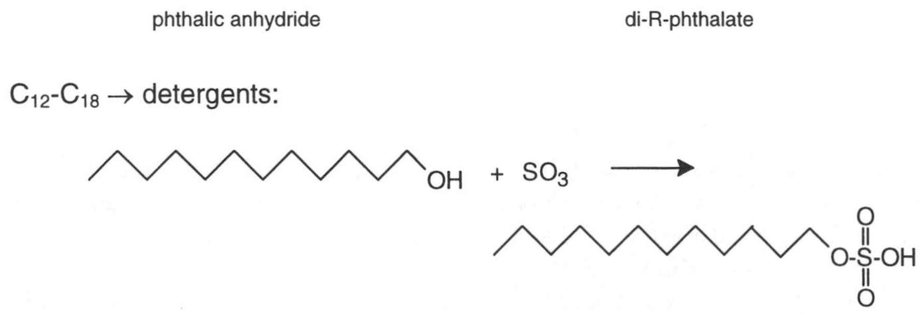

$C_6 - C_{11} \rightarrow$ plasticizers:

    phthalic anhydride   di-R-phthalate

C_{12}-$C_{18} \rightarrow$ detergents:

Figure 8.7 Applications of OXO-alcohols.

Table 8.7 Relative activities of metals for hydroformylation

Rh	Co	Ru	Mn	Fe	Cr, Mo, W, Ni
$10^3 - 10^4$	1	10^{-2}	10^{-4}	10^{-6}	0

8.4.2 Thermodynamics

Hydroformylation requires low temperature and elevated pressure as can be seen from Figure 8.8a, which shows the equilibrium conversion of propene to the corresponding aldehydes, at a typical inlet composition $H_2/CO/propene = 1:1:3$. As mentioned earlier, normal aldehydes are the preferred products. Figure 8.8b shows the equilibrium distribution of the normal- and iso-aldehyde as a function of temperature (pressure: 30 bar) in the hydroformylation of propene.

The iso-aldehyde is the thermodynamically favored product. During hydroformylation also other reactions can occur, in particular hydrogenation to propane. Thermodynamically, the latter reaction is the preferential reaction. Therefore, a catalyst is required which has a high selectivity.

8.4.3 Catalyst Development

The most important catalysts for hydroformylation are based on Rh and Co, which are introduced as carbonyls. Rh and Co are not the only metals showing catalytic activity but they are the most active ones. The metal activity for hydroformylation (relative units) is shown in Table 8.7. From this list it is understandable that in practice only Rh and Co are used.

A wide variety of ligands, in particular phosphines and phosphites, have been shown to influence activity, stability, and selectivity of the complexes (see Figure 8.9). Numerous patents have been filed showing that useful ligands have been found that provide increased selectivity to linear products, although the activity often is lowered.

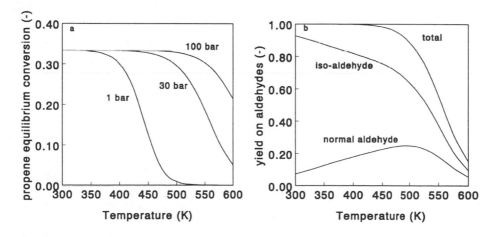

Figure 8.8 Hydroformylation of propene: (a) Propene equilibrium conversion; (b) normal/iso distribution; $p = 30$ bar; propene/syngas $= 3$ mol/mol.

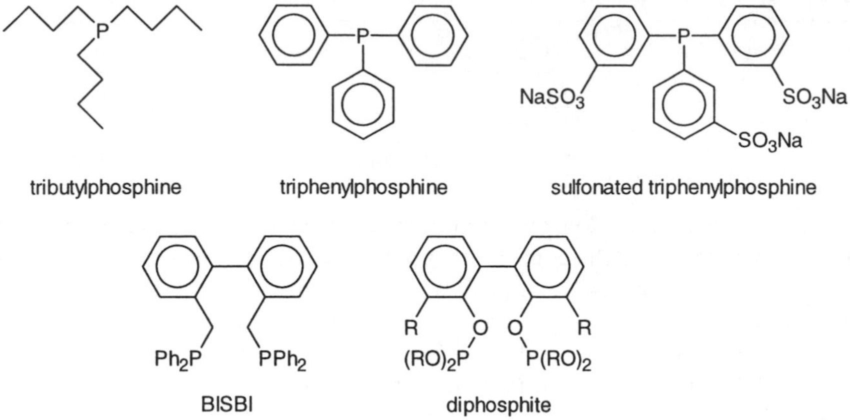

tributylphosphine   triphenylphosphine   sulfonated triphenylphosphine

BISBI   diphosphite

Figure 8.9   Examples of phosphine and phosphite ligands; Ph = phenyl.

Catalytic cycle

Scheme B.8.3 shows the catalytic cycle for hydroformylation of propene by
$HRh(CO)L_2$.

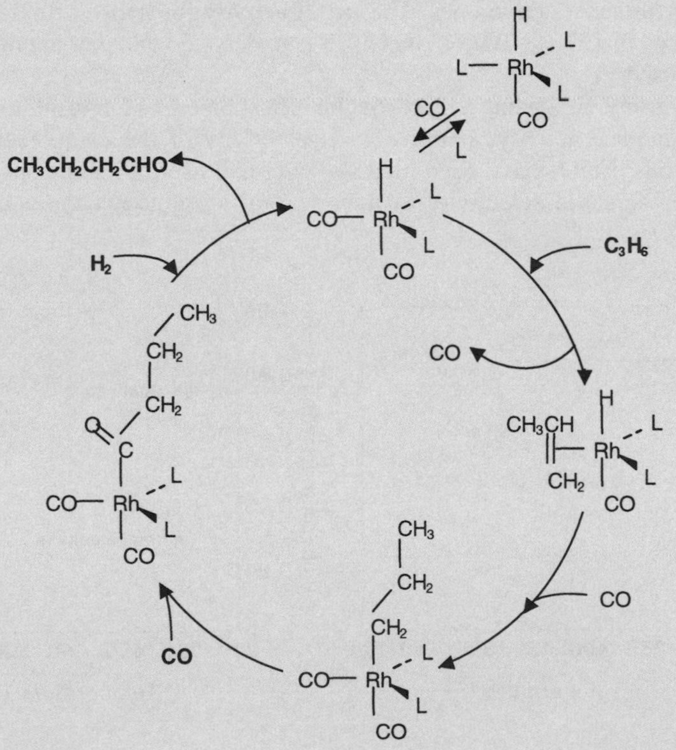

Scheme B.8.3   Catalytic cycle in hydroformylation of propene.

Kinetic studies have been performed for the hydroformylation reaction. For a simple system (the only ligands were CO) the following rate expression has been found:

$$r_{\text{hydroformylation}} = k[\text{Rh}][\text{H}_2][\text{CO}]^{-1} \tag{17}$$

For a Rh system containing triphenylphosphine ligands the following rate equation was reported:

$$r_{\text{hydroformylation}} = k[\text{Rh}][\text{PPh}_3]^{-0.7}[\text{alkene}]^{-0.6}[\text{H}_2]^{0.05}[\text{CO}]^{-0.1} \tag{18}$$

QUESTIONS:

> *Give an interpretation for these equations based on the mechanism. What are the consequences for reactor design?*

During hydroformylation side reactions occur, e.g. hydrogenation of alkenes to the corresponding alkanes, and double-bond isomerization (e.g. 2-hexene to 1-hexene). The extent of these side reactions depends on the metal and the ligand added. Compared to rhodium-based catalysts, cobalt-based catalysts strongly enhance double-bond isomerization. This is why cobalt-based processes are quite flexible in that they can be used for hydroformylation of a wide variety of alkenes. The selectivity to normal aldehydes of the rhodium catalyst, however, is much greater and, therefore, processes for hydroformylation of propene, the predominant feedstock, are more and more based on this type of catalyst.

QUESTIONS:

> *Hydroformylation in principle can be used for all alkenes. Why would α-alkenes (external double bond) react faster than internal alkenes?*
> *The major side reactions are hydrogenation of the alkenes to form alkanes and double-bond isomerization. Are these desired or undesired reactions?*

8.4.4 Processes

8.4.4.1 Hydroformylation of propene

Hydroformylation of lower alkenes, in particular propene, nowadays is usually based on rhodium catalysts. By adding the right ligands the selectivity towards linear products is high. The reaction takes place in a stirred-tank reactor at much lower pressure than the unmodified cobalt processes. The alkene and synthesis gas have to be purified thoroughly because the catalyst is extremely sensitive to poisons. Figure 8.10 shows a process scheme for the rhodium catalyzed hydroformylation of propene (the catalyst ligands are triphenylphosphines, approximately 75 (!) mol/mol Rh.). This process is often called the low-pressure OXO process (LPO).

The gaseous reactants are fed to the reactor through a sparger. The reactor contains

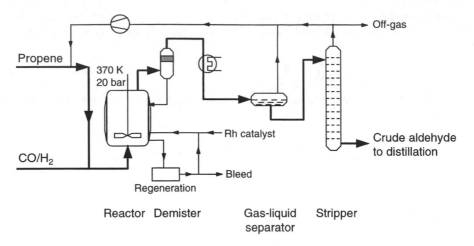

Figure 8.10   Flow scheme of rhodium-catalyzed hydroformylation of propene (LPO process).

the catalyst dissolved in product butyraldehyde and by-products (e.g. trimers and tetramers). Rhodium stays in the reactor, apart from a small purification cycle.

The gaseous reactor effluent passes through the demister in which fine droplets of catalyst that are contained in the product gases are removed and fed back to the reactor. The reaction products and unconverted propene (conversion per pass ca. 30%) are condensed and fed to the gas–liquid separator. Propene, apart from a small purge stream, is recycled to the reactor. The liquid reaction products are freed from residual propene in a stripping column and further purified by distillation.

A portion of the catalyst solution passes through a purification cycle to reactivate inactive rhodium complexes, and for removal of heavy ends and ligand decomposition products.

The demisting device is an essential part of the plant. It ensures that all rhodium remains in the reactor. In view of the high costs of rhodium, recovery should be as high as possible. In practice, the rhodium losses in the process are very small. The losses of phosphine ligands are more important and determine the economics of the process.

QUESTIONS:

> *What would the main impurity in propene be? Why is a purge needed? Why is this process scheme not suitable for hydroformylation of higher alkenes?*

An interesting novel industrial process is based on a *water-soluble* rhodium catalyst (Ruhrchemie/Rhône-Poulenc). It contains polar ligands (sulphonated triphenylphosphines) which are highly water-soluble. The advantage is that the catalyst does not dissolve in the organic phase. In fact, one could speak of a heterogeneous catalyst system. As a consequence, the rhodium losses are minimal. Figure 8.11 shows a flow scheme of the process.

In this process the reaction also takes place in a stirred-tank reactor which contains

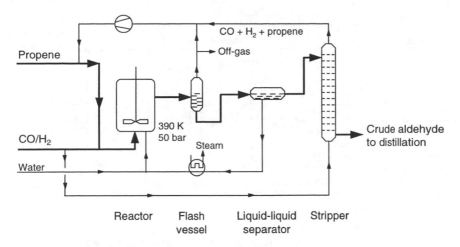

Figure 8.11 Hydroformylation of propene using a water-soluble rhodium catalyst.

the catalyst solution. Before entering the reactor, part of the syngas is first passed through a stripping column to recover the unreacted propene.

The liquid reactor effluent is flashed to remove dissolved gases, and then fed to a phase separator, where butyraldehyde is separated from the aqueous catalyst solution. The catalyst solution is returned to the reactor via the heat exchanger in which steam is generated. Some water is lost in the aldehyde phase and has to be replaced.

QUESTIONS:

Guess the composition of the off-gas. Why is a demister not needed in this case? What separation(s) would be performed in the distillation unit? Why has water been selected as the solvent instead of an organic solvent?

8.4.4.2 Hydroformylation of higher alkenes

The first industrial OXO-process was based on an unmodified cobalt catalyst: $HCo(CO)_4$. This catalyst system requires high pressures and temperatures. Hydroformylation of higher alkenes to produce plasticizer and detergent range alcohols still uses this type of catalyst.

To minimize catalyst consumption and to avoid problems in downstream processing (fouling of equipment, etc.) it is very important that cobalt is recovered. A process that offers an elegant solution to the problem of catalyst separation is the Kuhlmann process. This hydroformylation process is based on the following cobalt cycle:

$$2HCo(CO)_4 + Na_2CO_3 \rightarrow 2NaCo(CO)_4 + H_2O + CO_2 \tag{19}$$

The sodium salt is soluble in water. This enables recovery of Co catalyst by scrubbing with water. The aqueous solution containing Co as the sodium salt is transformed to the active catalyst by reaction with sulfuric acid:

$$2NaCo(CO)_4 + H_2SO_4 \rightarrow 2HCo(CO)_4 + Na_2SO_4 \tag{20}$$

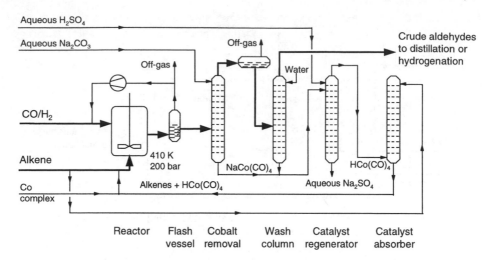

Figure 8.12   Kuhlmann hydroformylation process for the production of aldehydes.

This complex is soluble in alkenes and, as a consequence, it can be returned to the reactor dissolved in the reactant alkene flow.

Figure 8.12 shows a flow scheme of the Kuhlmann hydroformylation process for the production of aldehydes.

The alkene, with recycled and make-up catalyst, is fed to the hydroformylation reactor (usually a stirred tank) together with the synthesis gas. The crude product after passing the flash vessel, in which the gas is separated, is treated in countercurrent flow with an aqueous Na_2CO_3 solution. The Co-complex forms a water-soluble sodium salt. After scrubbing with water in the wash column, a virtually cobalt-free organic phase (crude aldehydes) and an aqueous phase containing the sodium salt remain. The crude aldehydes are sent to a distillation section (not shown). Here the excess alkenes are recovered and recycled to the reactor. The sodium salt is transformed to the original Co-complex ($HCo(CO)_4$) by adding H_2SO_4 in the regenerator. This complex is extracted by the alkene feedstock and recycled to the reactor.

8.4.4.3 Comparison of hydroformylation processes

Table 8.8 summarizes characteristics of the major hydroformylation processes. The rhodium catalyst is attractive for the conversion of propene to the linear aldehyde. The selectivity is approximately 95%. The branched product also has a market, although a much smaller one than the linear product, and its production is, therefore, desired in these quantities. Hydroformylation of higher alkenes with internal double bonds requires isomerization activity and, as a consequence, in that case cobalt catalysts are preferred. In all processes except one, aldehydes are the primary products. These can be hydrogenated to alcohols in a separate reactor. The modified cobalt catalyst is used for the direct production of alcohols rather than aldehydes, because this catalyst promotes hydrogenation.

Table 8.8 Comparison of hydroformylation processes (based on data from [12,14,15]

Catalyst	Rh/phosph.	Rh/phosph., water-sol.	HCo(CO)$_4$	Co/phosph.
Pressure (bar)	20	50	200	70
Temperature (K)	370	390	410	440
Alkene	Terminal C$_3$	Terminal C$_3$	Internal C$_3$–C$_{10}^+$	Internal C$_3$–C$_{10}^+$
Product	Aldehyde	Aldehyde	Aldehyde	Alcohol
Linearity (%)	70–95	95	60–80	70–90
Alkane by-product (%)	0	0	2	10–15
Metal deposition	No	No	Yes	Yes
Heavy ends	Little	Little	Yes	Yes
Poison sensitivity	High	High	Low	Low
Metal costs	1000	1000	1	1
Ligand costs	High	?	Low	High
Company	Union Carbide, Davy Powergas, Johnson Matthey	Ruhrchemie, Rhône-Poulenc	Kuhlmann	Shell

Two-stage low pressure Co-based process for direct alcohol production

Shell has commercialized a process based on a trialkylphosphine-cobalt catalyst that is much more stable than the unmodified catalyst, allowing operation at lower pressure (Figure B.8.3). This process produces the alcohol rather than the aldehyde, which is an added advantage since alcohols are usually the desired products of hydroformylation. A disadvantage is that the high hydrogenation activity also causes the formation of alkanes (up to 15 versus 2% for the non-modified catalyst). Therefore, operation may be carried out in two stages, with a lower hydrogen partial pressure in the first reactor to promote hydroformylation of the alkene rather than hydrogenation. Figure B.8.3 shows a simplified flow scheme of this version of the process.

The alkene feed is added to the first reactor together with make-up and recycle catalyst complex. Here, conversion takes place with a hydrogen poor synthesis gas to limit alkane formation. In the second reactor, hydrogen rich gas is added to ensure the hydrogenation of the aldehyde to the alcohol. Separation takes place by depressurization in a flash vessel and subsequent distillation. In the first column, unconverted alkenes are removed and recycled to the reactor. A purge stream is added in order to prevent the build-up of alkanes. The second column separates the catalyst complex and heavy by-products from the crude alcohols. By-products are formed by dimerization of aldehydes (see Section 8.4.1). Build-up of these aldols is prevented by purging part of the recycle stream.

Due to the high double-bond isomerization activity of the catalyst complex, this process is suitable for the production of primary alcohols from terminal alkenes as well as from internal alkenes. Although this process looks much simpler than the Kuhlmann hydroformylation process, it has two distinct disadvantages:

- thorough purification of the synthesis gas is very important because the ligand is very sensitive to oxidation;
- the activity of the modified cobalt catalyst is much lower, and therefore, the reactors need to be five to six times larger than in the conventional process for the same productivity.

Advantages, apart from the ones already mentioned are:

- higher linearity of alcohol product;
- possibility to separate the catalyst from the product by conventional distillation.

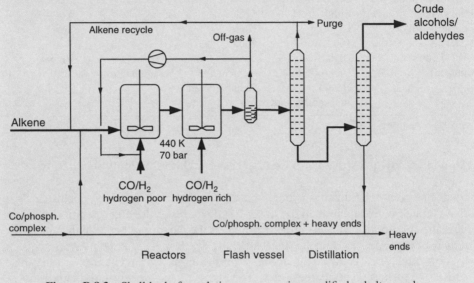

Figure B.8.3   Shell hydroformylation process using modified cobalt complex.

QUESTIONS:

> *Compare the four processes referred to in the table with respect to the phases present. Is mass transfer expected to be limiting? Compare the four processes in this respect also, and elaborate on the reactor choice.*

8.5 DIMETHYL TEREPHTALATE AND TEREPHTHALIC ACID PRODUCTION

Both terephthalic acid (TPA) and dimethyl terephthalate (DMT) are monomers for the production of poly(ethene terephthalate) by polycondensation with ethene glycol (see Scheme 8.3).

In the production of the polymer the acid has obvious advantages over the ester, because the yields are higher and no methanol recovery step is needed.

In most applications, and especially in the field of fibers, monomer purity is an essential factor in polymer quality. At present, manufacturers of poly-ethene terephthalate require a monomer purity of 99.6%. This is why in the past often the ester was preferred over the acid, which is more difficult to purify.

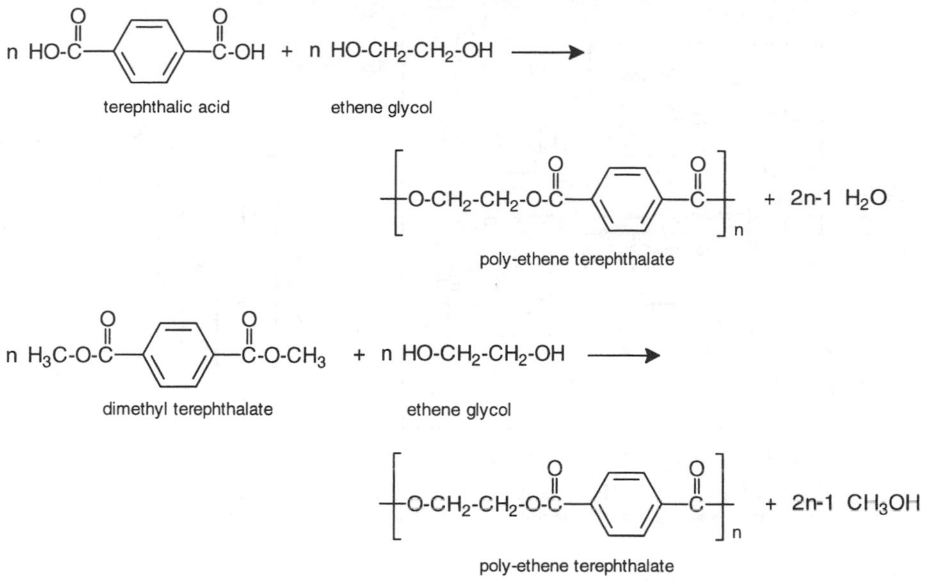

Scheme 8.3 Reactions for the production of poly-ethene terephthalate.

8.5.1 Background Information

The manufacture of terephthalic acid and dimethyl terephthalate is the major outlet for *p*-xylene. Oxidation of *p*-xylene with air to produces *p*-toluic acid (C_6H_4—$(COOH)(CH_3)$) is relatively easy. Further oxidation to terephthalic acid is difficult, however, due to the high stability of *p*-toluic acid. Therefore, the earliest process for terephthalic acid manufacture based on *p*-xylene involved oxidation by nitric acid, which is a very strong oxidant:

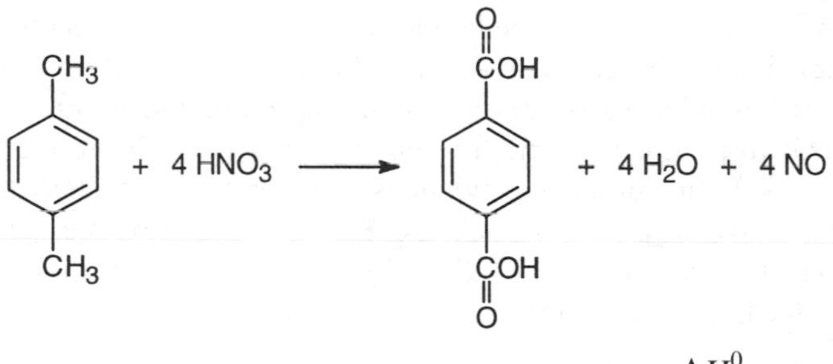

$$\Delta H^0_{298} = -750 \text{ kJ/mol} \qquad (19)$$

Not surprisingly, however, the product obtained by this reaction contains substantial amounts of impurities. Terephthalic acid is very difficult to purify (it is not volatile, insoluble in normal solvents, and does not melt). The solution found was the conversion of the acid into the corresponding methyl ester, viz. dimethyl terephthalate, which was then purified by vacuum distillation and/or recrystallization. The presence of nitrogen compounds, however, makes purification difficult and the process uneconomical.

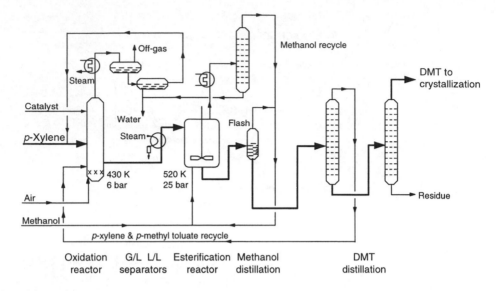

Figure 8.13   Simplified flow scheme for the production of dimethyl terephthalate (Witten).

8.5.2 Conversion of *p*-Toluic Acid Intermediate

The oxidation of *p*-xylene takes place by a radical mechanism with cobalt acetate or cobalt napthenate as the catalyst. The oxidation stops because of the high stability of *p*-toluic acid. Two methods are available to continue.

QUESTION:

> *Is the statement that the oxidation is a radical process in disagreement with the use of a catalyst?*

One method is to esterify *p*-toluic acid with methanol to *p*-methyl toluate, which is readily oxidized in air. Esterification proceeds in the liquid phase without a catalyst. During the 1950s and 1960s a number of processes combining oxidation in air and esterification were developed to produce dimethyl terephthalate, of which the most important is the Witten process. In this process *p*-xylene is converted in four alternate oxidation and esterification steps as shown in Scheme 8.4. The standard reaction enthalpy (298 K) for the oxidation reactions is -690 and -675 kJ/mol, respectively, and -30 kJ/mol for each esterification reaction.

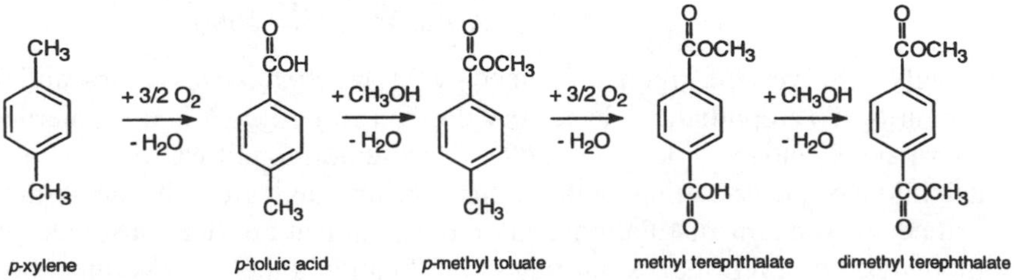

Scheme 8.4   Reaction sequence for the production of dimethyl terephthalate.

The second solution is the addition of extremely reactive radicals, such as Br atoms. These are formed by reaction of NaBr with acetic acid and oxygen. The Br radicals then attack *p*-toluic acid. Once the *p*-toluic acid radical has been formed, the reactions can proceed similarly to the oxidation of *p*-xylene, finally resulting in the formation of terephthalic acid. During the 1960s, a commercial process was developed based on this approach. The TPA produced is of a high purity. Since the discovery of this process, the Amoco process, there has been a trend towards the use of terephthalic acid rather than dimethyl terephthalate in polymerization processes. The overall reaction is as follows:

$$\Delta H^0_{298} = -1360 \text{ kJ/mol} \quad (22)$$

8.5.3 Processes

Figure 8.13 shows a simplified schematic of the Witten process for the production of dimethyl terephthalate. The heart of the process is formed by two reactors, one for oxidation and one for esterification. Fresh and recycled *p*-xylene, recycled *p*-methyl toluate, and catalyst (cobalt naphthenate) are fed to the oxidation reactor, in which mixing takes place by the upward movement of air. The temperature in the reactor is controlled by evaporation and condensation of excess *p*-xylene and formed water. Water is separated and *p*-xylene is returned continuously.

The product from the oxidation reactor is heated and fed to the esterification reactor. Excess evaporated methanol is sparged into the esterifier. *p*-Toluic acid and methyl terephthalate are transformed to their corresponding esters non-catalytically. Overhead vapors are condensed after which methanol is separated from the water and recycled. The liquid product is flashed to remove remaining methanol and water.

The esterification products are sent to a distillation column. *p*-Xylene and *p*-methyl toluate are collected overhead and returned to the oxidation reactor. The crude dimethyl terephthalate, after removal of heavy by-products and catalyst metal, is purified by crystallization. Although the cobalt catalyst used in this process is relatively cheap, recovery is necessary because heavy metals cannot simply be discarded. The residue can be mixed with water from the oxidation reactor in which the catalyst dissolves. After centrifugation the catalyst solution can be recycled.

QUESTIONS:

> *Where does the water come from? Why would the oxidation and esterification reactors operate at different pressures? What would be the composition of the off-gas?*

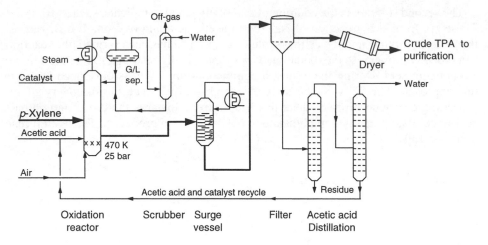

Figure 8.14   Simplified flow scheme for the production of pure terephthalic acid (Amoco).

The Amoco process produces fiber-grade terephthalic acid directly by the oxidation of *p*-xylene. Figure 8.14 shows a simplified process scheme. *p*-Xylene, acetic acid and catalyst (cobalt acetate, promoted by bromine) are introduced in the oxidation reactor. The reaction medium is agitated by introducing air at the bottom. Air is added in excess to maximize *p*-xylene conversion. Reaction takes place with acetic acid as the solvent for the catalyst. The heat generated is removed by vaporization of part of the solvent, and by condensation and reflux to the reactor. The generated steam can be used in the distillation part of the plant. A scrubber recovers acetic acid from the off-gases (nitrogen, unused oxygen, carbon dioxide resulting from overoxidation).

Because of its limited solubility in acetic acid (0.1 g/l at room temperature, 15 g/l at 470 K) most of the terephthalic acid crystallizes while it is being formed. A slurry forms, which is removed from the base of the reactor and sent to a surge vessel[3], which operates at lower pressure than the reactor. More terephthalic acid crystallizes. The suspension is filtered and the crude terephthalic acid is dried and sent to purification.

The mother liquor from the filter, containing acetic acid, water, the catalyst, some product and heavy by-products, and traces of *p*-xylene, is distilled to remove heavy residue and water. Acetic acid with dissolved catalyst is recycled to the reactor. The residue can be processed for additional catalyst recovery.

Prior to purification, the terephthalic acid purity is about 99.5 wt% (technical grade TPA). The most important impurity is 4-carboxybenzaldehyde (COOH—C_6H_4—COH), which is an intermediate in the oxidation reactions, and co-crystallizes with the terephthalic acid. The maximum allowable quantity is 25 ppm, because it will cause termination of the polymerizing chain in the production of poly-ethene terephthalate. An additional reason for removing the aldehyde is that it causes coloring of polymers.

The 4-carboxybenzaldehyde content may be reduced by selective catalytic hydro-

---

[3] Surge vessel: Vessel in which liquid is retained at a set level in order to prevent downstream or upstream equipment damages due to flow fluctuations.

genation. To this end the crystals are dissolved in water. 4-Carboxybenzaldehyde is converted to *p*-toluic acid over a palladium catalyst. The product has a much higher solubility and does not co-crystallize with TPA. Crystallization yields TPA with a 4-carboxybenzaldehyde content as low as 15 ppm.

Due to the presence of acetic acid and bromine, highly corrosive process conditions exist. Therefore, the use of special materials such as Hastelloy C or titanium is necessary in the reactor and some other parts of the process.

QUESTIONS:

> *What is the function of the heat exchanger at the top of the surge vessel? Why does complete oxidation to CO_2 and water not occur? Why is excess p-xylene used in the production of DMT, while excess air is used in the production of TPA?*

8.5.4 Process Comparison

The Amoco technology for the production of fiber-grade terephthalic acid accounts for nearly all new plants built. This is caused by its very high yield (>95%) accomplished in a single reactor pass, and the low solvent loss. Furthermore, as mentioned previously, polymerization from terephthalic acid has several advantages over polymerization from dimethyl terephthalate. The only important advantage of the Witten process for the production of dimethyl terephthalate is that no expensive construction materials are required because no very strong oxidants such as bromine radicals are used. Figure 8.15 shows the capacities for TPA and DMT since 1980, illustrating the increasing popularity of the Amoco process.

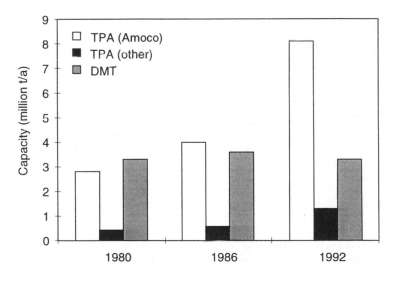

Figure 8.15 Capacity of TPA and DMT from 1980 to 1992 [4].

8.6 REVIEW OF REACTORS USED IN HOMOGENEOUS CATALYSIS

As we have seen in previous sections, in homogeneous catalysis usually a reaction takes place between a gaseous and a liquid reactant in the presence of catalyst that is dissolved in the liquid phase. The reaction generally takes place in the liquid phase. The contacting pattern between gas and liquid is generally one of the following:

- gas bubbles dispersed in continuous liquid phase:

    - bubble column (with gas-distribution device);
    - sparged stirred-tank (mechanical gas dispersion).

- liquid droplets dispersed in continuous gas phase:

    - spray column.

- thin flowing liquid film in contact with gas:

    - packed column;
    - wetted-wall column.

8.6.1 Choice of Reactor

The choice of reactor depends on factors such as the desired volume ratio of gas to liquid, the rate of the reaction (fast or slow) in relation to the mass-transfer characteristics, the kinetics (positive, negative, zero order), the ease of heat removal and temperature control, etc. Figure 8.16 shows examples of the reactor types mentioned above, while Table 8.9 shows some important parameters. The numbers in this table should be used with care because they depend on the physical and chemical properties of the reacting system and the detailed design of the reactor. In literature most data have been obtained for model systems (often air/water), and therefore large discrepancies are possible. They are meant to serve as a rough guide and to help in understanding the basics of reactor design.

QUESTIONS:

Compare the data of the reactors in Table 8.9. Are these data logical? Several factors are important in gas/liquid reactors, e.g. mixing or plug flow, fast or slow reaction, mass-transfer characteristics (e.g. main resistance in gas-film or in liquid-film). How do these factors influence the choice of reactor?

Table 8.9 Mass transfer parameters of various reactors. Adapted from [17,18]

Reactor type	Liquid holdup (%)	Gas holdup (%)	$k_1 a_1 (s^{-1})^a$	$\beta (m^3_{liquid}/m^3_{film})^b$
Stirred tank	>70	2–30	0.05–0.2	10^3–10^4
Bubble column	>70	2–30	0.05–0.15	10^3–10^4
Spray column	<20	>80	0.05–0.1	10–40
Packed column	5–15	50–80	0.01–0.8	10–40

$= $ mass-transfer coefficient based on unit volume of reactor ($m^3_{reactor}/(m^2_{interface}\cdot s)$); $a_1 = $ interarea ($m^2_{interface}/m^3_{reactor}$).

$= $ liquid phase volume/volume of diffusion layer within the liquid phase.

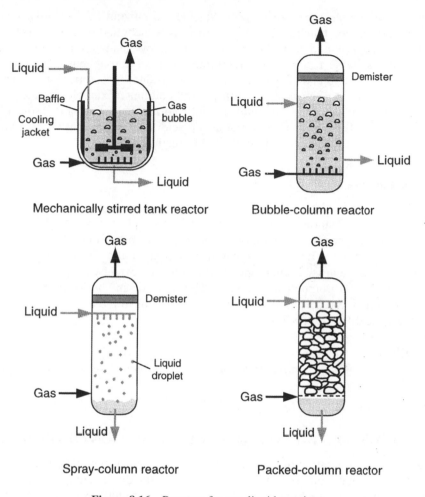

Figure 8.16 Reactors for gas–liquid reactions.

Bubble columns are most common, followed by sparged stirred tanks. Often both are used for a certain application. For instance, most processes discussed in the previous sections can use both reactor types. Stirred tanks and bubble columns have similar characteristics with respect to mass transfer (see Table 8.9). In both reactors the liquid is well mixed. The gas phase in the bubble column shows plug-flow behavior, while in stirred tanks it is well mixed.

The major advantage of the bubble column is the simple design of this type of reactor. Mixing of the liquid is achieved by the upflowing gas bubbles. The reactor shown in Figure 8.16 is an empty vessel. If a more uniform liquid distribution is required, a draft tube can be placed inside the column to enhance internal liquid circulation [16]. A disadvantage of bubble columns is that they have limited temperature and pressure ranges, while they are also unsuitable for processing viscous fluids. The major disadvantage of stirred tanks is that much backmixing occurs in both the liquid and the gas phase. A number of stirred tanks can be placed in series in order to narrow the residence time distribution. Mechanical stirring has the advantage that viscous fluids can be handled, but it increases investment and operating costs.

The spray column and packed column both have low gas-phase pressure drop, and are therefore suitable for processes requiring large gas throughputs. They are often used for gas treatment and off-gas scrubbing. For instance, absorption of CO_2 and H_2S from gas streams is often performed in packed columns. Examples of the use of a spray column are scrubbing of acetaldehyde from the reactor off-gas in the Wacker process (Section 8.2), and scrubbing of acetic acid from the off-gas of the oxidation reactor in the production of terephthalic acid (Section 8.5).

8.6.2 Exchanging Heat

Several measures can be taken to control the temperature in the reactor. For instance, cooling jackets or internal cooling coils can be used. External liquid recirculation through a heat exchanger is another possibility (see also Section 10.3.1). An elegant solution is to operate at the boiling point of the liquid mixture. In bubble columns and stirred tanks thus near isothermal operation can be achieved by evaporation of one or more of the liquid components present. This is done in the Wacker single-stage process (Section 8.2.2.1) and in the production of dimethyl terephthalate and terephthalic acid (Section 8.5.3).

QUESTION:

> *What are advantages and disadvantages of the above cooling procedures?*

8.7 REVIEW OF CATALYSIS/PRODUCT SEPARATION METHODS

The problem of separation of the catalyst from the product in homogeneous catalysis has been mentioned in the introduction as the main disadvantage of this type of catalysis. Subsequent sections showed how this problem has been tackled in various homogeneously catalyzed processes. In this section, we will summarize these solutions.

Complete catalyst recovery is important for several reasons:

- catalyst metal is expensive (especially rhodium);
- ligands are expensive (e.g. phosphines);
- catalyst metal or co-catalyst is hazardous to the environment (e.g. cobalt, MeI);
- metals act as oxidation catalysts;
- catalyst components are usually not allowed in the product.

Regeneration of the catalyst is also often required.

Product purification not only includes recovery of the catalyst. It also consists of removal of co-catalysts, decomposition products of the ligands, unconverted reactants, and by-products. The latter two, of course, are not specific for homogeneous catalysis. An example of a process that uses a co-catalyst, is the production of acetic acid. Methyl iodide acts as the co-catalyst. This substance is very toxic, and therefore must stay in the system.

Table 8.10 Classification of homogeneously catalyzed process based on separation principle

Separation principle	Process
No separation	Gas-phase polypropene process[a]
Solid product, catalyst in solution	Terephthalic acid
Gaseous product, catalyst in solution	Acetaldehyde (after flash in two-stage ethene oxidation)
	Hydroformylation of propene with conventional Rh complex
Distillation	Acetic acid
	Hydroformylation with modified Co complex
	Dimethyl terephthalate
Liquid-liquid separation	Hydroformylation of propene with water-soluble Rh complex
	Hydroformylation with unmodified Co complex

[a] Not treated in this chapter.

8.7.1 Current Separation Techniques

Table 8.10 shows separation techniques currently applied in homogeneous catalysis.

8.7.7.1 No separation at all

Very active, not too expensive, catalysts may be left in the product if they are used in such small quantities that they are not detrimental to the properties of the product. An example is the gas-phase polypropene process (not treated in this chapter).

8.7.7.2 Solid product, catalyst in solution

An example is the production of terephthalic acid. The acid crystallizes while being formed, whereas the cobalt acetate catalyst stays dissolved in the solvent.

8.7.7.3 Gaseous product, catalyst in solution

In the single-stage Wacker oxidation process, a two-phase system exists with ethene, oxygen, and the formed acetaldehyde in the gas phase (reaction mixture at boiling point of acetaldehyde), and the catalyst in an aqueous phase. The catalyst solution is separated from acetaldehyde and excess ethene by gas-liquid separation.

In the two-stage Wacker process the same situation is created after flashing of the product from the ethene oxidation reactor. In the catalyst oxidation reactor, two phases exist already, so gas-liquid separation is possible again.

The conventional rhodium catalyzed hydroformylation of propene also is a system with the catalyst in the liquid phase and the product in the gas phase. The catalyst stays in the reactor while the product leaves with the gases. However, some catalyst droplets may be entrained with the gas, so a demister is an essential part of this process.

8.7.7.4 Distillation of product from catalyst

The production of acetic acid by carbonylation of methanol is an example where the catalyst complex is separated from the product by distillation. Since rhodium complexes are relatively stable, distillation is not problematic in this case.

Separation in the hydroformylation of alkenes using a modified cobalt catalyst complex also involves distillation. The ligands are high-boiling so that they remain with the heavy ends when these are removed from the alcohol product.

Another example of separation by distillation is the production of dimethyl terephthalate. In this process, the catalyst is removed in the residue stream from the DMT column. Recovery of the catalyst takes place by dissolution in water.

8.7.7.5 Liquid–liquid separation

The hydroformylation of propene using a water-soluble rhodium catalyst is an example of a process in which the catalyst is separated from the product by liquid–liquid separation. The rhodium complex is in a different phase (water phase) than the aldehyde product (organic phase).

The Kuhlmann hydroformylation process is another process where catalyst separation is effected by liquid–liquid separation. In this case, however, a system consisting of two liquid phases is produced only after the reaction has taken place. This is done by reaction of the catalyst complex with a base. The now water-soluble catalyst complex is removed from the product by washing with water.

8.7.2 Future Developments

A somewhat different solution to the separation problem would be immobilization of the catalyst. No large-scale application is known so far, but research is underway. The catalyst complexes might be trapped in the pores of solid particles. A method can be to synthesize the complex inside the pores of a zeolite ('ship in a bottle'). Another way could be to trap catalyst complexes in porous materials and deposit a membrane at the outer surface. In essence, the catalyst becomes a heterogeneous one, eliminating the problem of separation of the catalyst from the product. Very promising are membranes that would allow permeation of the products but not of the catalyst. Homogeneous catalysis also offers potential in multifunctional reactors, e.g. in catalytic distillation.

References

1   Hagemeyer HJ (1991) 'Acetaldehyde' in: Kroschwitz JI and Howe-Grant M (eds) *Kirk-Othmer Encyclopedia of Chemical Technology*, vol. 1, 4th ed., Wiley, New York, pp. 94–109.

2   Wagner Jr FS (1991) 'Acetic acid' in: Kroschwitz JI and Howe-Grant M (eds.) *Kirk-Othmer Encyclopedia of Chemical Technology*, vol. 1, 4th ed., Wiley, New York, pp. 121–139.

3   Billig E and Bryant DR (1996) 'Oxo process' in: Kroschwitz JI and Howe-Grant M (eds.) *Kirk-Othmer Encyclopedia of Chemical Technology*, vol. 17, 4th ed., Wiley, New York, pp. 902–919.

4   Sheehan RJ (1995) 'Terephthalic acid, dimethyl terephthalate, and isophthalic acid' in: Gerhartz

W, et al. (eds.) *Ullmann's Encyclopedia of Industrial Chemistry*, A26, 5th ed., VCH, Weinheim, pp. 193–204.

5 James JW and Castor WM (1994) 'Styrene' in: Gerhartz W, et al. (eds.) *Ullmann's Encyclopedia of Industrial Chemistry*, A25, 5th ed., VCH, Weinheim, pp. 329–344.

6 Coty RR, Welch VA, Ram S and Singh J (1987) 'Ethylbenzene' in: Gerhartz W, et al. (eds.) *Ullmann's Encyclopedia of Industrial Chemistry*, A10, 5th ed., VCH, Weinheim, pp. 35–43.

7 Hammershaimb HU, Imai T, Thompson GJ and Vora BV (1992) 'Alkylation' in: Kroschwitz JI and Howe-Grant M (eds.) *Kirk-Othmer Encyclopedia of Chemical Technology*, vol. 2, 4th ed., Wiley, New York, pp. 85–112.

8 Czuppon TA, Knez SA and Rovner JM (1992) 'Ammonia' in: Kroschwitz JI and Howe-Grant M (eds.) *Kirk-Othmer Encyclopedia of Chemical Technology*, vol. 2, 4th ed., Wiley, New York, pp. 638–691.

9 Dever JP, George KF, Hoffman WC and Soo H (1993) 'Ethylene oxide' in: Kroschwitz JI and Howe-Grant M (eds.) *Kirk-Othmer Encyclopedia of Chemical Technology*, vol. 9, 4th ed., Wiley, New York, pp. 915–959.

10 English A, Rovner J and Davies S (1995) 'Methanol' in: Kroschwitz JI and Howe-Grant M (eds.) *Kirk-Othmer Encyclopedia of Chemical Technology*, vol. 16, 4th ed., Wiley, New York, pp. 537–556.

11 Masters C (1981) *Homogeneous Transition-Metal Catalysis – A Gentle Art*, Chapman and Hall Ltd., London.

12 Chauvel A and Lefebvre G (1989) *Petrochemical Processes 1. Synthesis Gas Derivatives and Major Hydrocarbons*, Technip, Paris.

13 Agreda VH (1993) *Acetic Acid and its Derivatives*, Marcel Dekker, New York.

14 Falbe J (1980) *New Syntheses with Carbon Monoxide*, Springer Verlag, New York.

15 Bahrmann H and Bach H (1991) 'Oxo synthesis' in: Gerhartz W, et al. (eds.) *Ullmann's Encyclopedia of Industrial Chemistry*, A18, 5th ed., VCH, Weinheim, pp. 321–327.

16 Zehner P and Kraume M (1992) 'Bubble columns' in: Gerhartz W, et al. (eds.) *Ullmann's Encyclopedia of Industrial Chemistry*, B4, 5th ed., VCH, Weinheim, pp. 275–330.

17 Trambouze P, Van Landeghem H and Wauquier JP (1988) *Chemical Reactors. Design/Engineering/Operation*, Chapter 8, Gulf Publishing Company, Houston, TX.

18 Krishna R and Sie ST (1994) 'Strategies for multiphase reactor selection' *Chem. Eng. Sci.* 49(24A) 4029–4065.

Chapter 9

Heterogeneous Catalysis – Concepts and Examples

9.1 INTRODUCTION

In previous chapters, implicitly, many processes based on heterogeneous catalysis have been discussed. This is no coincidence. Heterogeneous catalysis is a crucial technology. It is the workhorse in chemical reaction engineering. In this light, homogeneous transition-metal catalysis, as discussed in Chapter 8, can be considered the fancy horse. Usually its application is on a smaller scale and more specific than heterogeneous catalysis.

In industrial catalysis, it is generally preferable from an engineering perspective to use heterogeneous catalysts in continuous processes where possible. The main advantages of heterogeneous catalysis compared to homogeneous catalysis are:

- easy catalyst separation;
- flexibility in catalyst regeneration;
- less expensive.

It should be realized that these statements are general and should not be taken too literally. For instance, also very expensive solid catalytic structures are used in practice.

This chapter surveys the reactors commonly applied in heterogeneous catalysis, while attention is also paid to novel reactor types and special catalyst applications. The selection of a reactor configuration for heterogeneous catalysis is determined by many, often conflicting, factors. Firstly, there are the process needs such as safety and environmental requirements, and feasibility of scale-up to economical size. Secondly, there are the process 'wants' which usually include maximum conversion, maximum selectivity, high throughput, easy scale-up, and minimum overall process costs.

Some reactors are intrinsically safer than others. For example, in fixed-bed reactors hot spots can occur, possibly leading to runaways, whereas in fluidized-bed reactors temperature control is relatively easy and hot spots will not occur. However, fluidized-bed reactors are much more difficult to scale up, and catalyst particles have to be able to withstand the high mechanical stress of a fluid bed.

Often, with increasing conversion the selectivity decreases so one cannot have both maximum conversion and maximum selectivity. Both conversion and selectivity can be dependent on the degree of mixing of the phases present (from plug flow to completely mixed), the contacting pattern (co-current, counter-current, cross-current), the particle size, the temperature (distribution), etc.

Reference [1] provides a systematic approach to reactor selection, while references [2] and [3] for example, discuss design aspects of chemical reactors. Reactor selection is intimately linked with catalyst properties such as size, activity, etc. Therefore, we will first briefly discuss the influence of the various catalyst properties. A more detailed discussion is provided in reference [4]. Subsequent sections deal with conventional reactor types, and some unconventional techniques and applications. In Section 9.5 some processes that make use of the techniques discussed are treated in more detail.

9.2 CATALYST DESIGN

The choice of catalyst morphology (size, shape, porous texture, activity distribution, etc.) depends on intrinsic reaction kinetics as well as on diffusion rates of reactants and products. The catalyst cannot be chosen independently of the reactor type, because different reactor types place different demands on the catalyst. For instance, fixed-bed reactors require relatively large particles to minimize the pressure drop, while in fluidized-bed reactors relatively small particles must be used. However, an optimal choice is possible within the limits set by the reactor type.

9.2.1 Catalyst Size and Shape

The catalyst particle size influences the observed rate of reaction: the smaller the particle, the less time required for the reactants to move to the active catalyst sites, and for the products to diffuse out of the particle. Furthermore, with fast reactions in large particles the reactants may never reach the interior of the particle, thus decreasing the catalyst utilization. Catalyst utilization is expressed as the internal effectiveness factor η_i, which is defined as follows:

$$\eta_i = \frac{\text{observed reaction rate}}{\text{rate without internal gradients}} = \frac{r_{obs}(c,T)}{r_{v,chem}(c_s,T_s)} \qquad (1)$$

where:

$r_{obs}(c,T)$ = observed reaction rate, an average of the rates inside the particle at different concentrations c and temperatures T (mol/(s·m$^3_{particle}$))

$r_{v,chem}(c_s,T_s)$ = reaction rate at surface concentration c_s and temperature T_s (mol/(s·m$^3_{particle}$))

The internal effectiveness factor is a function of the generalized Thiele modulus (see for instance [1], [3], [4]). For an nth order reaction:

$$\phi_{gen} = \frac{V}{SA} \sqrt{\frac{k_v}{D_{eff}} \cdot \frac{n+1}{2} \cdot c_s^{n-1}} \qquad (2)$$

in which V/SA is the volume/external surface area ratio of the catalyst (for a sphere: $r_p/3$, in which r_p is the particle radius), k_v is the intrinsic reaction rate constant (per unit volume), D_{eff} is the effective diffusion rate of the molecule, and c_s the concentration at the catalysts surface.

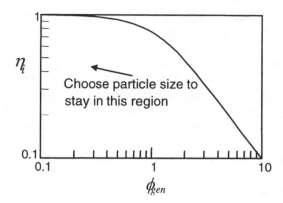

Figure 9.1 Internal effectiveness factor as a function of the generalized Thiele modulus for a first order reaction.

QUESTIONS:

> *According to Figure 9.1 $\eta_i \leq 1$. Is this always the case? (Hint: what happens when an exothermic reaction is carried out?)*
> *What are the dimensions of k_v?*

In order for diffusional limitations to be negligible, the effectiveness factor must be close to 1, i.e. nearly complete catalyst utilization, which requires that the Thiele modulus is sufficiently small (typically < 0.5), see Figure 9.1. It might be useful to remark that the physical meaning of the Thiele modulus is the ratio between catalytic activity and internal mass transfer. There is no doubt that catalyst activity can be improved enormously, while mass transfer is less accessible for fundamental improvements. Therefore, it is to be expected that mass transfer will become an increasingly important issue in catalyst development. The surface-over-volume ratio must be as large as possible (particle size as small as possible) from a diffusion (and heat-transfer) point of view. There are many different catalyst shapes that have different *SA/V* ratios for a given size. Figure 9.2 surveys the most common catalyst shapes.

Of course, the relative surface area is not the only factor influencing the perfor-

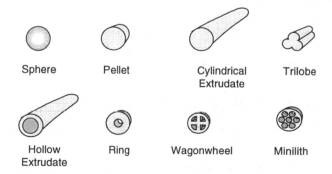

Figure 9.2 Various catalyst shapes.

mance of the catalyst. Smaller particles result in larger pressure drop over the reactor and in fixed-bed systems this is the main limiting factor. The catalyst shape also has effect on the pressure drop. In fixed-bed and moving-bed reactors the catalyst strength is also an important factor. The particles must be able to withstand the forces exerted by the bed above and by the pressure drop. In general, compact particles have a higher crushing strength than hollow ones, and the crushing strength increases with particle size. For the process economics it is also important that the manufacturing costs are within acceptable limits. The more exotic the catalyst shape, the more expensive its production will be.

With increasing particle size, the *SA/V* ratio, and hence the reaction rate, decreases, which is unfavorable, while the pressure drop, the crushing strength, and the manufacturing costs also decrease, which is favorable.

The definitive catalyst size and shape selection will be a compromise between high reaction rate (small particle, exotic shape), low pressure drop (large particle, exotic shape), large crushing strength (low porosity), and low manufacturing cost (large particle), within the limits set by the reactor type.

QUESTIONS:

What particle shape is most favorable with respect to catalytic activity? Is there a connection between the particle size and shape and the selectivity?
Does a relation exist between catalyst shape and reactor type?

9.2.2 Mechanical Properties of Catalyst Particles

One important aspect, which is crucial for reactor selection, has only been mentioned briefly. The mechanical transport properties required for a chosen reactor type play a decisive role in catalyst design. For instance, fluidized-bed and entrained-flow reactors require spherical particles in the range of 30–150 μm (microspheres) with a certain size distribution in order to obtain good fluidization behavior. Such microspheres can be produced by incorporating the catalyst material in a matrix of, for example, silica (see Sections 3.4.1 and 9.5.2). The advantage of the microspheroidal shape over other catalyst shapes is the mechanical attrition resistance. This is very important in entrained-flow reactors and in fluidized-bed reactors where catalyst particles are recirculated rapidly between the reactor and a regenerator (as in FCC).

Attrition resistance is also important for use of catalysts in moving-bed reactors. Usually, the particle size distribution must be narrow (2–6 mm) and no fines must be present. These fines can result from attrition and can clog the catalyst bed because they move in the spaces between the larger particles.

9.3 REACTOR TYPES AND THEIR CHARACTERISTICS

The most important factors for the choice and design of a reactor for heterogeneous catalysis are:

- The number and type of phases involved (G/S, G/L, G/L/S, L/L/S). Freedom exists in choosing the volume fractions of the various phases. Measures must be taken for appropriate mass and heat transport between the phases, e.g. determined by degree of mixing and contacting patterns (co-current, counter-current, cross-current).

- The kinetics of the main and side reactions. These determine the desired temperature and concentration profile, the residence time distribution (degree of mixing), etc.
- The heat of reaction. This is the most important parameter determining the measures to be taken for heat transfer.

Another issue that must be taken into consideration in reactor design is catalyst deactivation. The rate of catalyst addition and removal (every 2 years, continuously, etc.), and the need for regeneration are important design and engineering aspects.

For any particular situation, there usually are a large number of reactors to choose from, and the challenge is to make the optimal choice, keeping in mind the total process economics.

9.3.1 Reactor Types

Figures 9.3 and 9.4 give an overview of reactors for gas-solid catalyst reactions and for gas-liquid-solid catalyst reactions, respectively. Table 9.1 summarizes the applications of the reactors shown in Figures 9.3 and 9.4. Chapter 10 discusses some other three-phase reactors with moving catalyst particles used in the production of fine chemicals.

Numerous variations exist on each reactor type, such as the use of many small-diameter fixed-bed reactors, a different gas–liquid contacting pattern, and the manner of temperature control.

Occasionally, both gas-phase and liquid-phase operation are possible for single fluid phase reactions. In that case, flexibility exists and the question arises what is the best option. The liquid phase has the advantage of better heat-transfer properties and a high flexibility in concentration (by applying a solvent low concentrations are possible); a disadvantage is the low diffusion rate.

The most popular among the various reactor types is the adiabatic fixed-bed reactor. This is essentially a vessel, generally a vertical cylinder, in which the catalyst particles are randomly packed. It is the simplest reactor for use with solid catalysts, and it offers the highest catalyst loading per unit volume of reactor.

In principle, with one-phase flow the fluid can flow upward, downward or radially through the bed. Downward flow is most common, while radial flow is sometimes applied to decrease the pressure drop over the bed. Upward flow is hardly applied. A reason is that often flow rates are high and undesired movement of the particles might be the result. Similarly, with two-phase flow operation can be co-current downward, co-current upward or counter-current. Again, the first configuration, also called trickle flow, is used most frequently. The main example is hydrodesulfurization (HDS) of heavy oil fractions. Although counter-current operation is not industrially applied except in special cases (notably catalytic distillation, Section 9.4.3.3), it is often preferable from a theoretical point of view. An example is the HDS process mentioned above (see Section 9.5.3.2).

The moving-bed reactor is very similar to the fixed-bed reactor. However, the catalyst bed moves downward due to gravity forces. The fluid phase may flow from top to bottom or horizontally (axial flow). Complications are added by the need to

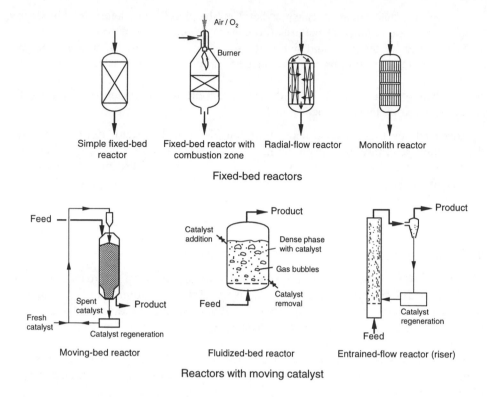

Figure 9.3   Reactors for gas-solid catalytic catalyst reactions.

circulate the catalyst, but these are offset by the possibility of continuous catalyst regeneration. Practical applications of this technology are limited. Examples are the continuously-regenerative reforming process (see Section 3.4.4) and the 'bunker' reactor in the Shell HYCON process for catalytic hydrogenation of residual oil fractions (see Section 3.5.2).

Monolith reactors are so-called structured reactors: they are ceramic or metallic blocks consisting of parallel channels covered with a catalyst layer. They are especially suited for gas-phase reactions requiring low pressure drop but are also investigated for application in three-phase processes.

In fluidized-bed reactors the catalyst particles move randomly due to the upward flow of gas, liquid or gas and liquid (also called ebullated-bed reactor). The advantages of this type of reactor are the excellent heat-transfer properties and the ease of catalyst addition and removal. Catalytic cracking (classical process), catalyst regeneration in catalytic cracking (a non-catalytic process – Section 3.4.1), and the LC-fining process for the catalytic hydrogenation of residues (Section 3.5.2) are examples of processes employing fluidized beds.

In a gas-fluidized-bed reactor, when the velocity of the gas is increased, at one point the catalyst particles become entrained by the gas flow. Then the term entrained-flow reactor or riser reactor applies. Such reactors are suitable for the use of very active catalysts that deactivate fast. The most important example is the FCC process using a zeolite catalyst (Section 3.4.1).

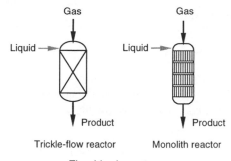

Fixed-bed reactors

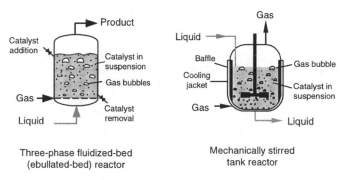

Reactors with moving catalyst

Figure 9.4   Reactors for gas–liquid–solid catalytic reactions.

Mechanically stirred tank reactors are not very common for solid-catalyzed bulk chemicals production. They are, however, the main reactors in the production of fine chemicals (see Chapter 10).

QUESTIONS:

> *Why would counter-current flow often be preferable in gas-liquid reactions such as HDS, provided that there are no practical constraints? For each process in Table 9.1 state why this type of reactor has been chosen? Could alternatives also be applied?*

9.3.2 Exchanging Heat

Fixed-bed reactors exhibit fairly poor heat-transfer characteristics. The particles are large (1–5 mm), resulting in a small surface-to-volume ratio and hence, a small heat-transfer area. Moreover, the catalyst material is usually a thermal insulator. Temperature control in fixed-bed reactors, therefore, is difficult and it is rule rather than an exception that a non-isothermal temperature profile in both axial and radial direction results with (dependent on the system and the conditions) the accompanying danger of local hot spots. Of course, also the phase (liquid or gas) plays a role. In the liquid

Table 9.1 Applications of different reactor types

Reactor type	Applications
Gas-phase reactions	
Simple fixed-bed reactor	HDS of naphtha
	Catalytic reforming (semi-regenerative)
	Steam reforming
	Water-gas shift
	Methanation
	Ammonia synthesis
	Methanol synthesis
Monolithic reactor	Exhaust gas cleaning
Fixed-bed with combustion zone	Autothermic reforming
Radial-flow reactor	Catalytic reforming
	Methanol synthesis
Moving-bed reactor	Catalytic reforming (continuous regenerative)
Fluidized-bed reactor	Classical FCC process
Entrained-flow reactor	Modern FCC process
Gas–liquid reactions	
Trickle-flow reactor	HDS of heavy oil fractions
Moving-bed reactor	Hydrogenation of residues (metal removal)
Three-phase fluidized-bed reactor	Hydrogenation of residues

phase, reaction temperature gradients are much smaller than those in gas-phase processes.

The adiabatic fixed-bed reactor is only suitable for reactions with small heat effect and for reactions in which the temperature rise due to reaction does not influence the selectivity. If in an adiabatic fixed-bed reactor the adiabatic temperature rise/fall would become too large (e.g. with large heats of reaction), a number of smaller adiabatic fixed-beds with intermediate cooling or heating can be placed in series. Figure 9.5 shows some possible configurations.

If the process imposes special requirements on temperature control, e.g. in the case of very exothermic or endothermic reactions, heat-transfer must take place throughout the reactor. Multi-tubular reactors are very suitable for such reactions. Usually hundreds of tubes surrounded by the heat-transfer medium are used. For highly exothermic reactions, the tubes are placed in a vessel with the heat-transfer medium (usually boiling liquids) flowing in the space surrounding the tubes (see Figure 9.6). The flow is directed back and forth across the tubes by baffles to provide good temperature control. Using this method near isothermal operation can be achieved. An example is the production of ethene oxide by partial oxidation of ethene (see Section 9.5.2.1).

Endothermic reactions requiring a large heat input at high temperature level can be carried out in tubes placed in a furnace (see Figure 9.6). The main example of this type of operation is steam reforming of methane or naphtha (see Chapter 5).

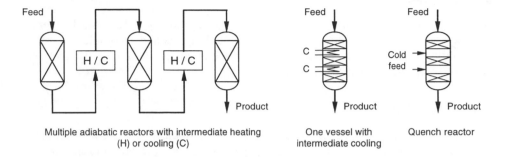

Multiple adiabatic reactors with intermediate heating (H) or cooling (C)

One vessel with intermediate cooling

Quench reactor

Figure 9.5 Adiabatic fixed-bed reactors with inter-stage heating/cooling.

Moving-bed reactors exhibit similar heat-transfer characteristics as fixed-bed reactors, but temperature gradients can be held within limits due to solids circulation.

Slurry reactors exhibit excellent heat-transfer properties. Due to the relatively small particle size, the heat-transfer area is relatively large, and high heat-transfer rates can be obtained. Backmixing of the catalyst results in a uniform temperature throughout the reactor. Additional heat-transfer area can be installed in the form of internal cooling/heating coils.

The heat-transfer characteristics of entrained-flow reactors are intermediate between those of fluidized-bed and moving-bed reactors.

In multiphase operation, excellent temperature control is possible by allowing for heat exchange by evaporation. This can be the reason to choose multiphase operation rather than, for instance, a single gas-phase reaction.

9.3.3 Role of Catalyst Deactivation

Catalyst deactivation, and in particular the time scale of deactivation, is an important issue in the choice of reactor type. Deactivation is the loss of catalyst activity with time, and can be either thermal (loss of surface area due to sintering) or chemical (fouling or poisoning) (see [4,5]).

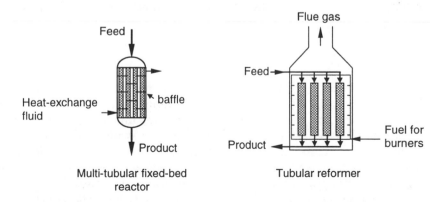

Multi-tubular fixed-bed reactor

Tubular reformer

Figure 9.6 Fixed-bed reactors with continuous heating/cooling.

9.3.3.1 Sintering

Sintering can occur when the temperature becomes too high. This can happen, for instance, as a result of maldistribution of the fluid phases and incomplete wetting of the catalyst particles in trickle-flow reactors (see Section 3.4.2). Locally, the heat is not removed efficiently leading to hot spots.

9.3.3.2 Poisoning

Poisoning of the catalyst occurs by the adsorption of impurities in the feed on specific catalytic sites, thus rendering them inactive for reaction. Common poisons are sulfur (e.g. in catalytic reforming (platinum catalyst, Section 3.4.4), steam reforming (nickel catalyst, Section 5.2)), CO and CO_2 (in ammonia synthesis (Section 6.1), iron catalyst), ammonia, and ethyne. Water can also have a negative effect because it accelerates sintering.

Poisoning can be limited by purification of the feed. For example, hydrodesulfurization (HDS) is applied to remove sulfur (Section 3.4.2). The removal of CO and CO_2 from synthesis gas for ammonia production is achieved by the water-gas shift reaction, subsequent absorption of CO_2, and a final methanation step (Section 5.4).

9.3.3.3 Fouling – Coke Formation

The most common cause of catalyst deactivation is fouling due to the occurrence of secondary reactions of reactants or products on the catalyst surface. The main example of this type of deactivation is coke formation. It occurs by condensation reactions (cycloaddition, dehydrogenation) of aromatic compounds on the catalyst surface to form large structures of low hydrogen content. Coke is often simply indicated as carbon, C. Optimal conditions for coke formation are a reducing environment, high temperature, and low hydrogen pressure. Therefore, it is logical that in previous chapters the problems associated with coke deposition were most pronounced in endothermic reactions proceeding at high temperature (>750 K), namely in fluid catalytic cracking (FCC, Section 3.4.1), catalytic reforming (Section 3.4.4), and steam reforming (Section 5.2). Coke is also formed in thermal (non-catalytic) processes such as steam cracking of ethane and naphtha (Chapter 4).

Coke formation can be limited by addition of hydrogen, either as H_2 or as H_2O:

$$C + 2H_2 \rightleftarrows CH_4 \tag{3}$$

$$C + H_2O \rightleftarrows CO_2 + 2H_2 \tag{4}$$

An alternative is frequent catalyst regeneration, by burning off the coke.

Table 9.2 illustrates the effect of catalyst deactivation time on the reactor choice.

In steam reforming, the use of excess steam not only limits coke formation but it also enhances the conversion of methane (or naphtha) to synthesis gas. In contrast, in the case of catalytic reforming operation takes place at conditions which are not optimal for the reforming reactions from a thermodynamic point of view (during reforming hydrogen is produced, so addition of hydrogen decreases the feed conversion). A compromise is necessary between conversion and catalyst deactivation. If one chooses to operate at high hydrogen pressure, the equilibrium conversion is relatively

Table 9.2 Effect of catalyst deactivation on choice of reactor; HC = hydrocarbon

Process	Measure to limit coke formation	Deactivation time	Reactor type
Steam reforming	Excess steam addition	2 years	Tubular fixed-bed reactor
Catalytic reforming			
Semi-regenerative	Large H_2/HC ratio	0.5–1.5 years	Fixed-bed reactor
Fully-regenerative	Moderate H_2/HC ratio	Days–weeks	Swing reactor
Continuous regenerative	Moderate H_2/HC ratio	Days–weeks	Moving-bed reactor
FCC	None (regeneration)	Seconds	Entrained-flow reactor

low, but simple adiabatic fixed-bed reactors can be used because of slow deactivation. One can also opt for a lower hydrogen pressure, but this adds to the investment costs due to the requirement of an additional reactor (swing-reactor) or the use of a moving-bed reactor, which is more cumbersome from the viewpoint of construction and operation. The advantages are a higher conversion and lower costs for the recycle of hydrogen.

9.3.4 Other Issues

9.3.4.1 Mixing of reactants

As has become clear from the previous section, reactors with a high degree of back-mixing, such as fluidized-bed reactors, have a more isothermal temperature profile and are better to control than other reactors (e.g. fixed-bed reactors). On the other hand, extensive backmixing is usually disadvantageous from the viewpoint of conversion and selectivity.

For first and higher order reactions, the reaction rate increases with increasing reactant concentration. Hence, in plug-flow reactors the reaction rate is high in the entrance zone of the reactor and decreases along the length of the reactor. A well-mixed reactor (e.g. CSTR) always operates at the lowest reactant concentration (which is the same throughout the reactor) and thus at the lowest reaction rate. Therefore, the maximum obtainable conversion in a plug-flow reactor is higher than that in a CSTR of the same size.

Consecutive reactions such as A $\rightarrow$ P $\rightarrow$ S, in which P is the desired product, pose another problem. In order to obtain maximum selectivity towards P plug-flow conditions should be aimed for. However, for highly exothermic reactions, from the point of view of avoiding large temperature gradients a thermally well-mixed system would be preferred. Provided they are very fast, consecutive reactions of the type

$$A \xrightarrow{B} P \xrightarrow{B} Q \qquad (5)$$

occasionally can best be performed in multiphase operation: the reaction will be limited to a thin diffusion layer around the gas bubbles, and the consecutive reaction will be eliminated. A monolith reactor can also be considered (see Section 9.4.2).

A solution to obtain a well-mixed temperature together with plug-flow concentration conditions can be to circulate the catalyst and remove the heat externally (see Section 9.5.2).

QUESTIONS:

> *Why, in the reaction given above, is plug flow preferable from a concentration point of view?*
> *Explain why consecutive reactions such as in Eqn. (5) can be carried out at relatively high selectivity in multiphase operation? How would you realize this idea in practice?*
> *Give advantages and disadvantages of temperature control by external heat exchange?*

9.3.4.2 Safety

Important safety-related hazards are high temperature excursions and runaways. Highly exothermal reactions can lead to such events when cooling is not sufficient. For instance, in a trickle-flow reactor the even distribution of the liquid over the catalyst bed is of decisive importance. In exothermic reactions, liquid maldistribution can lead to local hot spots, possibly leading to a temperature rise in the entire reactor, and, a total runaway.

9.3.4.3 Scale-up

The ease of scale-up of a chemical process, and with that the time and cost involved in going from the laboratory to a working industrial plant, depends on several factors. Some are intrinsically related to the reactor and process, for instance, the reactor hydrodynamics (gas, liquid, solid flow). Others depend on more or less external factors such as maximum size of equipment, proven versus new technology, experience of the company with the particular technology, etc. (see also Chapter 13).

It is much easier to scale up a fixed-bed reactor from laboratory scale to the full size plant than to scale up a fluidized-bed reactor. The reason for the difficulty in scaling up fluidized beds is that the fluidization characteristics of the bed not only depend on the local properties but also on the reactor size. For illustration, various fixed-bed reactors for industrial processes have been scaled up in one step from pilot reactors of 13–44 mm diameter to full-scale reactors of 3–9 m. In contrast, the scale up to fluidized-bed reactors with diameters of 7.5–17 m requires several steps, with each step only one or two orders of magnitude larger than the previous.

QUESTION:

> *Compare the various reactor types in terms of attainable conversion, relative reactor volume, safety, ease of operation and scale-up, and investment costs.*

9.4 NOVEL DEVELOPMENTS IN REACTOR TECHNOLOGY

Research and development activities in the past decades have produced a number of fascinating developments, both on the reactor and on the catalyst level. Many of these developments involve the use of multifunctional reactors. These reactors, for instance, combine reaction and heat transfer (Section 9.4.1), or more commonly, reaction and separation (Section 9.4.3). On the catalyst level, much progress has been achieved in the development of systems for emission control. In many applications monolithic reactors are used (Section 9.4.2). Due to their specific shape, such reactors are especially suited for applications requiring low pressure drop, e.g. automotive pollution control and the incineration of industrial off gases.

9.4.1 Adiabatic Reactor with Periodic Flow Reversal

Exothermic equilibrium reactions like ammonia and methanol synthesis suffer the disadvantage that in order to reach a favorable equilibrium between reactant and product the temperature must be as low as possible, while a sufficiently high temperature is required in order for the reaction to proceed at an acceptable rate. Furthermore, the temperature increases towards the exit of the catalyst bed due to the exothermicity of the reaction, lowering the obtainable conversion. Thus, one could say that the temperature profile is exactly the wrong way around. The consequence for process operation is that the feed must be preheated, while the product stream must be cooled. This is usually done in feed-effluent exchangers. In order to prevent the reactor temperature from increasing too much, heat must be removed between reaction stages (e.g. quench reactor). This obviously increases investment and operation costs.

An alternative, which makes use of the heat capacity of the catalyst bed, has been conceived by Boreskov et al. [6] and refined by Matros [7,8]: the adiabatic fixed-bed reactor with periodic flow reversal (RPFR). Figure 9.7 shows the principle of the RPFR. The main idea is to utilize the heat of reaction within the catalyst bed itself.

Feeding a hot catalyst bed with relatively cold gas will cool the inlet side of the bed; on the other hand, the temperature at the exit of the bed will increase due to the heat produced by the reaction. By reversing the direction of flow the heat contained in the catalyst bed will bring the cold inlet stream to reaction temperature. The part of the bed that has now become the outlet zone is relatively cold, which is favorable for the reaction equilibrium. After some time the inlet has cooled again, while the outlet has become warmer. Then the flow is reversed again and a new cycle begins. After a sufficient number of flow reversals, an oscillating but on average stationary state is attained.

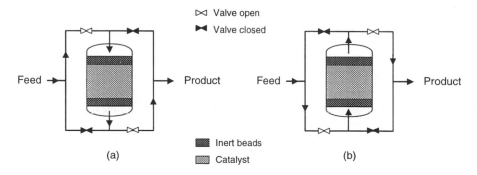

Figure 9.7   Principle of the RPFR; a: first half of the cycle; b: second half of the cycle.

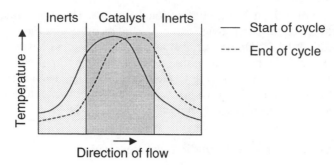

Figure 9.8   Temperature profile in RPFR at start and end of cycle.

By reversing the flow at the right time heat can be kept in the reactor, while the temperature in the middle zone will remain above the reaction ignition temperature. Once the process has been started up, the heat of reaction is sufficient to keep the process going. Figure 9.8 shows a typical temperature profile over the reactor as a function of time.

As only a sufficiently high temperature is maintained in the middle part of the reactor, part of the bed can consist of inert material (with high heat capacity and a large particle diameter). This lowers the cost while the conversion remains the same. An additional advantage is that the pressure drop is reduced.

Currently, the RPFR has three commercial applications. These are oxidation of SO_2 for sulfuric acid production, oxidation of volatile organic compounds (VOC) for purification of industrial exhaust gases, and NO_x reduction by ammonia in industrial exhaust gases. Other possible future applications are in steam reforming and partial oxidation of methane to produce synthesis gas, production of methanol and ammonia, and catalytic dehydrogenation [9].

QUESTIONS:

> *Reverse-flow operation has an obvious disadvantage. What is this disadvantage and how could it be solved?*
> *Why have the above processes been chosen for the application of the RPFR?*

9.4.2 Structured Catalytic Reactors

The use of structured catalysts in the chemical industry has been considered for years. In heterogeneous catalysis fixed-bed reactors with random catalyst packing are most commonly used. These reactors have some obvious disadvantages such as maldistribution of catalyst and fluids, high pressure drop, possible plugging by dust etc. Structured catalysts are promising with respect to solving these setbacks. Two basic types of structured catalysts may be distinguished:

- monolithic catalysts;
- arranged catalysts. Structured packings covered with catalytic material, analogous

to the packings used in distillation and absorption columns belong to this class. These will not be discussed here. Particulate catalysts arranged in arrays are also referred to as arranged catalysts.

9.4.2.1 Monoliths

Monoliths are continuous structures consisting of narrow parallel channels (typically 1–3 mm diameter). A ceramic or metallic support is coated with a layer of material in which active ingredients are dispersed (the washcoat). The walls of the channels may be either impermeable or permeable. In the latter case, the term 'membrane reactor' is used (see Section 9.4.3.1). Figure 9.9 shows an example of a monolith.

The shape of the channels differs widely: circular, triangular, square, hexagonal, etc. The shape of the entire monolithic block can be adapted to fit in the reaction chamber.

Monolithic reactors have recently found many applications in combustion, e.g. in automotive pollution control (see Section 9.5.3) and for the incineration of industrial off gases [11]. In both applications the primary advantage of monolithic reactors over conventional fixed-bed reactors is their low pressure drop. This is a result of the flow through straight channels instead of through the tortuous path in catalyst particles.

Other processes employing monolithic reactors, either already implemented, or in the development stage, are the catalytic combustion of fuels for gas turbines, oxidation of SO_2, oxidation of ammonia, and hydrogenation processes. The last category includes gas–liquid systems. In this case the hydrodynamics are very important, e.g. the feeding of the reactor has to be such that all channels are equally wetted [12].

9.4.2.2 Arranged catalysts – three-levels-of-porosity (TLP) reactors

Three-levels-of-porosity (TLP) reactors are alternatives for monolith reactors in certain applications. Conventional catalyst particles can be arranged in any geometric configuration. In such arrays, three levels of porosity can be distinguished: the pore space within the particles, the intraparticle space and the space between the arrays. An example of such a TLP reactor is the parallel-passage reactor (PPR) (see Figure 9.10 and [13]). The catalyst particles are confined between wire gauze screens that divide the reactor in a large number of catalyst layers with empty passages in between.

The gas flows *along* the catalysts layers, instead of through the bed as in a tradi-

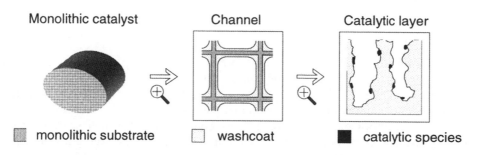

Figure 9.9 Example of a monolith with square channels. Adapted from [10].

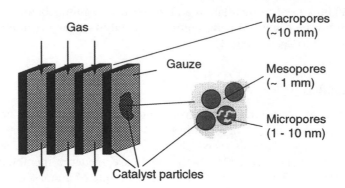

Figure 9.10   Example of TLP reactor: parallel passage reactor (PPR). Adapted from [13].

tional fixed-bed reactor. Because the gas flows through straight channels (ca. 10 mm wide), the pressure drop over the PPR is much lower than over a fixed-bed reactor. Reactants are transferred from the gas to the catalyst inside the gauzes mainly by diffusion. The PPR is very suitable for treating dust-containing gases, e.g. flue gases from power plants, as dust will not be collected on the catalyst particles as a result of the straightness of the gas passages.

9.4.3 Hybrid Systems

The potential of hybrid systems is drawing increasing attention. In hybrid systems different processes are coupled, e.g. reaction and separation or several separation processes.

An interesting example is a reactor in which reaction and separation take place simultaneously in such a way that yields and selectivities are enhanced. This type of reactor is a multifunctional reactor. Separation can take place in many ways, e.g. by application of membranes, adsorption, distillation, or by absorption in a solvent. The reaction will usually imply the use of a catalyst.

The most important benefit of incorporating two different functions in one piece of equipment, of course, is reduction of the capital investment. One process step is eliminated, along with the associated pumps, piping, and instrumentation. Other benefits depend on the specific reaction. In catalytic distillation, optimal heat integration is possible. Equilibrium-limited reactions are an obvious class of reactions that would benefit from the continuous removal of one of the products *in situ*.

9.4.3.1 Coupling membranes and reaction

Catalytic membranes offer the advantage of combining catalysis and separation in the same reactor. In principle a reaction can be carried out 'beyond its equilibrium' by continuous removal of one of the products. A module containing a catalytic membrane is an excellent example of a multifunctional reactor.

Catalytic membrane reactors are not yet commercial. A bench scale example, showing the potential, is a catalytic membrane reactor for the dehydrogenation of ethane:

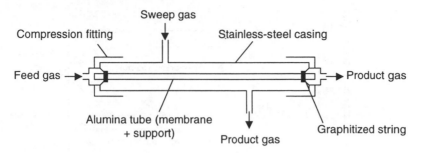

Figure 9.11 Membrane reactor for the dehydrogenation of ethane. Adapted from [14].

$$H_3C-CH_3 \xrightarrow{Pt} H_2C=CH_2 + H_2 \qquad (6)$$

The reactor consists of a porous alumina membrane tube covered with Pt crystallites showing good catalytic activity (see Figure 9.11).

The ceramic tube consists of a multi-layered composite porous alumina tube (inside diameter, 7 mm; outside diameter, 10 mm; length, 250 mm). The first layer (5 μm thick) has a unimodal pore structure (4 nm). Successive layers are thicker with progressively larger pores and are supported on a thick support layer (1.5 mm) with pores in the range of 10 μm [14].

For the gases playing a role in the reaction, Figure 9.12 shows the gas flow rate as a function of the pressure drop over the membrane. Clearly, the transport of H_2 is much faster than that of ethane and ethene. Hence, in principle it is possible to shift the equilibrium to the side of ethene by removal of H_2. Indeed, it has been shown that with this reactor conversions far beyond thermodynamic equilibrium can be realized (Figure 9.13).

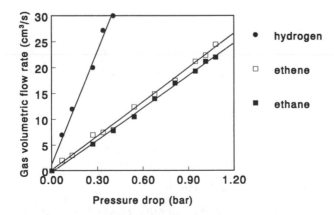

Figure 9.12 Permeation rate of hydrogen, ethane and ethene as a function of pressure. Adapted from [14].

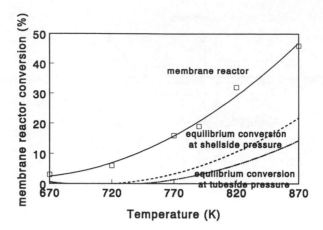

Figure 9.13   Conversion in a membrane reactor compared to thermodynamic equilibrium. Adapted from [14].

QUESTION:

Why are two equilibrium conversion curves shown in Figure 9.13?

Although membrane reactors certainly open opportunities, additional challenges have to be faced in the design and production of robust modules allowing high rates of mass (and heat) transfer.

Another example of the use of a membrane reactor is the production of fatty acids by fat hydrolysis (see Scheme 9.1). Usually, this reaction is carried out at high temperatures and sometimes in the presence of a catalyst. A major disadvantage of the process is the fact that the remaining water phase is heavily polluted thus making it uneconomical to recover glycerol.

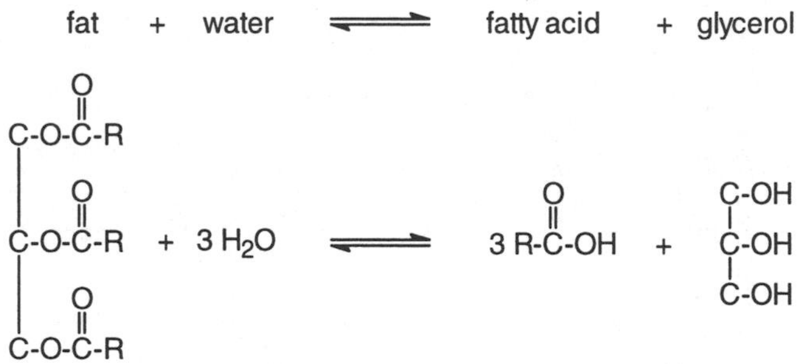

$$\text{fat} \ + \ \text{water} \ \rightleftharpoons \ \text{fatty acid} \ + \ \text{glycerol}$$

$$
\begin{array}{l}
\text{C-O-C-R} \\
\text{C-O-C-R} \ + \ 3\,H_2O \ \rightleftharpoons \ 3\,\text{R-C-OH} \ + \ \begin{array}{l}\text{C-OH}\\ \text{C-OH} \\ \text{C-OH}\end{array}
\end{array}
$$

Scheme 9.1   Production of fatty acids.

9.4.3.2 Coupling reaction and adsorption

An alternative multifunctional reactor combines reaction with adsorption. At laboratory scale, combination of reaction and adsorption has been applied for a few reactions, including the classical ammonia and methanol synthesis. Although these

Production of fatty acids by fat hydrolysis in a membrane reactor

A recent development at laboratory scale is the application of an enzyme (lipase) to catalyze the hydrolysis: water and fat are mixed at a low temperature (300 K) in continuous stirred-tank reactors (CSTR). The water phase contains the enzyme. A much purer glycerol solution is obtained than in the conventional process. The disadvantage is that the equilibrium is not favorable.

An elegant solution has been proposed based on a membrane reactor consisting of a module with hollow cellulose fibers (see Figure B.9.1) [15].

The enzyme is placed at the inner side of the fibers, to which the fat is fed. Water passes at the outside and diffuses through the membrane to react at the fat/lipase interface. The fatty acid formed stays in the oil phase, whereas the glycerol formed is transported through the membrane into the water phase. Laboratory studies show nearly complete conversions.

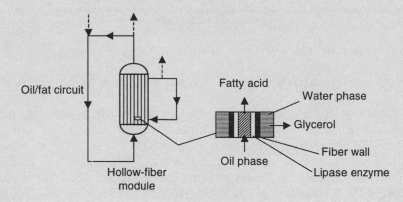

Figure B.9.1 Membrane reactor for the production of fatty acids. Adapted from [15].

processes have been optimized to a high degree of sophistication, process improvements are probably still possible. In particular, energy losses might be decreased further. The major energy losses in these processes are due to:

- condensation (inerts are present in the stream from which ammonia must be condensed, lowering heat-transfer rates);
- recycles (compressor, heating, etc.), due to the limited conversion per pass.

Moreover, energy loss occurs from a kinetic origin. This can be explained as follows: in the catalyst bed, over the length of the bed an increase of product concentration occurs, which causes a decrease of reaction rates (product inhibition by ammonia – see also [4]). Hence, longer beds are needed and, as a consequence, pressure drop over the catalyst bed increases.

It has been suggested that a multifunctional reactor be applied by adding a sorbent, which enables a higher conversion per pass. The reactor is called a gas–solid–solid trickle-flow reactor [16,17]. The reactor is a fixed-bed reactor, with the sorbent and the

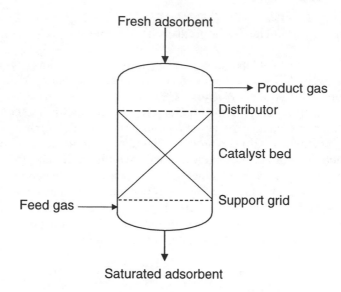

Figure 9.14   Gas–solid–solid trickle-flow reactor for methanol synthesis.

gas flowing counter-currently. Figure 9.14 shows such a reactor for methanol synthesis.

A porous powder 'trickles' over the catalyst pellets. It adsorbs the methanol produced and hence the equilibrium is shifted towards methanol. In this specific example an FCC catalyst appeared to be a useful, selective sorbent for methanol. In a laboratory-scale reactor it was confirmed that conversions exceeding 'equilibrium' values could be obtained. It should be remarked, however, that in most cases the adsorption of the product is a function of temperature. Therefore, a higher reaction temperature implies a higher flow rate of the adsorbent.

QUESTION:
   Give advantages but also disadvantages of this technology.

9.4.3.3 Catalytic distillation

Catalytic distillation (or reactive distillation) is another example of the use of a multi-functional reactor. It involves a combination of catalysis and distillation in a single column. Catalytic distillation is a two-phase flow process with gas and liquid flowing in counter-current mode. This requires special catalysts because with conventional fixed-bed catalysts (void fraction between particles ca. 0.3–0.4) flooding of the reactor would easily occur at high flow rates. For comparison, distillation column packings usually have a void fraction near 0.7.

In industry, catalytic distillation is already applied in a few cases, e.g. for the production of MTBE (methyl *tert*-butyl ether) (see Section 6.2.5), and more recently, cumene and ethylbenzene [18,19].

QUESTION:

> *When a membrane reactor or catalytic distillation is applied, it is often advantageous to combine a conventional reactor with the multifunctional reactor (see [20]). Why would this be the case?*

9.5 SELECTED EXAMPLES OF HETEROGENEOUS CATALYSIS

The previous part of this chapter has presented a conceptual approach to heterogeneous catalysis. In this part examples are given of some important industrial processes, which will elucidate the points discussed.

9.5.1 Ethylbenzene and Styrene Production

Both ethylbenzene and cumene (isopropylbenzene) are important chemicals. They are produced by alkylation of benzene with ethene and propene, respectively:

$$\text{benzene} + CH_2{=}CH_2 \rightleftharpoons \text{(ethylbenzene, } CH_2CH_3) \qquad \Delta H^0_{298} = -114 \text{ kJ/mol} \quad (7)$$

$$\text{benzene} + CH_3CH{=}CH_2 \rightleftharpoons \text{(cumene, } CH(CH_3)_2) \qquad \Delta H^0_{298} = -99 \text{ kJ/mol} \quad (8)$$

Cumene is mainly used for the production of phenol and acetone. Ethylbenzene is almost exclusively used as an intermediate for the production of styrene, one of the most important large-volume chemicals:

$$\text{(ethylbenzene, } CH_2CH_3) \rightleftharpoons \text{(styrene, } CH{=}CH_2) + H_2 \qquad \Delta H^0_{298} = 125 \text{ kJ/mol} \quad (9)$$

Styrene is mainly used as a monomer for polystyrene, which is used in the manufacture of all sorts of packaging materials. Other major applications are in ABS (acrylonitrile-butadiene-styrene) and SBR (styrene-butadiene rubber) and other copolymers.

The processes for ethylbenzene and cumene production have developed in a very similar way. Here we will limit ourselves to the production of ethylbenzene and its dehydrogenation to produce styrene.

9.5.1.1 Production of ethylbenzene

The production of ethylbenzene is an excellent example of the potential of heterogeneous catalysis for environmentally friendly processes. It also demonstrates the applicability of the shape selectivity of zeolites.

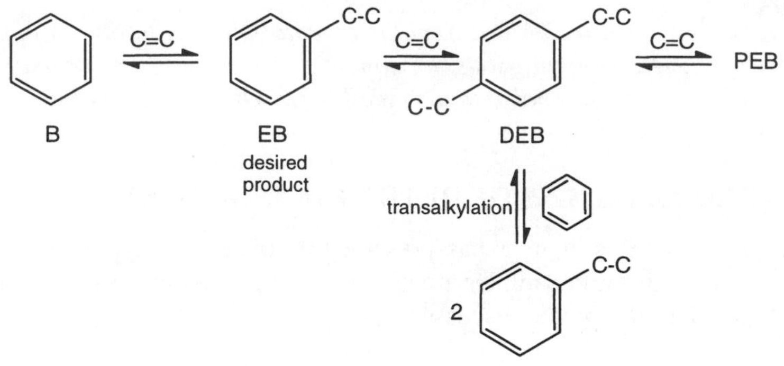

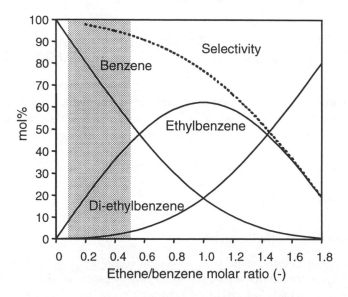

Scheme 9.2 Main reactions during alkylation of benzene with ethene; B = benzene; EB = ethyl-benzene; PEB = polyethylbenzene.

Reactions and thermodynamics

The production of ethylbenzene by alkylation of benzene with ethene is accompanied by various side reactions of which consecutive alkylations, leading to the formation of polyethylbenzenes, are the most important (see Scheme 9.2). These can be recycled and converted to ethylbenzene by transalkylation, and therefore are not by-products in the normal sense. The alkylation reactions are exothermic and thermodynamically favored by low temperature and elevated pressure. Transalkylation is virtually independent on temperature and pressure.

De-alkylation and isomerization also occur, as well as oligomerization of ethene (combination of two ethene molecules). The latter reaction is the first step in the formation of most of the heavy by-products (e.g. cumene, butylbenzenes, diphenyl compounds, and heavy aromatics and non-aromatics), which either end up as impu-

Figure 9.15 Effect of ethene/benzene mole ratio on the alkylation of benzene to ethylbenzene; $T =$ 500 K; $p = 1$ bar; the E/B molar ratio of industrial processes is indicated.

rities in the ethylbenzene product or as fuel. Control of ethene oligomerization has been the most important issue in catalyst development for ethylbenzene production.

Figure 9.15 shows the equilibrium yield on ethylbenzene and di-ethylbenzene as a function of the molar ethene/benzene ratio at 500 K and 1 bar. The conversion of benzene increases with increasing ethene/benzene ratio (= decreasing excess benzene), while the selectivity towards ethylbenzene (based on benzene) decreases. In practice, excess benzene is added to minimize the formation of polyalkylated compounds. Furthermore, a low concentration of ethene limits its oligomerization. The benzene to ethene mole ratio may vary from 2 to 16 depending on the process.

QUESTION:

A low ethene concentration will lower the rates of both desired and undesired reactions. Show that at the same yield of the desired product higher catalyst stability is expected at relatively low ethene concentration. Draw a block diagram for a possible process.

History of process development

The alkylation reaction is catalyzed by acid catalysts. Conventional alkylation catalysts are metal chlorides (BF_3, $AlCl_3$, etc.) and mineral acids (HF, H_2SO_4). The latter are used in, for example, the alkylation of isobutane with alkenes as discussed in Section 3.4.5. Traditional processes for the alkylation of benzene with ethene are based on an aluminium chloride catalyst. These processes take place in the liquid phase with the catalyst either in a separate fluid phase (e.g. Union Carbide–Badger – see [21]) or in homogeneous form (Monsanto–Lummus) as discussed below.

Although the conventional catalysts are effective for alkylation, they suffer some important disadvantages. Firstly, they are corrosive, which implies that the reactor section of the process must be constructed of special materials (glass-lined, brick-lined or Hastelloy). Secondly, the use of such catalysts results in waste disposal problems.

As a result of environmental pressures and the costs associated with waste disposal, several new processes have been developed during the 1970s and 1980s, all based on zeolite catalysts. Zeolites have the obvious advantages of being non-corrosive and harmless to the environment.

The first process to become commercially successful was a vapor-phase process with a ZSM-5 catalyst (1970s, Mobil–Badger). Nowadays, most plants built after 1980 use this process. Although the Mobil–Badger process has the advantages mentioned above, it has suffered the disadvantage of rather rapid catalyst deactivation (weeks) due to coke deposition. This problem has been overcome in later versions of the process.

Other zeolite-based processes have been developed since, amongst them one employing catalytic distillation (see [21]). Recently, a new liquid-phase alkylation process, based on a very selective zeolite called MCM-22, was developed, and commercialized in 1995 (Mobil). The catalyst is reported to be highly active for alkylation but inactive for oligomerization, permitting operation at low benzene/ethene ratios, while achieving the highest yield and product purity of all ethylbenzene

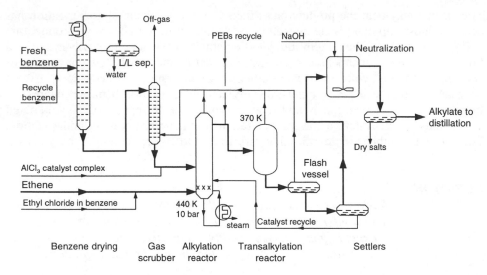

Figure 9.16   Monsanto–Lummus liquid-phase process for the production of ethylbenzene.

processes. However, experience with this process is limited, and information rather scarce.

The homogeneous liquid-phase process

The conventional liquid-phase alkylation process is based on a homogeneous AlCl₃ catalyst. Although new plants are not built anymore, still about 40% of ethylbenzene is produced by this process. Figure 9.16 shows a flow scheme. The catalyst system in this process consists of AlCl₃ and HCl (which acts as a catalyst promoter). HCl needs only be present in small quantities and is supplied as ethyl chloride.

Dry benzene, the catalyst complex, and ethyl chloride are fed to the alkylation reactor, which is an empty vessel with an internal lining of corrosion resistant brick. Ethene is sparged into the liquid in the reactor.

Alkylation takes place in the liquid phase. Excess benzene (2–2.5 mol/mol ethene) is used in order to limit oligomerization of ethene and the formation of polyethylbenzenes. The reactor effluent is mixed with recycle polyethylbenzenes and fed to the transalkylation reactor. In the transalkylation reactor the polyethylbenzenes, mainly di- and tri-ethylbenzene, are converted to ethylbenzene. The soluble catalyst complex is recycled to the alkylation reactor. The alkylate is treated with caustic to neutralize any HCl still present, remove residual AlCl₃, and to ensure that the alkylate is free of chlorinated compounds. The crude ethylbenzene product, still containing benzene and polyethylbenzenes, is then separated by distillation.

QUESTIONS:

> *How is the optimal ethene-to-benzene ratio in the reactor realized? Why is ethyl chloride added? How is the heat of reaction removed? What would be the composition of the off-gas? Suggest a distillation sequence and explain.*

Apart from the obvious disadvantages of corrosion, the need to remove the catalyst from the product, and the production of polluting waste, an additional disadvantage is that the benzene feed must be virtually free of water in order to prevent destruction of the catalyst complex, which requires an additional drying step. Furthermore, the catalyst complex needs to be prepared separately (not shown in Figure 9.16). Both steps add to investment and operation costs.

The zeolite-catalyzed vapor-phase process

The vapor-phase alkylation process (see Figure 9.17) is based on the zeolite ZSM-5. This catalyst contains pores of molecular dimensions, enabling shape selectivity (see Section 3.7). The ethylbenzene molecules can diffuse freely through the pores, whereas diffusion of the polyethylbenzene molecules is strongly inhibited (product selectivity).

The catalyst deactivates due to coke deposition, which is favored by high temperatures. An additional 'swing reactor' is present in the early plants to ensure continuous operation (catalyst life: weeks). However, improved technology has extended the catalyst life to 2 years eliminating the need for the 'swing reactor'. Regeneration of the catalyst takes place by burning off coke in oxygen-poor air or suitable N_2/O_2 mixtures. This operation takes about 24 h.

Benzene drying is not required since this catalyst is very resistant to water. A mixture of make-up and recycle benzene is preheated and fed to a multistage fixed-bed reactor. Part of the ethene required is added at this point. The benzene/ethene ratio at the entrance of the reactor is higher than in the $AlCl_3$-catalyzed liquid-phase process, namely 8–15 mol/mol. Cooling is performed by quenching with cold ethene after each catalyst bed in order to control the temperature rise resulting from the exothermic alkylation reaction.

Separation of the product stream takes place by distillation. Although the ZSM-5 catalyst is very selective for ethylbenzene, the formation of polyethylbenzenes cannot be prevented completely. The polyethylbenzene stream from the distillation section is

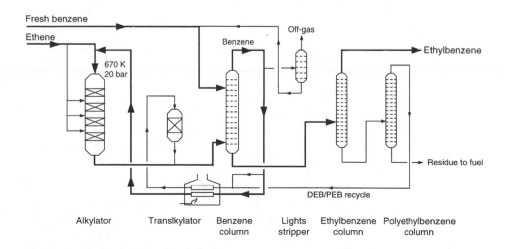

Figure 9.17 Mobil–Badger vapor-phase process for the production of ethylbenzene.

mixed with benzene and fed to a transalkylator where additional ethylbenzene is formed.

Heat is recovered at various points in the process (not shown) by generation of steam for use in the styrene plant.

Catalytic distillation in alkylation processes

Catalytic distillation (see also Section 9.4.3.3) is an alternative for the processes described previously. Figure B.9.2 shows a simplified flow-scheme based on a similar process for the production of cumene as described in [22].

The incentive for using a catalytic distillation process in alkylation reactions is to keep the ethene concentration low, thereby slowing down the oligomerization and limiting the amount of polyethylbenzenes (PEBs) produced. A process based on catalytic distillation inherently maintains a low concentration of ethene in the reactor, because unreacted benzene is separated from the liquid product mixture in the stripper section and returned to the reactor section.

The heat of the highly exothermic alkylation reaction is now used directly in the distillation of reactants and products: benzene evaporates from the liquid mixture. The vapor is then condensed in the overhead condenser and returned to the reactor. It is claimed that this process requires only half the amount of energy compared to other processes, while the investment costs are reduced by 25% [19]. Furthermore, due to the 'internal recirculation' of excess benzene, there is no need to recycle large amounts of benzene, reducing the operating costs of the process.

QUESTION:

Is counter-current contacting of ethene and benzene optimal for this reaction?

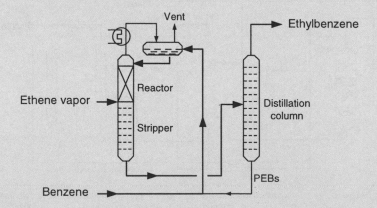

Figure B.9.2 Simplified flow-scheme of catalytic distillation process to produce ethylbenzene. Adapted from [22]

QUESTIONS:

> *Sketch the temperature profile along the reactor. The benzene/ethene ratio in the zeolite-based process is higher than that in the AlCl₃ catalyzed process. Is this logical in view of the shape selectivity of the zeolite?*
> *Recently, catalysts with higher selectivity have been developed. What would be the consequence for process design?*

The major advantages of the zeolite-based vapor-phase process are the absence of aqueous waste streams and the fact that no corrosive substances are used, so that high-alloy materials and brick linings are not required. Furthermore, equipment for catalyst recovery and waste treatment is eliminated. On the other hand, due to the larger benzene/ethene ratio, the costs for benzene recovery and recirculation are higher.

9.5.1.2 Production of styrene by dehydrogenation of ethylbenzene

Dehydrogenation of ethylbenzene leads to the formation of styrene:

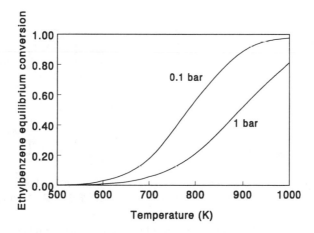

$$\Delta H^0_{298} = 125 \text{ kJ/mol} \quad (9)$$

The reaction is closely related to the primary dehydrogenation reactions that occur in steam cracking (see Chapter 4), but needs a catalyst because else side reactions such as isomerization and cracking with associated coke formation, would be excessive.

The forward reaction is highly endothermic and requires a high temperature as shown in Figure 9.18. Also shown is the effect of pressure. Low ethylbenzene partial pressures are clearly preferred. These two observations lead to the use of

Figure 9.18 Equilibrium ethylbenzene conversion versus temperature and pressure.

steam as a means of both supplying heat and lowering the partial pressure of ethylbenzene.

QUESTION:

> *Could there be another reason for adding steam?*

The intrinsic rate of styrene formation on an industrial catalyst (Shell 105, a potassium promoted iron catalyst) can be described by Langmuir–Hinshelwood kinetics (for situation far from equilibrium) [23]:

$$r_{styrene} = \frac{k \cdot K_{EB} \cdot p_{EB}}{1 + K_{H_2} p_{H_2} + K_{EB} p_{EB} + K_{ST} p_{ST}} \tag{10}$$

in which $r_{styrene}$ is the intrinsic rate of styrene formation (mol·m^{-3}·s^{-1}), k the reaction rate constant (mol·m^{-3}·s^{-1}) for the forward reaction (10), K_i the adsorption coefficient (bar^{-1}) and p_i the partial pressure (bar), respectively of component i (EB = ethylbenzene, ST = styrene).

QUESTION:

> *Knowing that the adsorption coefficient of hydrogen is low, while the adsorption terms for ethylbenzene and styrene are much larger than 1, rewrite Eqn. (10). How does the partial pressure of ethylbenzene influence the rate of reaction?*

Most processes for the production of styrene use adiabatic fixed-bed reactors [24]. Complete conversion of pure ethylbenzene under adiabatic operation would result in a very large temperature drop (see Figure 9.19), resulting in thermodynamic as well as kinetic limitations. Hence, steam is added to the feed for adiabatic fixed-bed reactors, and usually two or more reactors are placed in series as shown in Figure 9.19.

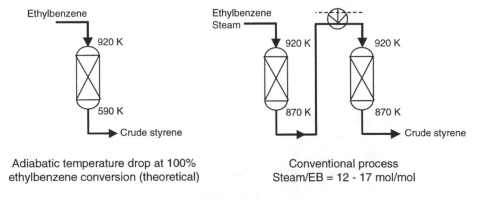

Figure 9.19 Temperature drop in adiabatic-fixed bed reactor (pure ethylbenzene, no steam) and conventional reactor configuration for styrene manufacture.

The steam/ethylbenzene ratio ranges from ca. 12 to 17 mol/mol in order to supply enough heat for the endothermic reaction, leading to high flow rates with associated large pressure drop [25].

QUESTION:

> *What can be done to limit the pressure drop in a fixed-bed reactor?*

Reverse-flow reactor for endothermic reaction: styrene production

The reverse-flow reactor (RPFR), as discussed in Section 9.4.1 for moderately exothermic reactions, could also be a good alternative for endothermic reactions such as the production of styrene. Figure B.9.3 shows a typical reactor set-up for styrene production [25].

Ethylbenzene and steam are introduced at one reactor end, while additional steam is fed co-currently at one or more downstream locations (one shown in Figure B.9.3). The flow direction of the streams is periodically reversed between the reactor ends. The inert beads serve as heat exchange medium and also prevent the occurrence of the reverse reaction, which would have taken place at the cooler reactor ends if catalyst material had been present.

The reverse-flow reactor operates at intermediate steam/ethylbenzene ratio (8–10 mol/mol). Near-isothermal operation at the desired temperature (ca. 900 K) can be obtained if steam is added at different points along the length of the reactor. Furthermore, the inlet and outlet streams are much cooler than in conventional operation (ca. 640 K) due to heat exchange with the inert material. Therefore, energy is used much more efficiently than in conventional operation.

QUESTION:

> *Sketch the temperature profile over the reverse-flow reactor before and after steady-state has been reached.*

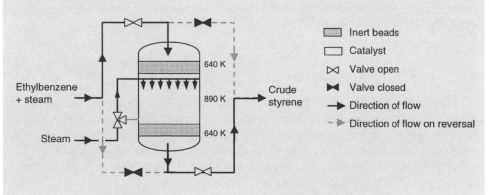

Figure B.9.3 Reverse-flow reactor for styrene production.

9.5.2 Selective Oxidations

Selective oxidation processes employing a solid catalyst are used for the synthesis of a variety of chemicals, the most important being formaldehyde, ethene oxide, maleic anhydride and phthalic anhydride. Catalytic selective oxidations constitute a technologically challenging class of processes because of the high exothermicity of the reactions, and the complex kinetic scheme leading to by-products and thus relatively low selectivities. Selective oxidation is also an area showing major innovations. Here, we will discuss the production of ethene oxide and maleic anhydride.

Selective oxidation of ethene to produce ethene oxide is interesting for several reasons. Silver is the only metal known to catalyze this reaction with sufficient selectivity, and this epoxidation reaction is unique to ethene [26]. No satisfactory catalyst has yet been found for the analogous reactions for the selective oxidation of propene and butene.

Maleic anhydride production is interesting from the point of view of both catalyst design and reactor technology.

9.5.2.1 Production of ethene oxide

Ethene oxide, because of its reactivity, is an important raw material for the production of a wide range of intermediates and consumer products. The main outlet for ethene oxide is the manufacture of ethene glycol, accounting for about 60%. Other uses are in surfactants, ethanolamines, etc. After polyethene, ethene oxide is the second largest consumer of ethene.

Background

As with acetaldehyde (Section 8.2), direct oxidation has largely replaced a more complex route to ethene oxide. Originally, ethene oxide was produced by indirect oxidation of ethene via ethene chlorohydrin:

$$2\,Cl_2 + 2\,H_2O \;\rightleftharpoons\; 2\,ClOH + 2\,HCl \tag{11}$$

$$2\,CH_2{=}CH_2 + 2\,ClOH \;\rightleftharpoons\; 2\,HOCH_2\text{-}CH_2Cl \qquad \Delta H^0_{298} = -220 \text{ kJ/mol} \tag{12}$$

$$2\,HOCH_2\text{-}CH_2Cl + Ca(OH)_2 \rightleftharpoons 2\,\underset{\displaystyle \diagdown O \diagup}{CH_2{-}CH_2} + CaCl_2 + 2\,H_2O$$

$$\Delta H^0_{298} = -7 \text{ kJ/mol} \tag{13}$$

Although this process provided good ethene oxide yields, most chlorine was lost to useless $CaCl_2$, and unwanted chlorine containing by-products were generated. This not only was inefficient, but also caused major pollution problems, resulting in the replacement of this route by direct oxidation of ethene with either oxygen or air.

The chlorohydrin route is still used in a number of plants for the manufacture of propene oxide from propene, mainly because attempts to achieve direct oxidation of propene to propene oxide failed. Many of these propene oxide plants are converted chlorohydrin based ethene oxide plants. Besides the chlorohydrin route, two major processes for propene epoxidation are based on epoxidation by organic hydroperoxides. Those processes are environmentally more benign but they produce two products, propene oxide and the product from the organic peroxide, viz. isobutene or styrene.

Reactions and kinetics

The basic reaction involved in the direct oxidation of ethene to ethene oxide is the following:

$$CH_2{=}CH_2 \;+\; 1/2\,O_2 \longrightarrow \underset{O}{CH_2{-}CH_2} \qquad \Delta H^0_{298} = -105 \;\; kJ/mol \quad (14)$$

The only significant by-products are carbon dioxide and water, which are either formed by complete combustion of ethene or by combustion of ethene oxide.

$$CH_2{=}CH_2 \;+\; 3\,O_2 \longrightarrow 2\;CO_2 \;+\; 2\;H_2O$$
$$\Delta H^0_{298} = -1324 \;\; kJ/mol \quad (15)$$

$$\underset{O}{CH_2{-}CH_2} \;+\; 5/2\,O_2 \longrightarrow 2\;CO_2 \;+\; 2\;H_2O$$
$$\Delta H^0_{298} = -1220 \;\; kJ/mol \quad (16)$$

All three reactions, and especially the latter two, are highly exothermic and thermodynamically complete at the operating conditions of ethene oxide synthesis. Moreover, the activation energies of the undesired reactions are higher than that of the desired reaction, which implies that even small changes in temperature can have a large effect on the selectivity towards ethene oxide. This temperature sensitivity demands very good temperature control.

QUESTION:

> *An important process parameter is the optimum ethene-to-oxygen ratio in the reactor. What is the optimum ratio (qualitatively)? Explain.*

A catalyst is needed which promotes the partial oxidation of ethene. To date, silver is the only metal known to catalyze the oxidation of ethene to ethene oxide with sufficient selectivity. All processes use a supported silver catalyst, although the catalyst has been much improved since its discovery. A promoter is added to increase the selectivity to ethene oxide by reducing the combustion reactions. This promoter is chlorine, introduced as ethyl chloride or vinyl chloride.

One of the numerous rate expressions available for the selective oxidation of ethene (25% Ag/2% Ba/SiO$_2$ at 333 K) is the following Langmuir–Hinshelwood equation [27,28]:

$$r_{EO} = \frac{kK_E p_E K_O p_O}{(1 + K_E p_E + K_O p_O)^2} \tag{17}$$

in which r_{EO} is the rate of formation of ethene oxide (mol·g^{-1}·h^{-1}), k is the reaction rate constant for reaction (14) ($15 \cdot 10^{-3}$ mol·g^{-1}·h^{-1}), K_E and K_O are the adsorption coefficients (10 bar^{-1} and 0.6 bar^{-1}) and p_E and p_O the partial pressures (bar) for ethene and molecular oxygen, respectively.

QUESTION:

> *Based on the reactions occurring and the rate expression, what would be the best reactor type, a plug-flow reactor or a CSTR, and how can the reactants best be fed (e.g., together at the reactor entrance, one at entrance, other distributed, etc.)?*

Process

The need for very effective temperature and selectivity control dominates reactor and process design in the selective oxidation of ethene to ethene oxide. Firstly, the ethene conversion per pass must be limited to ca. 7–15%, in order to suppress the consecutive oxidation of ethene oxide. Secondly, a multi-tubular reactor is preferred over other reactors because it allows better temperature control. The current large plant capacities (up to 150,000 t/a [29]) require large diameter reactors. However, the need to minimize the radial temperature gradients and the risk of hot spots calls for small diameter tubes. Therefore, ethene oxide reactors contain several thousand tubes (length, 6–12 m; internal diameter, 20–50 mm [30]) in parallel. The heat liberated by the reaction is removed by a high-boiling hydrocarbon or water, which flows around the reactor tubes. The process shown in Figure 9.20 employs a hydrocarbon-cooled reactor. Condensation of the vapors of this coolant in an external boiler is used to recover heat by steam generation.

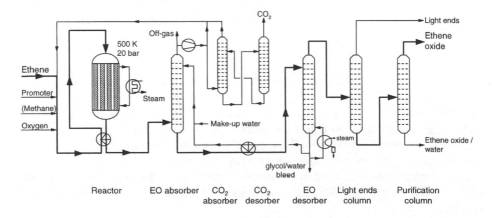

Figure 9.20 Production of ethene oxide by oxidation of ethene with oxygen.

In principle, a fluidized-bed reactor, possibly with internal heat transfer area, offers better temperature control than the multi-tubular reactor. However, a process based on fluidized-bed technology has not yet been commercialized, mainly due to problems in developing a catalyst that is resistant against attrition [30].

The temperature rise due to the exothermic reactions must be further controlled by adding inert diluents to the reactor feed. This suggests air as the source of oxygen rather than pure oxygen. However, as will become clear below, the use of air has disadvantages which are usually not compensated for by the lower cost of air. Therefore, nearly all modern plants employ oxygen as the oxidizing agent. Figure 9.20 shows a flow scheme of the oxygen-based ethene oxide process.

In this scheme, methane is used as the diluent, but other gases can also be used. The choice is based on the thermal properties of the gas. Methane has a higher heat capacity and thermal conductivity than nitrogen, which results in better isothermicity.

The preheated feed mixture consisting of ethene, oxygen and inerts (impurities in the reactants, added methane, and recycled CO_2) is fed to the reactor. A large excess of ethene is used to keep its conversion low, and the selectivity towards ethene oxide high. Although the pressure has been shown not to influence the conversion much at reaction temperatures, operations are conducted at elevated pressure to facilitate the subsequent absorption of ethene oxide in water. The gases leaving the reactor are cooled by heat exchange with the feed gas and sent to a column where ethene oxide (only present in a concentration of about 1–2 mol%) is absorbed in water. The aqueous solution rich in ethene oxide that leaves the bottom of the ethene oxide absorber, passes through a stripping column (EO desorber), separating ethene oxide at the top. The bottom stream, containing water and some ethene glycol ($HO-CH_2-CH_2-OH$), which is formed when ethene oxide comes into contact with water, is returned to the ethene oxide absorber. Two distillation columns separate ethene oxide from light ends (CO_2, acetaldehyde, other hydrocarbon traces) and residual water.

A small part of the gas leaving the top of the ethene oxide absorber is purged to prevent build-up of inerts (mainly CO_2, Ar, and methane). After compression, a side stream of the recycle gas goes to an absorber where CO_2 is removed, for instance by alkanolamine treatment, before being recycled to the reactor. This is necessary to keep the CO_2 concentration at an acceptable level.

Ethene and air in certain proportions form explosive mixtures. The feed to the reactor contains 20–40 mol% ethene and about 7 mol% oxygen to produce a reaction mixture that is always above the upper flammability limit. The presence of CO_2, which is formed in the combustion reactions, helps to reduce the flammability limit. Therefore, not all CO_2 is removed from the recycle stream. Depending on the amount of inerts recycled, methane can be added in order to reduce the oxygen concentration and thus the flammability zone.

QUESTION:

Which impurities would be present in the ethene and oxygen feed streams?

Air versus oxygen

The need to operate in the presence of diluents suggests the use of air instead of oxygen. However, in this case large amounts of nitrogen would end up in the recycle.

Therefore, a substantial part of the gas has to be vented, causing a significant loss of unconverted ethene (up to 5%). A secondary or purge reactor is needed with associated ethene oxide absorption system in order to improve the ethene conversion before venting. Therefore, additional investment is required. However, this is partially offset by the need for a carbon dioxide removal system in the oxygen process.

Also, the operating costs for the two processes differ. Especially the price of oxygen has a great bearing on the production costs. Furthermore, in the oxygen process considerable steam input is required in the carbon dioxide removal unit. The higher production rate per unit volume of catalyst and the less costly recycle of ethene when using oxygen may be a decisive advantage.

Alternative process: membrane reactor

A membrane reactor is not only suitable for removing one of the components selectively (e.g. hydrogen in ethane dehydrogenation, Section 9.4.3.1) it can also be applied to keep the reactants separated from each other, except on the catalytic sites. This application is especially interesting for partial oxidation of hydrocarbons, such as in the production of ethene oxide from ethene. The flammability limits restrict the feed composition, which in conventional reactors leads to the necessity of a large excess of the hydrocarbon. By adopting a multi-tubular cooled catalytic membrane reactor, oxygen and ethene can be kept separate, avoiding any flammability constraints. Figure B.9.4 illustrates this.

Oxygen diffuses through the membrane at a limited rate at various points along the reactor, so its concentration is kept low. Hence, besides avoiding flammability constraints, the selectivity of the reaction is also enhanced.

QUESTION:

> *Based on the kinetic scheme, could you design an optimal membrane reactor leading to maximum selectivity?*

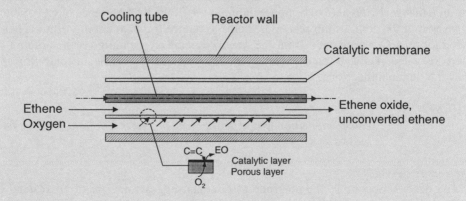

Figure B.9.4 Catalytic membrane reactor for the partial oxidation of ethene to ethene oxide.

Table 9.3 Comparison of oxygen- and air-based ethene oxide process. After [30]

Process	Oxygen based	Air based
Concentrations (vol%)		
C_2H_4	15–40	2–10
O_2	5–9	4–8
CO_2	5–15	5–10
Inert	Rest (mainly CH_4 and Ar)	Rest (mainly N_2)
Temperature (K)	570–630	570–630
Pressure (bar)	10–20	10–30
GHSV[a] ($m^3_{gas}/(m^3_{cat}\cdot h)$)	2000–4000	2000–4500
C_2H_4 conversion	7–15	20–65
Selectivity (%)	75–80	70–75

[a] GHSV = gas hourly space velocity.

QUESTIONS:

Why is a CO_2 removal system not required in the air-based process?
Estimate the residence time of the gas in the reactor.

Table 9.3 compares the operating parameters in oxygen- and air-based ethene oxide plants.

QUESTION:

Compare the data in Table 9.3 with the kinetic equation given earlier (Eqn. (17)). Check your answers with respect to the optimal reactor choice.

9.5.2.2 Production of maleic anhydride by selective oxidation of butane

Maleic anhydride is an important intermediate in the chemical industry. It is used in copolymerization and addition reactions to produce polyesters, resins, plasticizers, etc. The overall reaction, starting with butane is:

$$C_4H_{10} + 3.5\,O_2 \longrightarrow \text{maleic anhydride} + 4\,H_2O \qquad \Delta H^0_{298} = -1260 \text{ kJ/mol} \quad (18)$$

The commercial processes are catalytic and based on fluidized beds or multi-tubular fixed-bed reactor technology, as used in the production of ethene oxide. However, a fixed-bed reactor is not the optimal choice: the intermediates are more reactive than

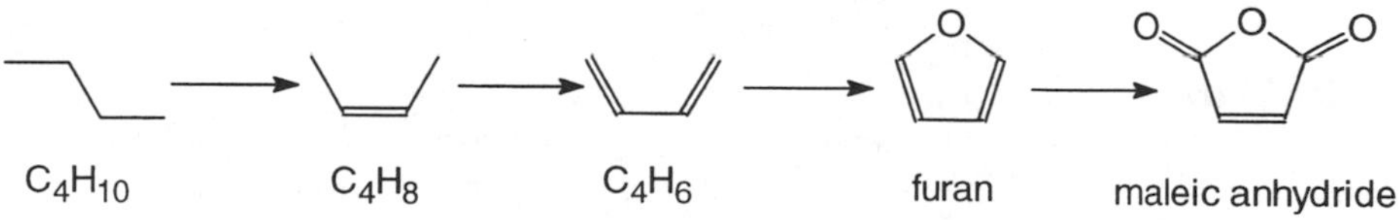

$$C_4H_{10} \longrightarrow C_4H_8 \longrightarrow C_4H_6 \longrightarrow \text{furan} \longrightarrow \text{maleic anhydride}$$

Scheme 9.3 Simplified reaction pathway for maleic anhydride production.

the starting material and the reactions are highly exothermic. Hot spots may easily occur leading to total combustion.

Note that the exothermicity of desired reaction is much larger than in the production of ethene oxide. This results in much lower single-reactor capacities for the production of maleic anhydride in fixed-bed reactors (up to 40 kt/a [31]).

Scheme 9.3 shows the simplified reaction network for the production of maleic anhydride. Just as in the selective oxidation of ethene to ethene oxide, by-products

Reaction network and kinetics

The reaction network for the production of maleic anhydride from butane can be represented as shown in Scheme B.9.1. Various intrinsic rate equations have been proposed to describe the reaction kinetics [31]. Here a Langmuir–Hinshelwood type adsorption model is presented.

Production of maleic anhydride in riser reactor

An interesting option for the production of maleic anhydride is being demonstrated by Monsanto and Du Pont. It involves the use of a riser reactor analogous to the reactor applied in the FCC process The catalyst played an important role in this development. A catalyst with the required high attrition resistance can be produced by incorporating the active ingredients in a matrix such as in the FCC process, but this decreases the activity and might lead to lower selectivity. A new technology, developed by Du Pont, encapsulates the active material in a porous silica shell. The pore openings of the shell permit unhindered diffusion of the reactants and products, without affecting conversion or selectivity. Figure B.9.5 shows the reactor configuration and a schematic of the conventional catalyst and the novel catalyst developed by Du Pont.

In the riser reactor, the oxidized catalyst transfers oxygen to butane to produce maleic anhydride. The reaction takes place at a temperature of about 770 K. The catalyst is separated from the product in a cyclone. Subsequently, the catalyst is reoxidized in a conventional fluidized-bed reactor.

Separation of the two steps enables fine-tuning of the reaction conditions [30]:

- The oxidation is performed without molecular oxygen, so the reaction mixture is never within the explosion limits;
- Residence times in the reactor and the regenerator can be chosen independently and, as a consequence, optimal conditions for both reactors can be chosen.

QUESTION:

> *Why is the lower activity due to incorporation in a matrix not a problem in FCC? Compare the different processes in view of the kinetics.*

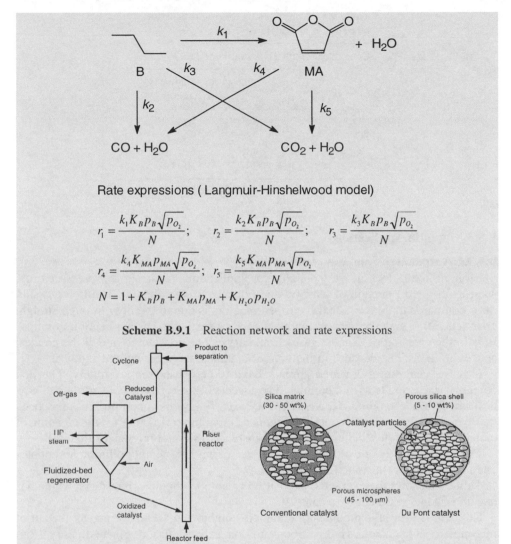

Scheme B.9.1 Reaction network and rate expressions

Figure B.9.5 Riser reactor and catalyst for catalytic oxidation of butane. Adapted from [31].

(CO_2, H_2O, and CO) are formed by complete or partial oxidation of the reactant butane and of the product maleic anhydride (see box).

Fluidized-bed reactors suggest themselves as an alternative to circumvent the problems encountered with fixed-bed reactors. They allow better temperature control and higher feed concentrations. However, catalyst attrition is a major problem. The world's largest plant is based on a fluidized-bed reactor [27].

QUESTION:

> *Summarize advantages and disadvantages of fluid-bed reactor technology for the production of maleic anhydride.*

Table 9.4 Contributions to various emissions by transport and stationary sources[a]

Component emitted	Contribution (%)	
	Transport	Stationary sources
NO_x	54	46
SO_2	8	92
HC	43	57
CO	90	10

[a] NO_x = NO, NO_2; HC = hydrocarbons.

9.5.3 Monolith Applications

9.5.3.1 Automotive emission control

Monolithic reactors are very suitable for applications requiring low pressure drop, such as in the conversion of harmful components in car exhausts. The low pressure drop compared to conventional fixed-bed reactors is due to the flow through straight channels rather than the tortuous paths in catalyst particles. The application of monoliths in the cleaning of exhaust gases from cars can be considered one of the greatest successes of the last decades in the area of chemical engineering and catalysis.

Gasoline and diesel powered engines have to be considered separately. The first category is referred to as otto engines. The success of the catalytic converter up to now is limited to otto engines. However, in the past few years much progress has been made in the development of catalytic filters of monolithic structure for the cleaning of exhaust gas from diesel engines (see, for instance, [32] and references therein).

The automobile is one of the major sources of environmental pollution and photochemical smog as illustrated in Table 9.4.

Table 9.5 shows the average composition of an otto engine exhaust gas, which has not undergone treatment by a catalyst.

CO_2 and H_2O are the products of complete combustion. CO is formed as a result of incomplete combustion. Hydrocarbons in the exhaust gas originate mainly from regions that are not reached by the flame (e.g. layers near the combustion chamber walls). Nitrogen oxides are formed by reaction of nitrogen and oxygen from the air when sufficient oxygen is present and the temperature is high. The small amount of sulfur compounds present in gasoline is oxidized largely to SO_2.

The exhaust gas composition depends on many parameters, the most important one being the air to fuel intake ratio, which is usually expressed using λ. $\lambda > 1$ implies a lean or oxidizing environment and $\lambda < 1$ a rich or reducing atmosphere. λ is defined as:

Table 9.5 Composition of a typical otto engine exhaust gas

Component	N_2	CO_2	H_2O	CO	O_2	NO_x	H_2	HC	SO_2
Vol%	74	12	10	2	1.0	0.5	0.4	0.1	0.006

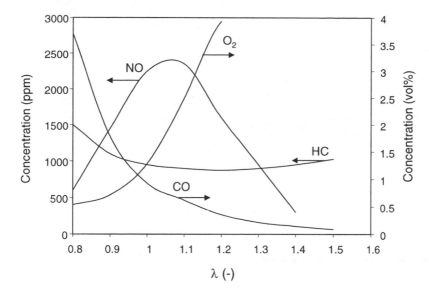

Figure 9.21 The concentration of CO, NO, HC (hydrocarbons) and O_2 emitted by a gasoline engine as a function of λ. Adapted from [32]

$$\lambda = (\text{fuel/air})_{\text{actual}}/(\text{fuel/air})_{\text{stoich}} \tag{19}$$

Figure 9.21 shows the influence of λ on the exhaust gas composition of a gasoline engine. With increasing λ the CO and HC (hydrocarbon) emissions decrease because of better combustion when more oxygen is present. The slight increase of CH at high λ is surprising at first sight. The explanation for this increase is that at lean conditions the temperature drops resulting in incomplete combustion. The NO production curve goes through a maximum. At rich conditions ($\lambda < 1$) the amount of air is low and as a result the emission of NO is low. At lean conditions the temperature drops, which causes a decrease in NO formation. The maximum amount of NO is formed near stoichiometric conditions.

Legislation requires the reduction of the CO, NO_x, and hydrocarbon emissions of automobiles. Basically two categories of emission reducing techniques exist: primary and secondary measures.

Primary measures

1. **Speed limitations.** It is well known that the emission of NO_x significantly increases with the driving speed. This increase is due to both a higher temperature and the operation in a relatively lean mixture at constant high speed.
2. **Fuel purification.** Hydrotreating of fossil fuels has a favorable influence on SO_2 emissions but only a small influence on NO_x emissions, since NO_x formation is mainly due to radical processes involving N_2 from the air.
3. **Engine modifications.** A combustion engine operating at for instance $\lambda > 1.3$ would result in a major decrease in CO and NO_x emissions. The limiting factor at high λ, due to the dilution of the fuel, is the ignition. In principle, this problem

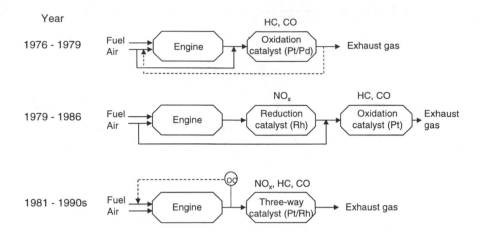

Figure 9.22 Development of emission control technologies for Otto engines; OC = oxygen control.

can be avoided by using a higher compression ratio. Additional advantages of this solution are a decrease in fuel consumption (estimated to be 15%) and up to 50% reduction in CO and NO_x emissions. These 'high compression lean burn engines', however, are still in the development stage, although the first commercial ones are currently being introduced into the market place at a large scale.

Secondary measures

Secondary measures to reduce exhaust emissions, also called end-of-pipe solutions all are based on catalytic conversion of the harmful emissions:

$$C_mH_n + (m + n/2)O_2 \rightarrow mCO_2 + n/2H_2O \qquad \Delta H^0_{298} < 0 \qquad (20)$$

$$CO + \tfrac{1}{2}O_2 \rightarrow CO_2 \qquad \Delta H^0_{298} = -288 \text{ kJ/mol} \qquad (21)$$

$$CO + NO \rightarrow \tfrac{1}{2}N_2 + CO_2 \qquad \Delta H^0_{298} = -373 \text{ kJ/mol} \qquad (22)$$

Figure 9.22 shows the evolvement of control technologies with time. Obviously, one reactor in which all three species are converted is preferable. However, the oxidation of hydrocarbons and CO requires other reaction conditions than the conversion of NO_x. Therefore, in the past (1976–1979) NO_x control was not attempted (regulations were not very strict then), and only CO and hydrocarbons were converted by an oxidation catalyst. Recirculation of exhaust gas in later designs helps to reduce NO_x emissions, due to a lower combustion temperature.

In later years (1979–1986) a dual catalyst system was applied. In the first reactor NO_x (and part of the HC and CO) were converted, while in the second reactor the remaining HC and CO were oxidized with a secondary supply of oxygen. A consequence of this arrangement is that the engine must operate at rich conditions in order for a reducing environment to prevail in the first reactor.

In 1981, emission standards became more stringent and the solution was found in

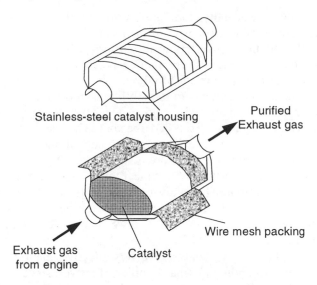

Stainless-steel catalyst housing

Purified Exhaust gas

Exhaust gas from engine

Catalyst

Wire mesh packing

Figure 9.23 Monolithic reactor and mounting system. Adapted from [33].

the application of the so-called 'three-way catalyst'. The name stems from the fact that the three components (NO_x, CO and HC) are removed simultaneously in one reactor. Improved catalysts and fuel/air control techniques have enabled this development.

In parallel with studies related to the active catalyst components (activity, stability), a number of engineering problems had to be tackled. Firstly, the pressure drop over the catalytic converter should be low and the converter should be as light as possible in

Table 9.6 Composition, performance of and conditions in a three-way catalyst[a]

Composition	
Carrier	Monolith: cordierite with 62 cells/cm^2
Washcoat	γ-Al$_2$O$_3$, CeO$_2$ (10–20%), La$_2$O$_3$ and/or BaO (1–2%)
Active phase	Pt + Rh: 1.2–1.4 g/l
Performance	
Controlled	$\lambda = 0.99 \pm 0.06$
	$X_{HC} > 80\%$
	$X_{NO,CO} > 70\%$
Uncontrolled	$\lambda = 1.05 \pm 0.2$
	$X_{HC} > 50\%$, avg. 70%
	$X_{CO} > 20\%$, avg. 55%
	$X_{NO} > 10\%$, avg. 50%
Operating conditions	
Temperature	570–1170 K
Space velocity	1–2×10^5 m$^3_{gas}$ m$^{-3}_{reactor}$ h^{-1}
Volume ratio	Catalyst/cylinder = 0.8–1.5

[a] X_i = conversion of component i.

view of fuel economy. Secondly, the catalyst should be resistant to the extreme temperatures and corrosive environment of the exhaust.

Not surprisingly, the first converters were reactors packed with catalyst particles, i.e. conventional fixed-bed reactors. The main concern with these reactors was attrition of the catalyst particles as a result of vibrational and mechanical stresses. Today the majority of catalytic converters are monolith reactors. These reactors combine a lower pressure drop with smaller size and weight, and hence provide better fuel economy than the conventional fixed-bed reactors. The catalyst is mounted in a stainless-steel container (the most difficult part of monolithic catalysts production) with a packing wrapped around it to ensure resistance to vibration (see Figure 9.23).

Table 9.6 shows the composition of a typical three-way catalyst, its performance, and some operation conditions. The catalysts, based on Pt and Rh, are very active and effective, provided they operate close to $\lambda = 1$. Deviation from this stoichiometric mixture will inevitably result in an overall lower conversion of the pollutants, as illustrated in Figure 9.24.

The presently used three-way catalysts are very effective due to their high activity combined with their outstanding thermal stability. However, since the active components are mainly Pt and Rh, costs and availability have become a point of concern. Indeed, today automotive catalysis already consumes 60% of the Pt and 70% of the Rh produced worldwide. The fast-growing European and Asian markets will further increase the Pt and Rh demands, and therefore, the already continuously increasing prices. Apart from the possible availability and price problems, the use of noble metal catalysts causes other problems. A reductive environment leads to reduction of SO_2 to H_2S, and at intermediate temperatures the formation of N_2O (ozone destruction in the atmosphere) is a problem. Research in industry and at universities is still ongoing to develop alternative and better catalytic systems.

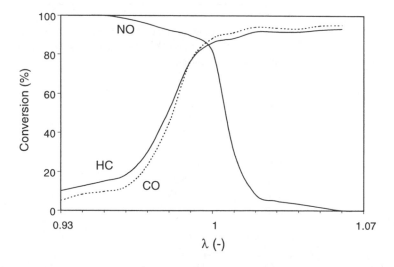

Figure 9.24 Efficiency of a three-way catalyst for the conversion of NO, CO and HC (hydrocarbons) as a function of λ. Adapted from [34].

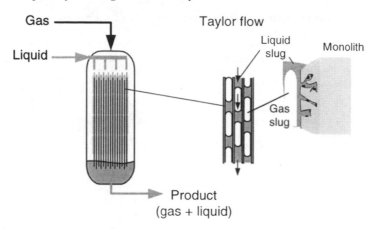

Figure 9.25   Monolith reactor with co-current gas–liquid flow and flow pattern.

QUESTIONS:

    Explain why H_2S can be formed. Compare the space velocity of catalytic converters with 'normal' chemical reactors.

9.5.3.2 Monolith reactors in gas–liquid reactions

The application of monolith reactors in two-phase flow operation is a relatively novel development. Currently, the interest in three-phase monolith reactors is focused on three main fields of applications: hydrogenations, oxidations, and biotechnology applications.

Flow pattern in monoliths

    In three-phase catalytic reactors, hydrodynamics play an important role in bringing about the contact between the phases. The performance of a monolith reactor is thus expected to be highly dependent upon the flow pattern in the monolithic channels. Many flow patterns have been described for gas-liquid flow through monolith channels (see [12,36–38]).

    The most important flow regime is so-called Taylor (or segmented) flow. This flow pattern consists of liquid slugs separated from each other by gas slugs (see Figure 9.25 [36]). The gas slugs are separated from the channel wall by a very thin liquid film. This thin liquid film provides low mass-transfer resistance from the gas/liquid interface to the channel wall. Mass transfer is further enhanced by the forced recirculation within the liquid slugs. Due to this flow pattern, plug-flow behavior is closely approached.

Comparison of three-phase monolith with conventional reactors

    For solid-catalyzed gas–liquid reactions monoliths compete with mechanically stirred tanks (mainly in fine chemicals), three-phase fluidized bed reactors and trickle-flow reactors (bulk chemicals and oil refining). Table 9.7 compares the latter two reactors with the three-phase monolith reactor. Characteristics of mechanically stirred tanks can be found in Table 10.6 in Chapter 10.

Table 9.7 Comparison of trickle-bed, three-phase fluidized-bed, and three-phase monolith reactor for catalytic gas-liquid reactions [2,36,39]

Characteristic	Three-phase fluidized bed	Trickle flow	Three-phase monolith
Particle diameter/wash coat thickness (mm)	0.1–3.0	1.5–6	0.01–0.2
Fraction of catalyst (m_{cat}^3/m_r^3)	0.1–0.5	0.55–0.6	0.07–0.25
Liquid hold-up (m_l^3/m_r^3)	0.2–0.8	0.05–0.25	0.2–0.8
Pressure drop (kPa/m)		10.0–50.0	0–3.0
Volumetric mass transfer coefficient			
Gas–liquid, $k_l a_l$ (s^{-1})	0.05–0.3	0.01–0.08	0.01–2
Liquid–solid, $k_s a_s$ (s^{-1})	0.1–0.5	0.06	0.05–0.06

The main advantage of the fluidized-bed reactor is that the relatively small particles result in good internal and liquid-solid mass-transfer characteristics leading to good catalyst utilization. Easy temperature control is another advantage. Its main disadvantages are catalyst attrition problems and the chaotic behavior of the reactor, which makes scale-up difficult.

For trickle-flow reactors these characteristics are reversed, i.e. scale-up is easier, while catalyst utilization is worse for conventional catalyst shapes. The latter is a result of a compromise between good catalyst utilization (small particles) and low pressure drop (large particles). In addition, temperature control is more difficult.

Three-phase monolith reactors combine the advantages of conventional reactors, while eliminating their main disadvantages. The main benefit results from the decoupling of catalyst utilization considerations (internal transport) from hydrodynamic considerations (pressure drop). Advantages of monoliths reactors compared to conventional reactors are:

- Active catalyst material can be tailored to minimize internal diffusion limitations, thereby mimicking a fixed catalyst with the diffusion properties of a small slurry catalyst, but without attrition and separation problems.
- Continuous operation is easy, unlike in mechanically stirred tank reactors (slurry reactors, see Chapter 10), where catalyst separation and recycle poses problems.
- Countercurrent contacting of gas and liquid is possible [37].
- There is a relatively large surface area for mass and heat transfer between the catalyst and the surrounding fluid. Trickle-flow reactors would require too small particles to achieve this.
- The pressure drop is much smaller than in the trickle-flow reactor.
- Scale-up should be straightforward.

Of course, monolith reactors also have disadvantages. They are more difficult to fabricate than conventional catalyst particles, and therefore are more expensive. Catalyst deactivation is as important an issue in monoliths as it is in trickle-flow reactors. Catalyst life should be sufficiently long. Although the scale-up of monoliths seems to

Monolith for countercurrent operation: hydrodesulfurization (HDS)

In general, hydrodesulfurization of heavy oil fractions (Eqn. (23)) takes place in trickle-flow reactors, in which liquid and gas flow co-currently downward.

$$S - compound + H_2 \leftrightarrow desulfurized\ compound + H_2S \tag{23}$$

A large proportion of the sulfur is removed in the top part of the reactor. This is a consequence of the different reactivities of the sulfur compounds present. The most reactive compounds are converted first, while conversion of the less reactive compounds occurs far more slowly in a later stage. Hence, hydrogen sulfide is generated in large quantities already in the top part of the reactor, and in co-current operation this hydrogen sulfide will pass through the remaining part of the reactor. Thus, a large part of the reactor operates under a relatively high hydrogen sulfide partial pressure. Since hydrogen sulfide suppresses the activity of the catalyst (product inhibition), the conditions are not optimal.

Counter-current operation with gas flowing upward and liquid flowing downward would be more efficient. The part of the catalyst bed where a high catalyst activity is required to convert the least active sulfur compounds operates under low hydrogen sulfide partial pressure. Only a small part of the bed, the part where the reactive compounds are converted, operates under high hydrogen sulfide partial pressure.

Figure B.9.6 shows the HDS and hydrogen sulfide profiles over the reactor for co-current and counter current operation [37].

Although counter-current operation would be preferable, in practice it is impossible in fixed-bed reactors at the gas and liquid flow rates of industrial interest. The reason is that due the relatively small catalyst particles (1–3 mm) and low bed voidage (ca. 0.4) the downward flow of liquid is impeded by the upward flow of gas, giving rise to flooding of the reactor.

For a given gas and liquid mass flow rate, the flooding criterion can be alleviated by reducing the particle surface-to-volume ratio (i.e. increasing its size) or by increasing the voidage of the catalyst bed. For instance, in packed distillation columns flooding is not a problem at the usual flow rates, because large particles (2.5–5 cm) with shapes resulting in a high voidage (ca. 0.9) are used.

However, in catalytic reactors, larger particles lead to poor catalyst utilization due to diffusion limitations (see Section 9.2.1). Particles in the mm range are required. The voidage of fixed beds can be increased by very loose packing of the catalyst particles. Both solutions result in poor usage of reactor volume, a significant cost factor for high-pressure processes such as hydrodesulfurization [37].

The monoliths used in gas-phase applications such as automotive exhaust control, with comparable surface-to-volume ratio and voidage as typical catalyst particles in fixed beds, are not suitable for counter-current operation either, because a phenomenon similar to flooding arises (see Figure B.9.7).

Because of the small channel diameter (ca. 1 mm), instead of flowing as a thin film down the channel wall, the liquid will bridge the channel and form a slug, which is then transported upward by the gas. Thus, instead of counter-current flow, an upward so-called segmented or Taylor flow is obtained.

Two solutions were found. First, larger diameter channels solve this problem but at the cost of a lower catalyst loading. This led to the development of the so-called internally-finned monolith, which has relatively large diameter channels (2.5–5 mm) with fins incorporated, which keeps the surface-to-volume-ratio the same as for small diameter channels (see Figure B.9.8). Second, the design of the outlet appeared to be essential. By the careful design of a special outlet, resulting in separate outlet flow lines for the liquid, flooding was greatly reduced.

The internally-finned monolith with counter-current flow could be a viable alternative for the trickle-bed reactor in the region of interest for hydrodesulfurization applications, especially when deep desulfurization is required.

QUESTIONS:

> *Why is simply increasing the channel diameter not an option?*
> *What are the advantages of the internally-finned monolith with counter-current flow over a trickle-bed reactor?*
> *What would be disadvantages of this type of monolith?*
> *What would be a disadvantage of fins for very fast reactions?*

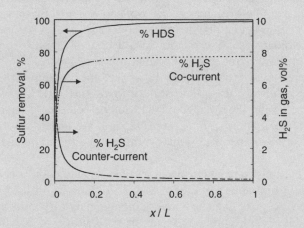

Figure B.9.6 HDS conversion and H_2S profile for co-current and counter-current operation. Adapted from [37].

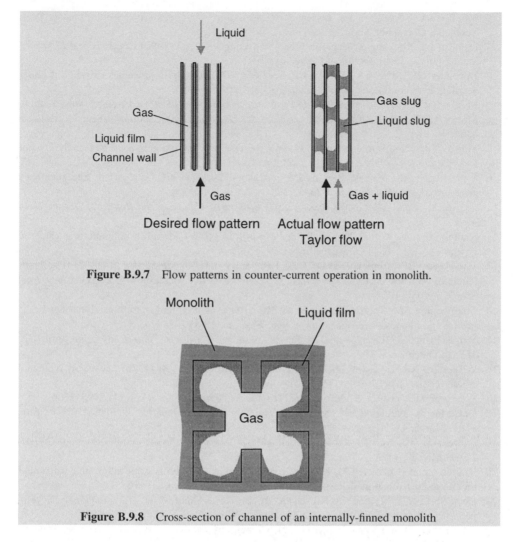

Figure B.9.7 Flow patterns in counter-current operation in monolith.

Figure B.9.8 Cross-section of channel of an internally-finned monolith

be straightforward, there are some questions as to how monolith blocks should be stacked and what the effect is on performance. Moreover, radial distribution should be good, because radial dispersion in the reactor does not occur. It is expected that monolithic reactors will play a major role in the Chemical Process Industry already in the near future.

References

1 Krishna R and Sie ST (1994) 'Strategy for multiphase reactor selection' *Chem. Eng. Sci.* 49(24A) 4029–4065.
2 Trambouze P, van Landeghem H and Wauquier JP (1988) *Chemical Reactors, Design/Engineering/Operation*, Gulf Publishing Corporation, Houston, TX.
3 Fogler SH (1986) *Elements of Chemical Reaction Engineering*, Prentice-Hall, London.

4 Moulijn JA, Kapteijn F and Jansen, JC (1999) *Catalysis Engineering*, lecture course, Delft Technical University.

5 Moulijn JA, Xiaoding X, Kapteijn F and Van Langeveld AD (1999) *Catalysis and Catalysts*, lecture course, Delft Technical University.

6 Boreskov GK, Matros Yu.Sh and Kiselev OV (1979) 'Catalytic processes carried out under nonstationary conditions' *Kinet. Katal.* 20 773–780.

7 Matros Yu.Sh (1985) *Unsteady-state Processes in Catalytic Reactors*, Elsevier, Amsterdam.

8 Matros Yu.Sh (1989) *Catalytic Processes Under Unsteady-state Conditions*, Elsevier, Amsterdam.

9 Matros Yu.Sh and Bunimovich GA (1996) 'Reverse-flow operation in fixed bed catalytic reactors' *Catal. Rev. – Sci. Eng.* 38(1) 1–68.

10 Van Diepen AE, Van de Riet ACJM and Moulijn JA (1996) 'Catalysis in fine chemicals production' *Rev. Port. Quím.* 3 23–33.

11 Cybyulski A and Moulijn JA (eds.) (1997) *Structured Catalysts and Reactors*, Dekker, New York.

12 Kapteijn F, Heiszwolf JJ, Nijhuis TA and Moulijn JA (1999) 'Monoliths in multiphase catalytic processes – aspects and prospects' *Cattech* 3(1) 24–41.

13 Van Hasselt BW, Lindenbergh DJ, Calis HP, Sie ST and Van den Bleek CM (1997) 'The three-levels-of-porosity reactor. A novel reactor for countercurrent trickle-flow processes' *Chem. Eng. Sci.* 52 3901–3907.

14 Champagnie AM, Tsotsis TT and Minet RG (1990) 'A high temperature catalytic membrane reactor for ethane dehydrogenation' *Chem. Eng. Sci.* 45(8) 2423–2429.

15 Van den Broek J (1985) 'Vette subsidie voor membraanreactor' *Chemisch Magazine* April 182–183 (in Dutch).

16 Westerterp KR, Bodewes TN, Vrijland MSA and Kuczynski M (1988) 'Two new methanol converters' *Hydroc. Process.* 67(11) 69–73.

17 Westerterp KR (1992) 'Multifunctional reactors' *Chem. Eng. Sci.* 47(9–11) 2195–2206.

18 DeGarmo JL, Parulekar VN and Pinjala V (1992) 'Consider reactive distillation' *Chem. Eng. Prog.* March, 43–50.

19 Podrebarac GG, Ng FTT and Rempel GL (1997) 'More uses for catalytic distillation' *CHEMTECH* May 37–45.

20 Van de Graaf JM, Zwiep M, Kapteijn F and Moulijn JA (1999) 'Application of a silicalite-1 membrane reactor in methatesis reactions' *Appl. Catal. A: General* 178 225–241.

21 Chen SS (1997) 'Styrene' in: Kroschwitz JI and Howe-Grant M (eds.) *Kirk-Othmer Encyclopedia of Chemical Technology*, vol. 22, 4th ed., Wiley, New York, pp. 956–994.

22 Shoemaker JD and Jones Jr EM (1987) 'Cumene by catalytic distillation' *Hydroc. Process.* 66(6) 57–58.

23 Kochloefl K (1997) 'Dehdrogenation of ethylbenzene' in: Ertl G, Knözinger H and Weitkamp J (eds.) *Handbook of Heterogeneous Catalysis*, vol. 5, VCH, Weinheim, pp. 2151–2159.

24 James JW and Castor WM (1994) 'Styrene' in: Gerhartz W, et al. (eds.) *Ullmann's Encyclopedia of Industrial Chemistry*, vol. A25, 5th ed., VCH, Weinheim, pp. 329–344.

25 Snyder JD and Subramaniam B (1994) 'A novel reverse flow strategy for ethylbenzene dehydrogenation in a packed-bed reactor' *Chem. Eng. Sci.* 49(24B) 5585–5601.

26 Satterfield CN (1991) *Heterogeneous Catalysis in Industrial Practice*, 2nd ed., McGraw-Hill, New York, pp. 279–285.

27 Farrauto R and Bartholomew H (1997) *Fundamentals of Industrial Catalytic Processes*, Blackie, London, Chapter 8.

28 Park DW and Gau G (1987) 'Ethylene epoxidation on a silver catalyst: unsteady and steady state kinetics' *J. Catal.* 105 81–94.

29 Mills PL, Harold MP and Lerou JJ (1996) *Industrial Heterogeneous Gas-phase Oxidation*

Processes, Niok Special Course, Advanced Catalysis Engineering, vol. D, August 19–September 6.

30 Rebsdat S and Mayer D (1994) 'Ethylene oxide' in: Gerhartz W, et al. (eds.) *Ullmann's Encyclopedia of Industrial Chemistry*, vol. A10, 5th ed., VCH, Weinheim, pp. 117–135.

31 Lerou JJ and Mills PL (1993) 'Du Pont butane oxidation process' in: Weijnen MPC and Drinkenburg AAH (eds.) *Precision Process Technology*, Kluwer, Dordrecht, pp. 175–195.

32 Van Diepen AE, Makkee M and Moulijn JA (1999) 'Emission control from mobile sources: otto and diesel engines' in: Janssen FJJG and Van Santen RA (eds.) *Environmental Catalysis*, Imperial College, London, pp. 257–291.

33 Kieboom APG, Moulijn JA, Sheldon RA and Van Leeuwen PWNM (1999) in: Van Santen RA, Van Leeuwen PWNM, Moulijn JA and Averill BA (eds.) *Catalysis: an Integrated Approach to Homogeneous, Heterogeneous and Industrial Catalysis*, 2nd ed., Elsevier, Amsterdam, Chapter 2.

34 Lepperhoff G, Pischinger F and Koberstein E (1985) in: Gerhartz W, et al. (eds.) *Ullmann's Encyclopedia of Industrial Chemistry*, vol. A3, 5th ed., VCH, Weinheim, pp. 189–200.

35 Kieboom APG, Moulijn JA, Van Leeuwen PWNM and Van Santen RA (1999) in: Van Santen RA, Van Leeuwen PWNM, Moulijn JA and Averill BA (eds.) *Catalysis: an Integrated Approach to Homogeneous, Heterogeneous and Industrial Catalysis*, 2nd ed., Elsevier, Amsterdam, Chapter 1.

36 Irandoust S, Cybulski A and Moulijn JA (1997) 'The use of monolithic catalysts for three-phase reactions' in: Cybulski A and Moulijn JA (eds.) *Structured Catalysts and Reactors*, Dekker, New York, Chapter 9.

37 Sie ST and Lebens PJM (1997) 'Monolithic reactors for countercurrent gas-liquid operation' in: Cybulski A and Moulijn JA (eds.) *Structured Catalysts and Reactors*, Dekker, New York, Chapter 11.

38 Lebens, PJM, Heiszwolf JJ, Kapteijn F, Sie ST and Moulijn JA (1999) 'Gas-liquid mass transfer in an internally finned monolith operated countercurrently in the film flow regime' *Chem. Eng. Sci.* 54 5119–5125.

39 Ramachandran PA and Chaudari RV (1983) 'Three phase catalytic reactors' in Hughes R (ed.) *Topics in Chemical Engineering*, Vol. 3, Gordon and Breach, New York.

Literature

Chauvel A and Lefebvre G (1989) *Petrochemical Processes 1. Synthesis Gas Derivatives and Major Hydrocarbons*, Technip, Paris.

Chauvel A and Lefebvre G (1989) *Petrochemical Processes 2. Major Oxygenated, Chlorinated and Nitrated Derivatives*, Technip, Paris.

Ertl G, Knözinger H and Weitkamp J (eds.) (1997) *Handbook of Heterogeneous Catalysis*, vol. 1–5, VCH, Weinheim.

Henkel KD (1992) 'Reactor types and their industrial applications' in: Gerhartz W, et al. (eds.) *Ullmann's Encyclopedia of Industrial Chemistry*, vol. B4, 5th ed., VCH, Weinheim, pp. 87–120.

Thoenes D (1994) *Chemical Reactor Development: from Laboratory Synthesis to Industrial Production*, Kluwer, Dordrecht.

Van Santen RA, Van Leeuwen PWNM, Moulijn JA and Averill BA (eds.) (1999) *Catalysis: an Integrated Approach to Homogeneous, Heterogeneous and Industrial Catalysis*, 2nd ed., Elsevier, Amsterdam.

Chapter 10

Fine Chemicals

Traditionally, chemical engineers have been mainly involved in bulk chemicals production. Numerous novel processes have been put into practice, and the chemical engineers have evidently been successful. Also in environmental catalysis the chemical engineering community has contributed visibly, a noticeable example being the catalytic converter used in the cleaning of car exhaust gases. Less attention has been devoted to fields where the production volumes are smaller, and the chemical complexity is higher, e.g. in the production of pharmaceuticals and agrochemicals. However, also in these areas the chemical reaction engineering approach is useful.

10.1 INTRODUCTION

In the chemical industry, usually a distinction is made between commodities (or bulk chemicals), fine chemicals and specialties.

As with bulk chemicals, fine chemicals are identified according to specifications (what they are). In contrast, specialties are identified according to performance (what they can do). Fine chemicals include advanced intermediates, bulk drugs and bulk pesticides, active ingredients, bulk vitamins, and flavor and fragrance chemicals. Some examples of specialty chemicals are adhesives, diagnostics, disinfectants, pesticides, pharmaceuticals, photographic chemicals, dyestuffs, perfumes, and specialty polymers. Table 10.1 shows a small selection, illustrating the difference between bulk, fine, and specialty chemicals.

Fine chemicals differ from bulk chemicals in many respects, as shown in Table 10.2 (the limits are slightly arbitrarily). Specific for fine chemistry is the formation of relatively large amounts of by-products. Table 10.3 illustrates this.

QUESTIONS:

Try to find the volumes produced per year of the major bulk chemicals. Compare them with the production volumes of some fine chemicals. Also find price information and compare.

Why is the energy consumption per unit product in the production of bulk chemicals usually much lower than in the production of fine chemicals?

In fine chemistry, catalysis does not play the important role it does in the production of bulk chemicals. Multistep synthesis reactions are common and they usually consist of a number of stoichiometric reactions rather than catalytic reactions. These reactions

Table 10.1 Position of fine chemicals in the chemical industry; some examples[a]

Bulk		Fine	Specialty
Toluene	(1) →	Isobutylbenzene (i) (3) → (5) Ibuprofen (ai)	Pharma-ceuticals
Trans-2-butene Acetic acid	(3) →	C₅ aldehyde intermediate (I) +	Additive in feed industry,
Isobutene Formaldehyde	(4) →	Citral (i) Intermediate (i) (5) → Viitamin A acetate (i) ↓ Vitamin A (bv)	Pharma-ceuticals
Phenol	(2) →	*p*-Hydroxy-benzaldehyde (i, ff)	Perfumes, Flavoring, Agro-chemicals,
p-Cresol	(1) ↗ (2) →	*p*-Anisaldehyde (i, ff)	Pharma-ceuticals

[a] ai, active ingredient; i, intermediate; bd, bulk drug; bv, bulk vitamin; ff, flavor and fragrance chemical; (number) = number of steps; X = Cl, Br, I, F; PPh₃ = triphenylphosphine.

Table 10.2 Fine versus bulk chemicals

	Fine chemicals	Bulk chemicals
Price	>5$/kg	<5$/kg
Volume	<10 kt/a	>10 kt/a
Product variety	High	Low
Chemical complexity	High	Low
Added value	High	Low
Applications	Limited number (often one)	Many
Synthesis	Multi step	Few steps
	Various routes	One or few routes
Catalysis	Exception	Often
Reactant and product phase	Liquid and often solid	Usually gas or liquid
Raw materials consumption (kg/kg)	High	Low
Energy consumption (kg/kg)	High	Low
By-products (kg/kg)	High	Low
Toxic compounds	Often (e.g., phosgene, HCN)	Exception
Plants	Often MPP[a]	Dedicated
	Usually batch	Often continuous
Investment		
($)	Low	High
($/kg)	High	Low
Labor	High	Low
Market fluctuations	High	Low
Producers	Limited number	Many

[a] MPP, multi-product or multi-purpose plant.

Table 10.3 Production and by-product formation in various sectors of industry [1]

Industry segment	Production (metric ton/year)	By-product formation (kg by-product/kg product)
Oil refining	10^6–10^8	0.01–0.1
Bulk chemicals	10^4–10^6	<1–5
Fine chemicals	10^2–10^4	5–50
Pharmaceuticals	10–10^3	25– >100

often result in the formation of large amounts of by-products, predominantly inorganic salts. The following examples illustrate this.

Hydrogenation is often carried out with a mixture of a metal and an acid (Fe/HCl), producing stoichiometric amounts of iron chloride as by-product. So, even at 100% selectivity the process is environmentally not attractive. A similar reasoning holds for oxidation with permanganate or dichromate, halogenations, sulfonations, nitrations, etc.

Friedel–Crafts reactions are other obvious examples. Scheme 10.1 shows the Friedel–Crafts alkylation of benzene with 2-propylchloride to yield cumene.

The reaction scheme shows that the Cl atom in 2-propylchloride ends up as HCl. Furthermore, equimolar quantities of $Al(OH)_3$ are formed. Catalytic processes do not produce so many by-products. Perfect examples in bulk chemistry are the homogeneously catalyzed hydroformylation, Wacker oxidation, and methanol carbonylation (Chapter 8), which all have an atom utilization[1] of 100% based on a theoretical (100%) yield. Not surprisingly, catalysis is and will be more and more applied in fine chemistry. In that respect, fine chemistry can benefit from the experience in bulk chemistry.

QUESTIONS:

Give some more examples of 100% atom-utilization reactions in previous chapters. Do reactions with lower atom-utilization also take place in catalytic processes?
What is the atom utilization of the cumene production process of Scheme 10.1?
Suggest an alternative process for the production of cumene.

A trend in fine chemistry is the use of reactants that do not produce by-products with unfavorable properties. For instance, oxidation with permanganate is unattractive in this respect. Examples of attractive reactions are oxidations carried out with O_2, H_2O_2, and organic peroxides, e.g. *tert*-butyl-hydroperoxide: $(CH_3)_3$—COOH. These oxidation reactions often have to be catalyzed.

---

[1] Atom utilization or atom selectivity: ratio of the molecular weight of the desired product to the total molecular weights of all materials produced.

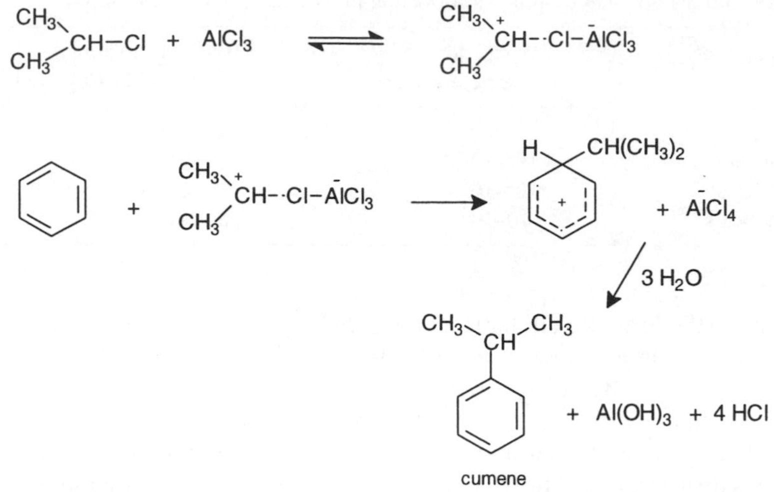

Scheme 10.1 Cumene production by Friedel–Crafts alkylation of benzene with 2-propylchloride.

QUESTIONS:

Why are these oxidation reactants attractive? Why is O_2/air used in particular in the production of bulk chemicals?

Another trend in fine chemistry is the increasing interest in the production of compounds with high purity, especially optically pure compounds. With pharmaceuticals, it is often not safe to apply mixtures of the optical isomers (enantiomers). It is possible that only one enantiomer is active for the desired purpose while the other enantiomer is inactive or even has a damaging effect. An example is the Softenon disaster in the beginning of the sixties. Softenon used to be the trade name for thalidomide, which was typically given as a painkiller to pregnant women. However, one of the enantiomers of thalidomide (Figure 10.1) appeared to lead to disfigured children. As a result of this affair, at present in the EU and Japan, manufacturers have to prove that enantiomers present in a product do not show undesired phenomena. This makes it very attractive to exclusively produce the desired enantiomer.

Homogeneous transition-metal catalysts can provide the selectivity needed for the manufacture of chiral compounds. In practice, however, whereas homogeneous catalysis is widespread in bulk chemicals production (see Chapter 8) only a few processes in fine chemicals and pharmaceuticals production apply soluble transition-metal catalysts.

One of the first commercial catalytic asymmetric syntheses is the Monsanto process for the production of L-dopa ((L-amino-(3,4-dihydroxyphenylanine) propanoic acid, Figure 10.1), which is used in the treatment of Parkinson's disease [2]. Eqn. (1) shows the key step, viz. stereoselective hydrogenation (with Ac = $CH_3(CO)$—), which takes place in the presence of a soluble bisphosphine-modified rhodium catalyst [3]. The rhodium complex solution is mixed with a slurry of the reactant in aqueous ethanol.

Example: production of hydroquinone

The production of hydroquinone presents an illustrative example of the replacement of stoichiometric reactions using salt-producing reagents with catalytic reactions (see Scheme B.10.1).

Hydroquinone was traditionally manufactured by oxidation of aniline with stoichiometric amounts of manganese dioxide to produce benzoquinone, which was then reduced with iron and hydrochloric acid (this latter step was later replaced by a catalytic hydrogenation). The aniline was produced from benzene via nitration and reduction with Fe/HCl. The overall process generated over 10 kg of organic salts ($MnSO_4$, $FeCl_2$, Na_2SO_4 and $NaCl$) per kg hydroquinone. In contrast, a more modern route to hydroquinone, involving the catalytic oxidation of *p*-di-isopropylbenzene, followed by acid-catalyzed rearrangement of the bishydroperoxide, produces less than 1 kg of organic salts per kg hydroquinone. In this process only acetone is formed as by-product.

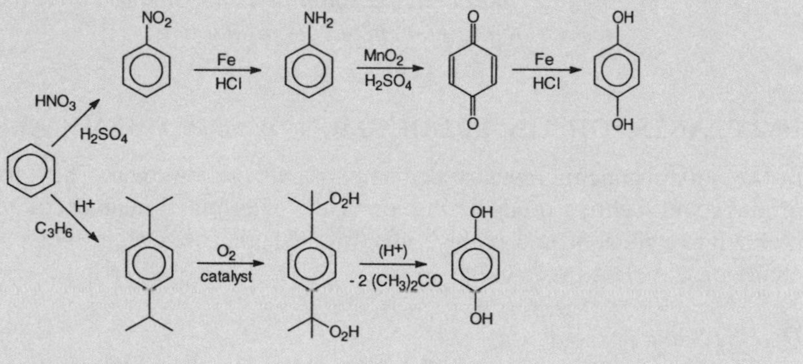

Scheme B.10.1   Two routes to hydroquinone. From [1].

During hydrogenation the reactant dissolves while the product precipitates. The product, essentially optically pure, is collected by filtration. The catalyst and racemic product (mixture of the two possible optical isomers) remain in solution. Note that this is an example of the separation principle discussed in Section 8.7, where the catalyst stays in solution while the formed product is a solid.

Figure 10.1   Molecular structure of thalidomide (left) and L-Dopa (right); C* is the chiral center.

Some other examples of the use of homogeneous catalysis in fine chemicals production are steps in the production of vitamin A (rhodium-catalyzed hydroformylation of diacetoxybutene intermediate) [3], agrochemicals, flavors, and fragrances [4].

The use of catalysis, including biocatalysis, in the fine chemicals industry has already resulted in much improved process efficiency, but further progress is possible and needed [5].

QUESTION:

> *What could be the reason that homogeneous catalysis is not applied more often in fine chemicals production?*

10.2 PLANTS FOR THE PRODUCTION OF FINE CHEMICALS

In the bulk chemicals industry dedicated plants predominate. They are optimized to produce well-defined products in a precisely determined manner. In a sense they are both highly efficient and highly inflexible. Major changes are not possible without changing the plant profoundly.

QUESTION:

> *Illustrate the inflexibility regarding the incorporation of new findings (e.g. the development of a revolutionary catalyst) for ammonia and methanol plants.*

In fine chemistry usually reactions are carried out batch-wise rather than continuously. Typical plant types are the so-called Multi-Product Plant and Multi-Purpose Plant (MPP). In Multi-Product-Plants, each product may go through the same pieces of equipment (stages), and in the same order. In Multi-Purpose-Plants, different equipment is required for different products and different products may follow different routes through the equipment. A typical MPP consists of:

- stirred stainless-steel and glass-lined batch reactors provided with reflux condensers;
- feed systems for gaseous, liquid, and solid reactants;
- feed systems for blanketing with inert gases;
- equipment for separation and purification, e.g.,

    - filters, centrifuges (most fine chemicals are solids);
    - fluidized-bed driers, tray driers, rotary driers;
    - distillation equipment;

- facilities for recovery of solvents;
- storage facilities;
- effluent treatment installations:

  - sewage treatment;
  - liquids and solids incinerators;
  - off-gas treatment;

- utilities.

Also in MPPs the reactors form the heart of the plant. The most commonly used reactors are mechanically stirred batch reactors. Table 10.4 shows the main characteristics of commercially available batch reactors. For most equipment the volume-price dependency can be described by:

$$\text{Cost} = A \cdot \text{Volume}^B \tag{2}$$

For glass-lined reactors $B - 0.32\text{--}0.42$ and for stainless-steel reactors $B = 0.75$.

QUESTIONS:

> *Obviously, the larger the reactor, the lower the cost per unit volume. Why are the volumes applied nevertheless limited? Why is the exponent in the cost estimation formula higher in the case of stainless-steel reactors?*

Table 10.5 gives a rough impression of the investments in MPPs. The reactor accounts for 20–40% of the total costs. Glass-lined reactors are used extensively, although they are more expensive than stainless-steel reactors. Of course, the data

Table 10.4 Main characteristics of standard reactors available on the market. After [6]

Capacity (m^3)	Diameter (m)	Weight (t)	Area[a] m^2	Type	Pressure (bar)	Material[b]
8	2	9.5	18.6	DJ	15	GS
14	2.5	14	26.6	DJ	15	GS
25	2.8	21	40.0	DJ	15	GS
32	3.1	24	45.6	DJ	15	GS
10	2.4	11	19.8	DJ	6	GS
20	2.8	18	33	DJ	6	GS
25	3.0	22	38.8	DJ	6	GS
1	1.2	1.8	4.45	DJ	6	SS
4	1.8	2.5	10.9	DJ	6	SS
10	2.4	5.2	11.2	EC	6	SS
25	3.0	7.9	19.5	EC	6	SS
0.7	0.9	0.8	3.4	DJ	1.5	G

[a] Area available for heat exchange: DJ, double jacket; EC is welded external coil.
[b] GS, glass-lined steel; SS, stainless steel; G, glass.

Table 10.5 Investments in MPPs; investment in 1000 US $/m^3 (1982)

	Glass lined		Stainless steel	
	1 m^3	4 m^3	1 m^3	4 m^3
Reactor system	70	32	50	25
Solids & liquid recovery	29	23	29	23
Drying, size reduction	16	14	16	14
Liquid storage	7	7	7	7
Analytical instruments	2	1	2	1
Utilities	6	6	6	6
Effluent treatment	22	22	22	22
Building & civil	18	18	18	18
Total	170	125	150	118

given are an underestimation of the present costs. In particular, more sophisticated and expensive analytical instruments are used nowadays, and together with increased automatization, this leads to higher investment costs. On the other hand, operating costs are lowered because less labor is required.

In fine chemistry also dedicated plants exist. For example, aspirin is produced in such a dedicated plant (see Figure 10.2 [7–9]); in fact, the production volume in this case is tremendous (about 35 kt/a). It is questionable whether this process fits the category 'fine chemistry'.

Salicylic acid and acetic anhydride react according to Eqn. (3).

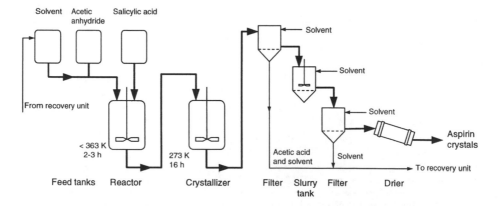

$$\Delta H^0_{298} = -84 \text{ kJ/mol} \tag{3}$$

Figure 10.2 Simplified scheme of aspirin production (batch process).

The reaction takes place in a glass-lined or stainless-steel batch reactor, in which the temperature must be kept below 363 K. At the end of the reaction period (2–3 h), the liquid product mixture is pumped into a crystallizer where it is cooled to 273 K. The aspirin crystallizes from the acetic acid/acetic anhydride mother liquor. The resulting suspension is transferred to a filter for removal of acetic acid and solvent (often acetic acid, but carbon tetrachloride and hydrocarbons are also used). After washing with solvent, the aspirin crystals are slurried and filtered again. The aspirin crystals are then dried and sent to sifting, granulation, and tableting. Acetic acid, which is formed as a by-product, is recovered. Solvent and possibly unconverted anhydride are recycled to the reactor.

QUESTION:

> *What are the consequences of the different time scales for reaction and crystallization?*

Also examples exist where dedicated plants have been designed for purely technical reasons. An example is the oxidation reaction represented by Eqn. (4).

$$(4)$$

The reaction is carried out at low temperature at a residence time of a few seconds. Longer residence times lead to further oxidation. Accordingly, a continuous process using a tubular reactor is the only viable option[2].

There is an increasing tendency to build dedicated plants for fine chemicals that used to be manufactured in MPPs. This is mainly due to rising requirements towards product purity; perfect cleaning of equipment before the next product is manufactured either is impossible or very expensive. Moreover, more catalytic processes are implemented in fine chemicals manufacture because of environmental needs. Heterogeneous catalytic processes are less flexible and often require dedicated plants, or at least a set of dedicated equipment items.

QUESTION:

> *The economic figures in Table 10.5 are from 1982. Would they still apply? What are the major differences between then and now?*

---

[2] Attainable residence times for the three basic reactor types are approximately: 15 min to 20 h for a batch reactor, 0.5 s to 1 h for a tubular reactor (PFR), and 10 min to 4 h for a single CSTR [6].

10.3 BATCH REACTOR DESIGN

This section treats the design of batch (and semi-batch) reactors. The reactor usually is the heart of the process. Therefore, the design and scale-up of reactors is very important. Reactor design requires information on chemical kinetics, heat and mass transfer, and reactor hydrodynamics. In fine chemistry, reaction systems are quite diverse. They can be classified based on the phases present, just as in bulk chemicals, except that liquid phase reactions dominate. When a reaction has been studied in the laboratory, in process development it has to be decided how to design the industrial reactor. Up till now, the majority of the reactors applied are mechanically stirred tank reactors used in the batch mode. The most important choices in the design of batch reactors concern:

- reactor volume;
- selection of the mixer;
- speed of the agitator;
- geometry of the tank, including baffles;
- heat-exchange area (internal and external).

It will be illustrated later that heat transfer is fundamental in design and operation.

10.3.1 Mechanically Stirred Batch Reactors

Figure 10.3 shows some of the batch reactor systems most often encountered in practice. Double-jacket reactors are most easily built, but the heat exchange area is limited. Exchange area may also be installed inside the reactors but its presence will hinder stirring and cleaning. Moreover, no glass-lined reactors equipped with internal cooling systems are available on the market. Therefore, when a large heat exchange area is needed, an external heat exchanger is usually required. A reflux condenser can be used when a reaction is carried out at boiling point. Alternatively, part of the liquid can be circulated through an external heat exchanger. In this case, the reactor is not mechanically agitated (obviously) but mixing is provided by liquid circulation.

QUESTIONS:

Give the advantages of a configuration where the reaction is carried out in a boiling solvent. How is the desired temperature realized?

If one reactant is gaseous, it can be supplied through a sparger at the bottom of the reactor (see Figure 10.4). This type of operation is termed semi-batch: one reactant, in this case a liquid, is loaded in the reactor, while the other, in this case a gas, is introduced continuously. (Note: when the liquid is also supplied continuously, we have a CSTR.) Other possible reactor types for gas-liquid reactions are bubble-column reactors (empty, packed or with trays), spray columns, etc. [2,5] (see Figure 8.16 in Chapter 8). These reactors can also be used if mechanical stirring is not desired, for whatever reason. When no gas is involved in the reaction, an inert gas, usually nitrogen, can be used to promote the agitation and mixing of two fluid phases [2].

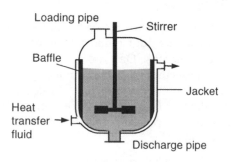
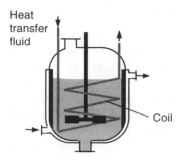

Batch reactor with double jacket

Batch reactor with double jacket and internal coil

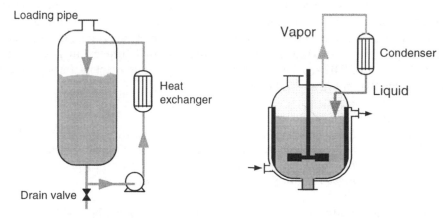

Batch reactor with external heat exchanger on recirculation loop

Batch reactor with cooling by vapor phase condensation

Figure 10.3 Batch reactor systems.

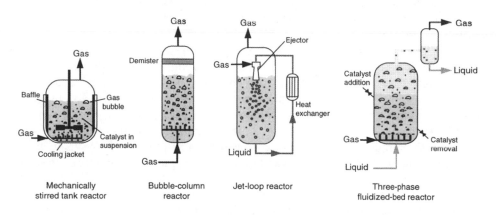

Mechanically stirred tank reactor

Bubble-column reactor

Jet-loop reactor

Three-phase fluidized-bed reactor

Figure 10.4 Reactors with moving catalyst particles.

10.3.2 Batch Reactors for Gas–Liquid–Solid Systems

In fine chemicals production gas–liquid–solid catalyst systems are common for catalytic hydrogenations, oxidations, hydrodesulfurizations, and reductive aminations. The reasons that a liquid phase is present are numerous. One reason can be low volatility of the generally bulky molecules, which together with their limited thermal stability (due to the often complex structure) prohibits operation in the gas phase (which would require relatively high temperatures). Furthermore, the selectivity might decrease with increasing temperature. It is also possible that it is desired to surround the catalyst particles with a liquid layer in order to

- avoid deactivating deposits and thus ensure higher catalyst effectiveness;
- achieve better temperature control due to the higher heat capacity of liquids;
- modify the active catalyst sites to promote or inhibit certain reactions.

In these instances, a liquid phase, often a solvent, is added on purpose. However, besides the positive effects mentioned above, a disadvantage is also introduced, i.e. an extra barrier is introduced between the gaseous reactant (hydrogen, oxygen, etc.) and the catalyst (see box: Mass transfer and reaction of hydrogen in gas–liquid–solid catalyst systems). The mass-transfer rate of the gaseous reactant through the liquid film is often, but not always, the limiting step.

QUESTION:

Explain why in hydrogenation in multiphase reactors hydrogen transfer from the gas bubbles to the liquid is often rate determining. Comparing water with an organic solvent, in which solvent do you expect the highest rate of hydrogenation?

Several gas–liquid–solid reactors are available, which can be divided in fixed-bed reactors (usually co-current trickle-flow), and reactors in which the catalyst particles move. Trickle-flow reactors, i.e. fixed-bed reactors with co-current downflow of gas and liquid, have large-scale applications in oil refining processes such as hydrotreating (Section 3.4.2) and catalytic hydrogenation of residues (Section 3.5.2). Despite their continuous operation, trickle-flow reactors also have significant potential in fine chemicals production, because they are well suited for high pressure operations. The main difficulty with trickle-flow reactors, as discussed in Section 3.4.2, is the wetting of the catalyst particles.

Figure 10.4 shows some examples of reactors with moving catalyst particles. The most commonly used reactors in fine chemicals production are still suspension (or slurry) reactors. The stirred-tank reactor, bubble-column, and jet-loop reactor are all suspension reactors, in which very fine catalyst particles (1–200 μm) are distributed throughout the volume of the liquid. Many variations exist for each reactor type. For instance, stirred tanks may have various types of agitators, cooling jackets or cooling coils, or both. Bubble columns may be empty, packed, or fitted with trays. Note that all three reactors can also be used for gas-liquid systems, then of course without a solid catalyst.

In all reactor types presented in Figure 10.4, good mixing is important to aid in the transport of hydrogen from the gas phase to the catalyst. The mechanically stirred tank reactor is most commonly used in batch processes. The catalyst particles are suspended in the liquid, which is almost perfectly mixed by a mechanical agitator. Cooling is usually accomplished by coils within the reactor or by a cooling jacket. Another option is circulation of the liquid/solid slurry over external cooling elements.

In bubble columns, agitation of the liquid phase and hence suspension of the catalyst is effected by the gas flow. Gas recycle causes more turbulence and thus better mixing. Often, circulation of the liquid is required to obtain a more uniform suspension. This can either be induced by the gas flow (airlift loop reactor) or by an external pump. In the latter case, it is possible to return the slurry to the reactor at high flow rate through an ejector (venturi tube). The local under-pressure causes the gas to be drawn into the passing stream, thus providing very efficient mixing. This type of reactor is called a jet-loop or venturi reactor. Jet-loop reactors tend to replace stirred-tank reactors in the most recent fine chemical hydrogenations. The external heat exchanger on the liquid circulation loop enables a high heat-removal capacity, which is a great advantage in highly exothermic reactions. A limitation on the use of jet-loop reactors is that the catalyst must be compatible with the pump, i.e. possess low hardness and high attrition resistance.

The three-phase fluidized-bed reactor (also called ebullated-bed reactor) differs from the suspension reactors in the use of larger catalyst particles (0.1–3 mm) and the formation of a well-defined agitated catalyst bed. Whereas the suspension reactors can operate in both batch mode and continuously with respect to the liquid phase (and catalyst), the ebullated-bed reactor (see also Section 3.5.2) only operates in the continuous mode, and hence is generally not the appropriate choice for fine chemicals.

QUESTION:

Compare the fixed-catalyst with the moving-catalyst reactors. What are advantages and disadvantages? Compare the moving-catalyst reactors.

Table 10.6 compares characteristics of the mechanically stirred tank, the bubble-column and the jet-loop reactor (see Table 9.7 for data on trickle-flow, three-phase monolith, and three-phase fluidized-bed reactors).

Two examples of gas–liquid reactions taking place in the presence of a solid catalyst will be discussed, which clearly illustrate the trend towards cleaner and less risky processes. The first example is the production of dimethyl carbonate (see box), which in itself is hardly a fine chemical (plant capacity: ca. 10,000 t/a), but is an interesting chemical for fine chemistry with respect to both waste reduction and the replacement of toxic substances with safer ones. The second example is the production of the drug ibuprofen (see box).

Table 10.6 Comparison of mechanically stirred tank, bubble-column, and jet-loop reactor for catalytic gas-liquid reactions [2,6,12–14]

Characteristic	Stirred tank	Bubble column	Jet loop
Particle diameter (μm)	1–200	1–200	1–200
Fraction of catalyst ($m_{cat}^3\ m_r^{-3}$)	0.001–0.01	0.001–0.01	0.001–0./
Liquid hold-up ($m_l^3\ m_r^{-3}$)	0.8–0.9	0.8–0.9	0.8–0.9
Mass transfer parameters			
Gas-liquid, $k_l a_l$ (s^{-1})	0.05–0.2	0.05–0.15	0.2–1/
Liquid-solid, $k_s a_s$ (s^{-1})	0.1–0.5	$\approx$0.25	0.1–/

Mass transfer and reaction of hydrogen in gas–liquid–solid catalyst systems

Figure B.10.1 shows mass transfer and reaction of hydrogen in a gas–liquid–solid catalyst system according to the film model. The gas phase consists of pure H_2, which is the limiting reactant. The reaction taking place is first order in hydrogen and irreversible. The reactor is isothermal and at steady state.

Mass transfer from the gas phase to the surface of the catalyst particle, i.e. external mass transfer, takes place by a number of steps in *series*. In contrast, diffusion inside the particle (internal mass transfer) and reaction occur *simultaneously*.

External mass transfer
In the absence of internal diffusion limitations (i.e. diffusion inside the catalyst particle is much faster than reaction and $c = c_s$), the observed rate of reaction equals:

$$r^{obs} = k_1 a_1 (c_i - c_1) = k_s a_s (c_1 - c_s) = k c_s \tag{5}$$

and is determined by the slowest step. $p = H_2$ pressure (bar); $c = H_2$ concentration inside particle, dependent on distance from surface (mol m_l^{-3}); $c_i = H_2$ concentration at interface, c_1 in bulk liquid, c_s at catalyst surface (mol m_l^{-3}); r^{obs} = observed rate of reaction (mol m_l^{-3} s^{-1}); k_1 = mass-transfer coefficient from liquid film to bulk liquid ($m_r^3 m_i^{-2} s^{-1}$); a_1 = gas–liquid interfacial area per unit volume of reactor ($m_i^2 m_r^{-3}$); k_s = mass-transfer coefficient from bulk liquid to catalyst surface ($m_r^3 m_s^{-2} s^{-1}$); a_s = liquid–solid interfacial area per unit volume of reactor ($m_s^2 m_r^{-3}$); k = reaction rate constant for first-order reaction (s^{-1}).

QUESTIONS:

> Derive an expression for r_{obs}, which does not contain c_i and c_s. Why is this useful? What expression do you obtain for very slow reactions? What is the maximum observable rate for very fast reactions?
>
> Relate the rate of reaction as defined above to the rate of reaction per unit mass of catalyst in a reactor.

10.4 BATCH REACTOR SCALE-UP[3]

This section focuses on the scale-up of homogeneous stirred-tank batch reactors. In general, the histories of temperature, concentrations, and mixing conditions in a full-scale reactor differ significantly from those in a laboratory reactor. Therefore, it is not surprising that selectivities usually change upon scale-up.

---

[3] A. Cybulski has made a large contribution to this section.

Internal mass transfer

Inside the catalyst pellet diffusion occurs simultaneously with reaction. This leads to an internal concentration profile: at larger penetration depths the concentration of the reactants becomes lower due to partial conversion to reaction products. For positive reaction orders, this lower concentration results in a less efficient utilization of the catalyst particle, i.e. to so-called internal diffusion limitations. A measure for the degree of internal diffusion limitations is given by the effectiveness factor, η, see also Section 9.2.1.

The simplest expression for the internal effectiveness factor results from irreversible first-order kinetics and a slab geometry for the pellet:

$$\eta = \frac{\tanh\phi}{\phi} \qquad \text{and} \qquad \phi = L\sqrt{\frac{k}{D_{\text{eff}}}} \qquad (6)$$

with ϕ the Thiele modulus, L the length of the catalyst slab (m_p), and D_{eff} the effective diffusion coefficient ($m_p^2\,s^{-1}$).

It follows from Eqn. (6) that the importance of internal diffusion limitations only depends on the Thiele modulus, i.e. on the ratio and not on the individual values for the reaction rate constant and the diffusion coefficient.

QUESTIONS:

What is the value of η for very large and very small ϕ? What is the physical meaning of these values? Sketch the concentration profiles in a catalyst particle for various values of these variables.

A more elaborate treatment of this subject can be found in numerous texts, e.g. [6,10,11].

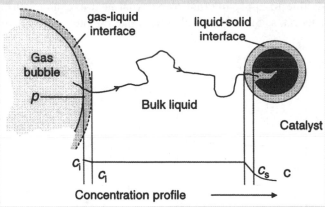

Figure B.10.1 Mass transfer and reaction of hydrogen in a gas/liquid/solid catalyst system according to the film model.

10.4.1 Temperature Control

Batch reactors are usually operated as shown in Figure 10.5. At first, the temperature of the heat transfer medium (T_h) is high in order to warm up the reaction mixture to reaction temperature (T_r). When an exothermic reaction starts, the temperature of the heat transfer medium is switched down, often to the lowest possible level. After the reaction is under control (case a), the temperature may be raised to keep the reaction mixture at the optimal reaction temperature to get sufficiently high reaction rates. If a reaction is faster and/or more exothermic, it might be impossible to keep the temperature under control even with maximal cooling (case b). It is obvious that such an event is undesired: a so-called runaway takes place. Batch reactors are unsuitable for such processes, so other procedures are called for. Often, so-called semi-batch reactors are used (see Section 10.4.3).

10.4.2 Heat Transfer

Usually, the stirred reactors in the fine chemicals industry are of the jacketed type. As a consequence, the heat-transfer area per unit volume usually decreases with increasing reactor volume (see Table 10.4). In principle, heat transfer can be improved by more intensive mixing. In the average commercial reactor, however, this does not result in drastically increased heat-transfer rates. This can be understood from the relationship for the overall heat-transfer coefficient (for a jacketed stirred-tank reactor (see Table 10.7)).

Heat transfer may also be improved by using a lower coolant temperature, but:

- use of media at very low T costly;
- components from reaction mixture might precipitate on wall.

It is also possible to apply internal or external heat-exchange surfaces that allow large heat transfer rates at relatively mild cooling temperature. Moreover, if necessary,

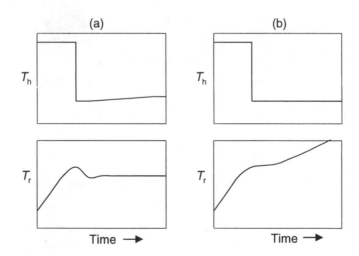

Figure 10.5 Control of reactor temperature (T_r) by manipulating the temperature of the heat transfer medium (T_h); (a) reaction under control; (b) reaction out of control.

heat production rates are lowered by reducing concentrations of reactants or catalysts. Another option is semi-batch operation with controlled dosing of (one of the) reactants, see next section.

QUESTIONS:
> *Derive the formula for the overall heat-transfer coefficient. If α_i could be increased by 100%, what is the effect on U?*

10.4.3 Example of Scale-Up of a Batch and Semi-batch Reactor

The influence of heat transfer on yield and selectivity in scaling up batch and semi-batch reactors will be illustrated by simulation using a series reaction. This reaction is composed of two irreversible elementary steps, both exothermic and both with first-order kinetics:

$$A \xrightarrow{k_1} P \xrightarrow{k_2} S \tag{11}$$

A is the reactant, P the desired product, and S an undesired by-product. k_1 and k_2 are the reaction rate constants (s^{-1}). Tables 10.8 and 10.9 show the required mole and energy balance equations and the accompanying data, respectively.

The overall heat-transfer coefficient, U, is evaluated from Eqn. (10), while the individual coefficients α can be calculated from the appropriate expressions (see [15]). The system of differential Eqns. (12)–(14) is conveniently solved numerically using the Runge–Kutta method.

Figure 10.6 shows the results of calculations for a small and a large reactor. Clearly,

---

Production of dimethyl carbonate

The traditional route to dimethyl carbonate involved reaction of methanol with phosgene:

$$2CH_3OH + Cl-CO-Cl \rightarrow (CH_3O)_2CO + 2HCl \tag{7}$$

Apart from the very dangerous effects of phosgene (it was used as a poison gas in World War I), its industrial application leads to problems concerning the removal and disposal of polluting by-products (HCl, chlorinated hydrocarbons). A more recently developed process is based on the oxidative carbonylation of methanol:

$$2CH_3OH + CO + 0.5O_2 \rightarrow (CH_3O)_2CO + H_2O \tag{8}$$

The reaction takes place in a suspension reactor in the presence of copper salts as the catalyst. This reaction carries much less risk than the phosgene route and no HCl is produced. The technology is evolving rapidly because of the great advantage of dimethyl carbonate over phosgene and other toxic compounds as an intermediate in the production of fine chemicals and pharmaceuticals, and as a specialty solvent.

Production of ibuprofen

Scheme B.10.2 shows the classical route by Boots (the inventor of the drug), and a new route developed by Hoechst Celanese.

Both routes share the intermediate, *p*-isobutylacetophenone, but the classical route involves a further five steps with substantial inorganic salt formation, while the alternative requires only two steps, one of which is a catalytic hydrogenation. The other step involves catalytic carbonylation. Both of these catalytic steps are 100% atom selective, and no waste is produced.

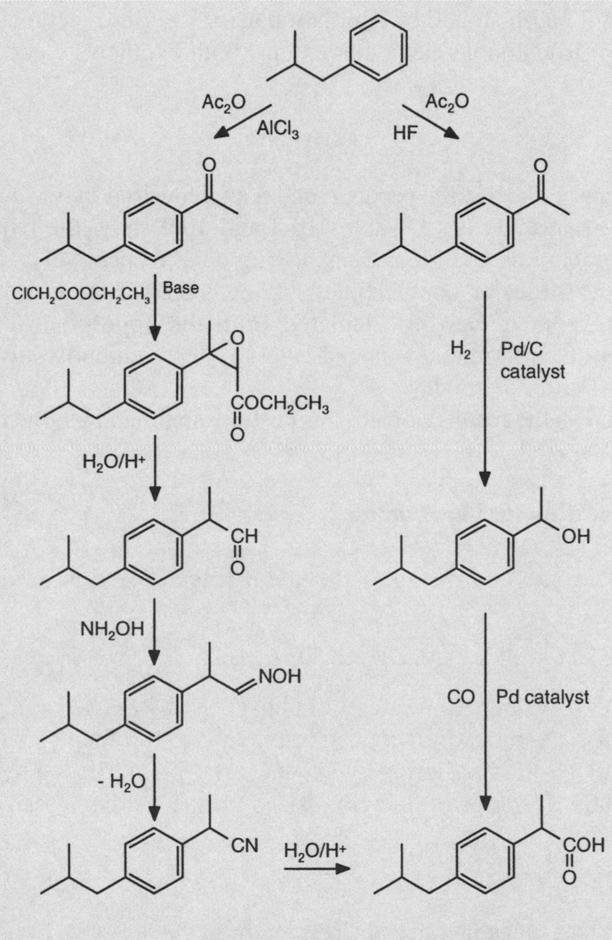

Scheme B.10.2   Two routes to ibuprofen; Boots (left) and Hoechst Celanese (right) [1].

the two reactors behave quite differently as function of the residence time. At large residence time the selectivity of the process in a large-scale reactor is significantly lower than that in a pilot plant reactor. Imagine what the operator will think when in the laboratory it was found that after 5000 s the product concentration was 400 mol/m^3

and in the commercial reactor at identical conditions the yield is close to zero! His/her first action would probably be to ask for a new catalyst.

Nothing surprising has happened, however. The reason for the unsatisfactory yield is that the A_h/V ratio of the large-scale reactor is much smaller than that of the small reactor, which results in a temperature excursion ('runaway'). Hence, the rate of the second reaction increases, consequently resulting in decreased selectivity. Obviously, the procedure followed in this case is far from optimal. It is not advisable to apply a step shaped temperature profile. It is common practice to switch off the heat supply after initiating the reaction and start cooling intensively. When the temperature peak has passed, the reaction mixture is cooled moderately to maintain the temperature at the desired level.

When the temperature rise is difficult to control, semi-batch operation is generally used. At the start, only a small amount of A is loaded into the reactor and when the desired temperature is reached, more A is steadily added over time. By carefully dosing A, the rate of heat evolution is controlled and a runaway can be prevented.

In the simulation of the semi-batch mode, the process conditions are the same as in the batch reactor except for the dosing scheme for A. The initial concentration of A is lower, and the remaining amount of A is dosed with a rate r_d starting at time t_d. It was assumed, for purposes of simplicity, that the volume of the reaction mixture and its physical properties did not change during dosing. Eqns. (12)–(14) describe the course of the process before dosing of the additional amount of A starts. The mole balance equation for the period of dosing A is modified as shown in Table 10.10. This table also gives additional data. Figure 10.7 shows the results of calculations for a large semi-batch reactor.

Table 10.11 compares the results of computations for both batch and semi-batch reactors. The presented data clearly show that even in the case of an inappropriate cooling policy, the yield and selectivity in a large semi-batch reactor are better than

Table 10.7   Heat transfer in a stirred-tank reactor

Heat flow in a stirred-tank reactor:

$$Q = U \cdot A_h \cdot \Delta T \tag{9}$$

Q = amount of heat transferred (W)
U = overall heat-transfer coefficient (W m^{-2} K^{-2})
ΔT = temperature difference between reaction mixture and coolant (K)
A_h = heat-transfer area (m^2)

Overall heat-transfer coefficient:

$$\frac{1}{U} = \frac{1}{\alpha_i} + \frac{\delta A_i}{\lambda A_m} + \frac{A_i}{\alpha_e A_e} \tag{10}$$

A_i, A_e = internal, external heat-transfer area (m^2)
A_m = logarithmic heat-transfer area ($-$)
δ = wall thickness (m)
λ = wall thermal conductivity, typically 30–100 W m^{-1} K^{-1}
α_i = internal heat-transfer coefficient, typically 200–700 W m^{-2} K^{-1}
α_e = external heat-transfer coefficient, typically up to 1000 W m^{-2} K^{-1}

Table 10.8 Mole and energy balance equations for reaction in a batch reactor

Mole balance equations:

$$\frac{dC_A}{dt} = r_1 = -k_1 C_A \tag{12}$$

$$\frac{dC_P}{dt} = r_2 - r_1 = k_1 C_A - k_2 C_P \tag{13}$$

Energy balance equation:[a]

$$\rho c_p \frac{dT}{dt} = (r_1)(-\Delta H_1) + (-r_2)(-\Delta H_2) + U(T_h - T)A_h/V \tag{14}$$

Initial conditions: $t = 0$; $C_A = C_{A0}$; $C_P = C_{P0}$; $T = T_0$

[a] The reactant and products have all been dissolved in water. They are supposed not to influence the heat capacity and density of the mixture.

those in an even smaller batch reactor. It is also clear that this type of modeling greatly helps in the design and operation of (semi-) batch processes.

Semi-batch operation is also often applied when two or more different reactants are present. Initially, one reactant and additives are loaded into the reactor and when the desired temperature is reached the other reactant(s) and additives are added with time. Figure 10.8 illustrates semi-batch operation for the exothermic reaction:

$$A + B \rightarrow C \tag{16}$$

Initially the reactor contains only A, possibly dissolved in, or mixed with an inert solvent. After A has reached a predetermined temperature, B is fed continuously or in portions while the temperature of the heat transfer medium is decreased. By appropriately adjusting the feed rate of B, the reactor stays far from its runaway temperature. The temperature of the heat-transfer medium may be increased to keep the reaction mixture at the optimal temperature.

Table 10.9 Explanation of symbols and reaction data

$-r_1 =$	Rate of reaction of A (mol m^{-3} s^{-1})	$k_{10} = 0.5$ s^{-1}
$-r_2 =$	Rate of reaction of P (mol m^{-3} s^{-1})	$k_{20} = 10^{11}$ s^{-1}
$k_i =$	$k_{i0} \exp(-E_i/RT)$ (s^{-1})	$E_1 = 20$ kJ mol^{-1}
$k_{i0} =$	Pre-exponential factor (s^{-1})	$E_2 = 100$ kJ mol^{-1}
$E_i =$	Activation energy for reaction i (J mol^{-1})	$\Delta H_1 = -300$ kJ mol^{-1}
$R =$	Gas constant (8.314 J mol^{-1} K^{-1})	$\Delta H_2 = -250$ kJ mol^{-1}
$\Delta H_i =$	Reaction enthalpy (kJ mol^{-1})	
$c_p =$	Heat capacity (kJ kg^{-1} K^{-1})	$c_p = 4$ kJ kg^{-1} K^{-1}
$\rho =$	Density (kg m^{-3})	$\rho = 1000$ kg m^{-3}
		$U = 500$ W m^{-2} K^{-1}
$C_i =$	Concentration of i (mol m^{-3})	$C_{A0} = 1000$ mol m^{-3}, $C_{P0} = 0$ mol m^{-3}
$T_r =$	Temperature of reaction mixture (K)	$T_{0(r)} = 295$ K
$T_h =$	Temperature of heat transfer medium (K)	$T_h = 345$ K, $0 < t < 3.6$ ks
$A_h =$	Area available for heat transfer (m^2)	295 K, $3.6 < t < 5.4$ ks
$V =$	Reactor volume (m^3)	$A_{h, small} = 9.5$ m^{-1}, $A_{h,large} = 2.6$ m^{-1}
		$V_{small} = 0.063$ m^3, $V_{large} = 6.3$ m^3

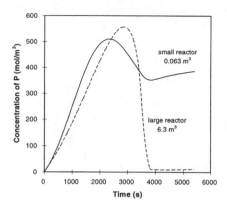 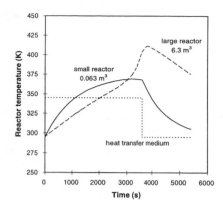

Figure 10.6 Concentration of P and reactor temperature versus time in batch reactor (data: Tables 10.8 and 10.9).

QUESTION:

> *When a runaway is possible, semi-batch operation coupled with a low coolant temperature can lead to safe operation. However, when the temperature is too low, a dangerous situation may exist, whereas at higher temperatures this would not be the case. Explain this.*

Sometimes continuous operation is desired, even for small-scale production (see Section 10.2). For instance, batch reactors are unsuitable for very fast reactions that require reaction times in the order of minutes or even seconds.

10.5 SAFETY ASPECTS OF FINE CHEMICALS

Illustrative for the risks involved in fine chemicals production, is the large amount of incidents with batch processes. According to Rasmussen [16] 57% of all incidents in the chemical process industries originated from batch processes and another 24% from storage, which can be considered as a batch operation also. Most of these incidents resulted from insufficient information about the process and from design errors. A

Table 10.10 Modified mole balance and data for semi-batch operation

Modified mole balance from start of dosing: $\dfrac{dC_A}{dt} = r_1 + r_d$ (15)

Initial conditions for this period: $t = t_d$; $C_A = C_{Ad}$; $C_P = C_{Pd}$; $T = T_d$
$C_{A0} = 460$ mol m^{-3}
r_d = dosing rate of A $= 0.3$ mol m^{-3} s^{-1}
t_d = time of beginning of dosing $= 1800$ s
C_{Ad} = concentration of A (mol m^{-3}) at t_d
C_{Pd} = concentration of P (mol m^{-3}) at t_d
T_d = temperature (K) at t_d

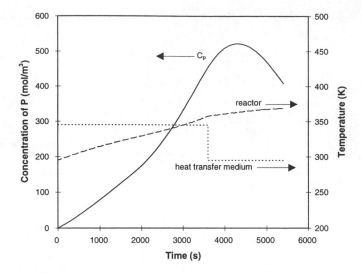

Figure 10.7 Concentration of P and reactor temperature versus time in semi-batch reactor (based on data in Tables 10.8–10.10).

large part of the incidents leading to runaways is directly related to the intended reactions.

10.5.1 Thermal Risks in The Production of Chemicals

Undesired reactions or poorly controlled desired reactions may lead to thermal runaways [17]. For example, in storage tanks heat dissipation is limited and, as a consequence, even slowly occurring exothermic reactions can cause a thermal runaway. In a reactor, loss of control of the desired reactions can occur due to the fact that heat removal typically shows a linear and heat production an exponential relationship with temperature. It is therefore necessary to determine a range of stable reactor operation conditions. Furthermore, a runaway of the desired reactions can result in secondary reactions that cause the formation of undesired products.

Semi-batch operation is not always safe either. For instance, if the temperature in a semi-batch reactor is chosen too low, the rates will be low and, consequently, it is possible that large amounts of intermediates or raw materials accumulate. When they participate in highly exothermic primary or secondary reactions, a runaway might occur. So, a runaway can be caused by too low an initial temperature or coolant temperature!

Table 10.11 Batch and semi-batch reactor results (after 5400 s)

Reactor	V (m³)	A_h/V (m⁻¹)	Conversion (%mol)	Selectivity (%mol)	Yield of P (%mol)
Batch	0.063	9.5	92.5	41.6	38.5
Batch	6.3	2.6	97.8	1.2	1.2
Semi-batch	6.3	2.6	85.6	47.2	40.2

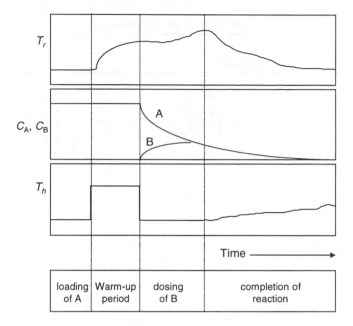

Figure 10.8 Semi-batch operation of a stirred-tank reactor with reactants A and B; T_r = temperature of reaction mixture; T_h = temperature of heat transfer medium; C_A, C_B: concentration of A and B.

QUESTION:

Various heat-transfer systems may be chosen: electrical heaters, steam heaters or heating by a circulating liquid. Which operation is safest?

10.5.2 Safety and Process Development

Safety is the major factor when working with hazardous materials and/or highly exothermic reactions. Usually, purely economic considerations are less important, although the safest possible process in the long run is also the most profitable. Three basic routes exist for protection against chemical process hazards [18,19]. These routes are schematically depicted in Figure 10.9.

Containment usually is an expensive means of reducing damage resulting from what has already happened. Installation of on-line systems for the detection of process deviations associated with trip systems for corrective actions adds to the initial investment, but decreases the risk. The objective of trip systems is to stop the process driving to explosion. Trip systems are becoming cheaper due to the progress in electronics. The third route, i.e. the design of an inherently safe process, requires the deepest understanding of the process and hence the highest research expenditure. However, the risk of hazardous situations is much reduced. A process is inherently safe if no fluctuation or disturbance can cause an accident. Inherent safety [18,19] is served by:

- identification of all reactions that may result in dangerous materials or overpressure, and establishing the process conditions disfavoring these reactions;

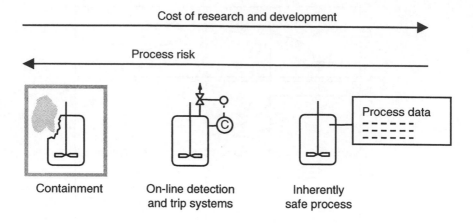

Figure 10.9 Three routes for protection against chemical process hazards.

- minimization of the amounts of hazardous materials;
- avoidance of external sources that could trigger runaways, e.g. by utilizing fluids which do not react in a dangerous way with the reaction mixture;
- operation of the process far within the region of stability.

The choice of reactor is a major decision. Batch reactors are only suitable for non-hazardous processes of low or moderate rate. If batch operation is too risky, semi-batch operation is preferred. Continuous reactors are the best solution for fast and highly exothermic reactions.

10.5.3 Summary of Scale-up of Batch Reactors

A laboratory unit can be transposed to an industrial unit based on the reaction time alone, provided the temperatures and the concentration of all components involved are identical in both units. However, in practice this is often difficult to achieve. One reason is that for processes involving highly exothermic reactions maintaining the same temperature in a large unit as in the small laboratory unit is difficult. Further-more, mixing and hydrodynamic conditions in a large tank are different from those in a small tank resulting in concentration differences. Especially when the reaction networks are complex, unexpectedly low yields and/or selectivities can be the result.

Fast changes in market demand often require quick development of a new process. A traditional approach towards reactor scale-up is then usually preferred: The process is tested in semi-technical and/or pilot reactors whose size is not less than 10% of the size of the industrial reactor. Ranges of safe operation and satisfactory yields are determined on the intermediate scales without knowledge of the process kinetics. This procedure may lead to a process that is not optimal. Therefore, if there is enough time and money, a chemical reaction engineering approach based on kinetic models, including mass and heat transfer phenomena, is recommended.

References

1 Sheldon R (1993) 'The role of catalysis in waste minimization' in: Weijnen MPC and Drinkenburg AAH (eds.) *Precision Process Technology; Perspectives for Pollution Prevention*, Kluwer, Dordrecht.
2 Mills PL, Ramachandran PA and Chaudhari RV (1992) 'Multiphase reaction engineering for fine chemicals and pharmaceuticals' *Rev. Chem. Eng.* 8(1–2), 5–176.
3 Parshall GW and Nugent WA (1988) 'Making pharmaceuticals via homogeneous catalysis, Part 1' *CHEMTECH* 3, 184–190.
4 Parshall GW and Nugent WA (1988) 'Making pharmaceuticals via homogeneous catalysis, Part 3' *CHEMTECH* 6, 376–383.
5 Bruggink A (1998) 'Growth and efficiency in the (fine) chemical industry' *Chimica Oggi* Sept.
6 Trambouze P, van Landeghem H and Wauquier JP (1988) *Chemical Reactors, Design/Engineering/Operation*, Technip, Paris.
7 Thomas MR (1997) 'Salicylic acid and related compounds' in: Kroschwitz JI and Howe-Grant M (eds.) *Kirk-Othmer Encyclopedia of Chemical Technology*, vol. 21, 4th ed., Wiley, New York, pp. 601–626.
8 Lowenheim FA, Moran MK (1975) *Faith, Keyes, and Clark's Industrial Chemicals*, 4th ed., Wiley, New York, pp. 117–120.
9 Anon. (1953) 'Taking headaches out of aspirin production' *Chem. Eng.* 6, 116–120.
10 Kapteijn F, Marin GB and Moulijn JA (1999) in: Van Santen RA, Van Leeuwen PWNM, Moulijn JA and Averill BA (eds.) *Catalysis: an Integrated Approach to Homogeneous, Heterogeneous and Industrial Catalysis*, 2nd ed., Elsevier, Amsterdam, Chapter 8.
11 Fogler SH (1986) *Elements of Chemical Reaction Engineering*, Prentice-Hall, London.
12 Irandoust S, Cybulski A and Moulijn JA (1997) 'The use of monolithic catalysts for three-phase reactions' in: Cybulski A and Moulijn JA (eds.) *Structured Catalysts and Reactors*, Dekker, New York, Chapter 9.
13 Van Dierendonck LL, Zaharadník J and Linek V (1998) 'Loop venturi reactor – a feasible alternative to stirred tank reactors?' *Ind. Eng. Chem. Res.* 37, 734–738.
14 Nardin D (1995) 'Trends and opportunities with modern Buss loop reactor technology' presented at the *Chemspec Europe 95 BACS Symposium*.
15 Green DW, Maloney JO and Perry RH (eds.) (1997) *Perry's Engineers' Handbook*, 7th ed., McGraw-Hill, New York.
16 Rasmussen B (1987) *Unwanted Chemical Reactions in the Chemical Process Industry*, Report of Risoe Natl. Lab. Denmark, Risoe-M-2631.
17 Gygax R (1988) 'Chemical reaction engineering for safety' *Chem. Eng. Sci.* 43(8), 1759–1771.
18 Regenass W, Osterwalder U and Brogli F (1984) 'Reactor engineering for inherent safety' *Inst. Chem. Eng.*, Publ. Ser. No. 37, Edinburgh, 364.
19 Regenass W (1984) 'The control of exothermic reactors', *Proceedings of the Symposium on Protection of Exothermic Reactions*, Chester, April.

Chapter 11

Polymerization – Production of Polyethene

11.1 INTRODUCTION

Humans have used natural polymeric materials, such as wood and horn, since prehistoric times. During the last century modified forms of natural polymers have been produced (e.g. vulcanized rubber, modified forms of cellulose such as cellulose nitrate and later cellulose acetate). The first fully synthetic polymers were the phenol–formaldehyde resins, which were developed in the beginning of the 20th century. Since then, the polymers industry has seen a spectacular growth, as shown in Figure 11.1.

Nearly 100 million metric tons of synthetic polymers are produced per year worldwide. Table 11.1 indicates the contributions of the various types of synthetic polymers to the total polymer production in the USA. World figures will show a similar trend.

Polymers of the largest class, the thermoplastics, become flexible solids above a particular temperature, and become rigid again upon cooling below this temperature. This flexible/rigid cycle can be repeated on reheating and cooling again. Thermoplastics, when flexible, can be molded into shapes that are preserved in the rigid state after cooling. Thermoplastic fibers can be drawn (pulled out) into strands with considerable tensile strength and durability. The non-fibers cannot undergo such processing. Thermoset resins (network copolymers) do not become flexible at all until the temperature becomes so high that thermal decomposition occurs. Synthetic rubbers, or elastomers,

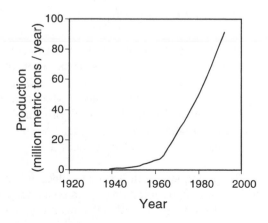

Figure 11.1 World production of plastic materials. Adapted from [1].

Table 11.1 Main components of total polymer production (1990) in USA [2]

Type	Fraction of total (%)	Main specific polymers (% of total)
Thermoplastics (non-fibers)	68	Polyethene (23%)
		Poly(vinyl chloride) (13%)
		Polypropene (11%)
		Polystyrene (8%)
Thermoplastics (fibers)	14	PET (5%)
		Nylon 6,6 (4%)
		Poly(alkenes) (2%)
Thermoset resins		Phenol–formaldehyde (5%)
		Urea-formaldehyde (2%)
Synthetic rubbers	10	SBR (2%)
	8	Poly(butadiene) (1%)

can be deformed quite severely by a small stress but regain their original shape on removal of the stress. It is evident that a small number of polymers dominate the production statistics, and of these polyethene (PE) is the largest volume polymer. This chapter is certainly not intended to cover the whole field of polymerization and wll focus on the production of PE.

11.2 POLYMERIZATION REACTIONS

11.2.1 Types of Polymerization

Polymerization can proceed according to two different mechanisms, referred to as chain-growth and step-growth polymerization. In chain-growth polymerization (also called addition polymerization) reaction occurs by successive addition of monomer molecules to the reactive end (e.g. a radical end) of a growing polymer chain.

The most important group of chain-growth polymerizations is polymerization of vinyl monomers such as ethene, propene, styrene, and vinyl chloride:

$$nCH_2=CHX \rightarrow -(-CH_2-CHX)_n- \tag{1}$$

in which X = H, CH_3, C_5H_6, or Cl, respectively.

An initiator or a catalyst is usually required to start the chain-growth reaction.

Step-growth polymerization involves the reaction between the functional groups (HO—, HOOC—, etc.) of any two molecules (either monomers or polymers). By repeated reaction long chains are gradually produced. Most commonly, the reactions are condensation reactions resulting in the elimination of a small molecule like water or methanol. Examples are the production of poly(ethene terephthalate) by reaction of ethene glycol with either terephthalic acid or dimethyl terephthalate (see Scheme 11.1).

Scheme 11.2 shows the principle reactions taking place when two different functional groups are present on the same molecule (indicated as AB, e.g. in the production of caprolactam), and when two different molecules each contain two identical functional groups (indicated as AA and BB, e.g. in production of PET).

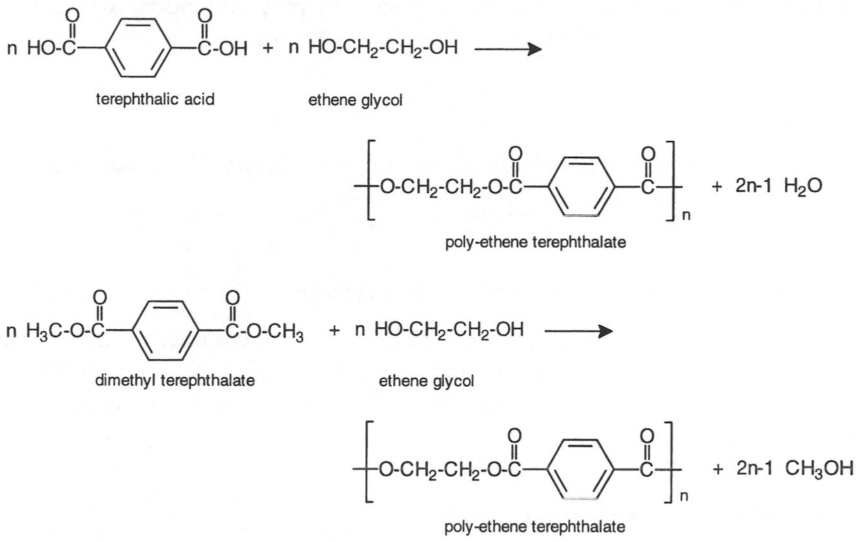

Scheme 11.1 Reactions for the production of polyethene terephthalate (PET).

In chain-growth polymerization, a high molecular weight product is produced right from the start, while the monomer quantity decreases slowly with time. In contrast, in step-growth polymerization there is a slow increase in average molecular weight of the product. The molecular weight is usually not as high as in chain-growth polymerization, and relatively small amounts of unreacted monomer are present after the start of the reaction.

Chain-growth polymerization generally is fast, irreversible, and moderately to

Polymerization of bifunctional monomer (A, B: two different functional groups):

$$
\begin{array}{lll}
AB & + \ AB & \rightarrow \quad ABAB \ (or \ (AB)_2) \\
AB & + \ (AB)_2 & \rightarrow \quad (AB)_3 \\
AB & + \ (AB)_3 & \rightarrow \quad (AB)_4 \\
(AB)_2 & + \ (AB)_2 & \rightarrow \quad (AB)_4 \\
\cdots \\
(AB)_r & + \ (AB)_s & \rightarrow \quad (AB)_{r+s}
\end{array}
$$

Polymerization of two monomers with different functional groups:

$$
\begin{array}{llll}
AA & + \ BB & \rightarrow & AABB \\
AABB & + \ AA & \rightarrow & AABBAA \\
AABBAA & + \ BB & \rightarrow & AABBAABB \\
AABB & + \ AABB & \rightarrow & AABBAABB \\
Etc.
\end{array}
$$

Scheme 11.2 Reactions in step-growth polymerization.

highly exothermic. On the other hand, step-growth polymerization is usually slow, equilibrium-limited, and isothermal to slightly exothermic.

QUESTION:

> *Explain the differences mentioned between chain- and step-growth polymerization.*

11.2.2 Mechanisms of Chain-Growth Polymerization

Chain-growth polymerization can be classified (in order of commercial importance) as radical, coordination, anionic, or cationic polymerization, depending on the type of initiation. The next two sections will briefly discuss radical and coordination polymerization, respectively.

11.2.2.1 Radical polymerization

Scheme 11.3 outlines a typical reaction scheme for radical polymerization. Most radical polymerizations need an initiator to produce the first radical and thus start the chain of addition reactions. The most common initiation reaction is the thermal decomposition of molecules containing weak bonds, e.g. peroxides ($-O-O-$) or azo compounds ($-N=N-$). The formed radicals then react with the monomers. Once initiated, a chain will grow by repeated additions of monomer molecules with simultaneous creation of a new radical site. This propagation is very fast, so very long polymer chains will form already in the earliest stage of the reaction.

Termination can occur by disproportionation or recombination. In the first case, the

Initiation: $R\text{-}R \quad \rightarrow 2\,R^\bullet \qquad k_i \approx 10^{-4} - 10^{-6}\,s^{-1}\,(300 - 350\,K)$
                $R^\bullet + M \rightarrow RM_1^\bullet$

Propagation: $RM_1^\bullet + M \rightarrow RM_2^\bullet \qquad k_p \approx 10^2 - 10^4\,m^3\,kmol^{-1}\,s^{-1}\,(300 - 350\,K)$
                $RM_2^\bullet + M \rightarrow RM_3^\bullet$
                Etc.
                $RM_{n-1}^\bullet + M \rightarrow RM_n^\bullet$

Termination
by disproportionation: $RM_m^\bullet + RM_n^\bullet \rightarrow RM_m^= + RM_n$

Termination
by recombination: $RM_m^\bullet + RM_n^\bullet \rightarrow RM_m\text{-}M_nR$

Transfer to solvent: $RM_n^\bullet + S \rightarrow RM_n + S^\bullet$

Transfer to monomer: $RM_n^\bullet + M \rightarrow RM_n + M^\bullet$

Scheme 11.3   Steps in radical polymerization; M_1, M_n, number of monomers in chain.

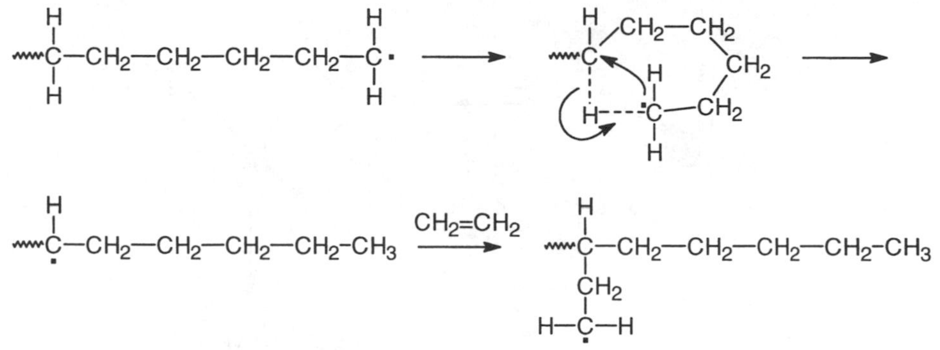

Figure 11.2 Backbiting in the synthesis of LDPE by radical polymerization.

final polymers on average have the same length as the growing chains. Termination occurs by transfer of a hydrogen atom from one of the radicals to the other, leading to unsaturation at one chain end. Recombination results in polymers with on average double the length of the growing chains.

Chain transfer is common in many radical polymerization processes. It involves the transfer of the radical end of a growing polymer chain to another species, for instance a monomer or a solvent. Chain transfer reduces the average polymer size and molecular weight. It is possible to add a special agent, a modifier, in order to control the average degree of polymerization [2].

In the production of low-density polyethene (LDPE), side chains are generated by internal chain transfer, in which the end of the chain abstracts a hydrogen atom from an internal —CH₂— group, a process termed backbiting. Figure 11.2 illustrates this process. A branch starts to grow from the internal carbon radical. In this branch, backbiting is also possible resulting in widely branched chains. Backbiting has a significant influence on the structure, and hence, the properties of the final polymer.

11.2.2.2 Coordination polymerization

In coordination polymerization, usually transition-metal catalysts are involved. Figure 11.3 indicates the main features of chain propagation in the coordination polymerization of ethene. A growing polymer chain is coordinatively bound to a metal atom that has another coordinative vacancy (partially empty d-orbitals). A new ethene molecule is inserted by the creation of bonds between one of its carbon atoms and the metal and between the other carbon atom and the innermost carbon atom of the existing chain. Branching will not occur through this mechanism since no radicals are involved; the active site of the growing chain is the carbon atom directly bonded to the metal. High-density polyethene (HDPE) is produced by this type of polymerization.

When higher 1-alkenes are added, the resultant polymer chain will have a few short branches. These are all of the same length since they are simply the rest of the 1-alkene molecule. Figure 11.4 shows the incorporation of 1-butene in a growing polyethene chain. The ethene copolymer known as linear-low-density polyethene (LLDPE) is produced by coordination polymerization of ethene and 1-butene or 1-hexene.

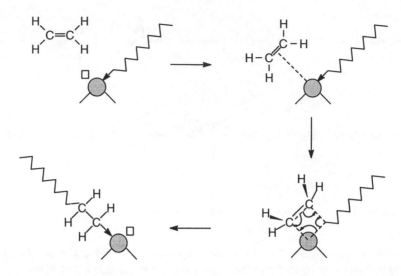

Figure 11.3   General mechanism of propagation in coordination polymerization of ethene; ○ represents metal atom and □ indicates vacancy for two electrons. Adapted from [2].

The most important catalysts for coordination polymerization are so-called Ziegler–Natta or Ziegler catalysts, and 'Phillips' catalysts, both discovered to be effective for alkene polymerizations in the 1950s. Ziegler catalysts combine transition-metal compounds such as titanium and vanadium with organometallic compounds. An important property of these catalysts is that they yield stereoregular polymers when higher alkenes are polymerized, e.g. polymerization of propene produces isotactic[1] polypropene with high selectivity. On Ziegler catalysts, the polymer chains grow to a very long length. Therefore, a chain-transfer agent is added to the system. Most commonly H_2 is used, which donates a hydrogen atom to terminate and detach the chain from the metal atom [2].

In catalyst systems developed at Phillips Petroleum Co., chromium oxide supported on silica is the most important constituent. Phillips catalysts are less active than

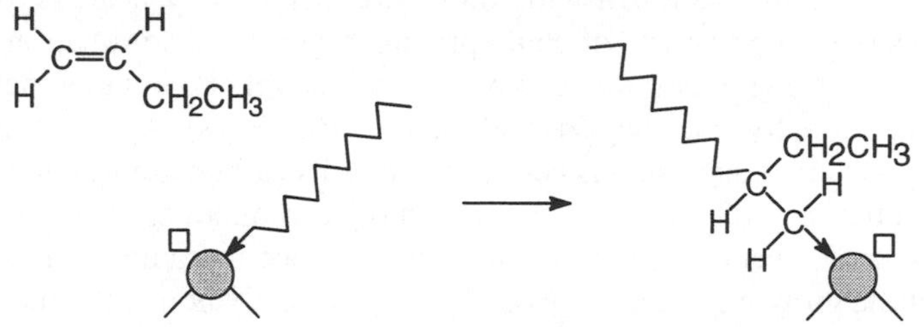

Figure 11.4   Incorporation of 1-butene in polyethene to produce a short branch.

---

[1] In isotactic polypropene all methyl groups point into the same direction when the backbone (the carbon atoms in the chain) is stretched. In contrast, in syndiotactic polypropene the methyl groups alternate along the chain, while the atactic polymer lacks regularity.

Ziegler catalysts, and therefore are limited to the production of polyethene. They produce a polymer with a large molecular-weight distribution, with chains varying in molecular weight from 10^3 to 10^6 g/mol.

It is not surprising that polymerization catalysis draws a lot of attention, and many new systems have been discovered. Even plastic materials that only melt at temperatures of over 770 K (!) can be produced [3].

11.3 POLYETHENES – BACKGROUND INFORMATION

11.3.1 Catalyst Development

The chance discovery (1935, ICI) that ethene can be polymerized, at very high pressure (over 2000 bar), to produce a high molecular weight semicrystalline material, was the starting point of the manufacture of polyethene. After the discovery of higher density polyethene materials, it was called low-density polyethene (LDPE). LDPE was first produced commercially in the United Kingdom in 1938.

In the 1950s, Ziegler and Natta discovered catalysts capable of polymerizing ethene at lower pressures and temperatures than those used in the production of LDPE. These catalysts contain titanium halides and alkylaluminium compounds. The polymer formed by this new process had a higher density (a property that could easily be measured) than LDPE, and thus was called high-density polyethene (HDPE). At about the same time, researchers at Phillips Petroleum Co. discovered that catalysts of chromium oxide on silica were also capable of producing HDPE at moderate temperatures and pressures. Commercial production of HDPE started in the late 1950s, using both types of catalyst systems. HDPE produced with these catalysts is linear without branching.

Since the introduction of the original Ziegler catalysts, high-activity catalysts based on titanium supported on magnesium dichloride and various other supports (among which polyethene) have been developed. Similarly, numerous modifications of the Phillips catalyst have been made by the introduction of various additives such as titanium alkoxides. These developments in catalyst composition have made possible the control of physical properties and molecular weight.

The addition of 1-alkenes during catalytic polymerization of ethene results in polymers with a density similar to LDPE or lower. A small degree of branching of these polymers is deliberately introduced and controlled by the type of comonomer added. Commercial production of such polymers started in 1968 by Phillips Petroleum Co. The most important of these catalytically produced ethene copolymers is linear low-density polyethene (LLDPE).

A new family of catalyst was discovered in the 1970s. These catalysts (Kaminsky catalysts) contain a metallocene complex, usually based on zirconia and methylaluminoxane (see Figure 11.5) [4]. Metallocene catalysts enabled the synthesis of ethene copolymers with highly uniform branching.

11.3.2 Classification and Properties

In the previous section, we have already encountered the three most important polyethene classes, namely LDPE, HDPE, and LLDPE. In 1994 the polyethene world production exceeded 40 million metric tons, divided mainly amongst LDPE,

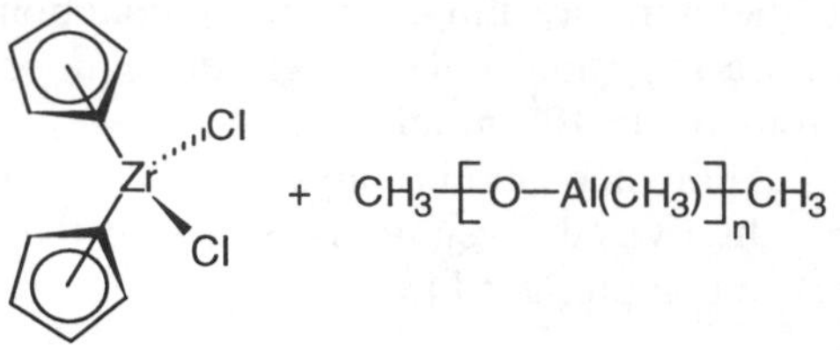

Figure 11.5   Metallocene catalyst (Kaminsky type).

LLDPE and MDPE, and HDPE (17, 13, and 14 million metric tons per year, respectively). Table 11.2 gives a commercial classification of the various polyethenes.

The polymer density (or crystallinity) depends on the degree of branching as well as on the molecular weight. Linear unbranched polymers have a more densely packed molecular structure, and hence contain larger portions of crystalline (ordered) material than branched polymers. This is due to the interference of the side branches with the alignment of the polymer chains. Figure 11.6 summarizes the chain structure of the main polyethenes. HDPE has no branches, LLDPE has short and uniform branches, while LDPE has branches of more random length. The difference is a result of the different operating conditions, catalysts, and comonomers employed in the production of these polymers.

In the case of LDPE production, which takes place by radical reactions at extremely high pressure and relatively high temperature (see Section 11.4.3), polymer density is controlled by the operating temperature. A higher temperature results in more side reactions causing branching, and hence a lower density.

In the low-pressure catalytic processes to produce HDPE and LLDPE (see Section 11.4.4), the density is controlled by the type and amount of comonomer added to the polymerization reactor. Common comonomers are 1-alkenes ranging from 1-butene to 1-octene. The amount of comonomer present varies from 0 mol% in the highest density HDPEs to about 3 mol% in LLDPE. In VLDPE over 4 mol% comonomers of different types are present. Due to the presence of side-chains of variable length, this polymer has an even lower density than LDPE.

In general, polymer density decreases with increasing molecular weight of the polymer chains. The reason is that with longer chains there is more entanglement, which makes alignment of the chains difficult.

With increasing density, certain polymer properties increase, e.g. yield strength,

Table 11.2   Commercial classification and properties of polyethenes

Acronym	Designation	Density (kg/m^3)
HDPE	High-density polyethene	>940
UHMWPE[a]	Ultrahigh-molecular-weight polyethene	930–935
MDPE	Medium-density polyethene	926–940
LLDPE	Linear low-density polyethene	915–925
LDPE	Low-density polyethene	910–940
VLDPE	Very-low-density polyethene	880–915

[a] Linear polymer with molecular weight of over $3 \cdot 10^6$ g/mol.

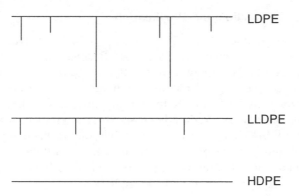

Figure 11.6 Simple representation of the backbone structures of LDPE, LLDPE, and HDPE.

stiffness, and impermeability to gases and liquids, while others decrease, e.g. transparency and low temperature brittleness resistance. Therefore, the density to a large extent determines the possible applications of a specific polymer.

11.3.3 Applications

LDPE has a wide variety of applications of which those requiring processing into thin film take up the largest segment (60–70%). Polyethene film applications include food and nonfood packaging, carry-out bags, and trash can liners. Other applications of LDPE include extrusion coating (e.g. for packaging of milk), injection molding for the production of toys, and sheathing for wires and cables [4].

LLDPE is replacing LDPE in film applications, which is now the largest outlet for LLDPE (66–70%). Injection molding and sheathing for wires and cables are other important applications [5]. LLDPE is less likely to replace applications requiring extrusion, because of its lower flexibility and lubricity. Neither is LLDPE used in applications requiring high clarity.

The major application of HDPE is in blow molding (40%) to produce amongst others milk bottles, containers, drums, fuel tanks for automobiles, toys, and housewares. Other large outlets are films and sheet (e.g. for wrapping, carrier bags), extruded pipe (e.g. water, gas, irrigation, cable insulation), injection molding (e.g. crates, containers, toys) [5].

UHMWPE is used in special applications requiring high abrasion resistance such as bearings, sprockets, and gaskets.

11.4 PROCESSES FOR THE PRODUCTION OF POLYETHENES

Before discussing polymerization processes for the production of LDPE in Section 11.4.3 and for the production of HDPE and LLDPE in Section 11.4.4, brief attention will be paid to monomer production and purification.

11.4.1 Monomer Production and Purification

Most ethene for the production of polyethene is produced by cracking of ethane or naphtha in the presence of steam (see Chapter 4). Besides ethene, the cracked gas also contains ethane, and other compounds in smaller quantities (e.g. methane, hydrogen, ethyne). Table 11.3 gives typical maximum amounts of these and other impurities required for polymer-grade ethene. Note that the purity is specified on a ppm level.

Downstream processing of crude ethene starts with removal of sour gases (CO_2, H_2S) and water are removed by absorption and drying, respectively. Then ethene is further separated and purified by cryogenic distillation. Ethyne is one of the most harmful compounds, and its presence must be limited to below 1 ppm [6]. This is achieved by selective hydrogenation of the triple bond.

11.4.2 Polymerization – Exothermicity

The main difference between polymerization processes and processes for the manufacture of smaller molecules lies in the fact that polymer systems are highly viscous. In chain-growth polymerization an extra problem is the often high exothermicity of the reaction (see Table 11.4). Therefore, adequate temperature control is essential for an economical and safe process.

As can be seen from Table 11.4, the heat of reaction for polymerization of ethene is one of the largest, with an accompanying excessive adiabatic temperature rise (ΔT_{ad} = temperature rise if no cooling would be applied = 1610 K!!). In addition, at temperatures above about 570 K ethene decomposition into carbon, hydrogen and methane poses a serious threat. This reaction is highly exothermic ($\Delta H^0_{298} = -120$ kJ/mol), and once started is difficult to control, leading to a thermal runaway. Hence, very special attention has to be paid to temperature control in the polymerization of ethene, both to ensure polymer quality and safe operation of the process.

QUESTION:

> *According to a rule of thumb hydrogenation reactions are exothermic. Does this agree with the exothermicity of ethene decomposition?*

Table 11.3 Specification for polymer-grade ethene

Component	Value
Ethene, minimum mol%	99.95
Methane + ethane	Balance
Other impurities, mol ppm max	
Ethyne	1.0
Oxygen	1.0
Carbon monoxide	1.0
Carbon dioxide	1.0
Water	2.0
Sulfur	2.0
Hydrogen	5.0
Propene	10.0
C_4^+	10.0

Table 11.4 Typical enthalpies of chain-growth polymerization [7]

Polymer	ΔH^0_{298} (kJ/mol monomer)	$\Delta T_{\text{adiabatic}}$ (K)[a]
Polyethene	− 95.0	1610
PVC	− 95.8	730
Polystyrene	− 69.9	320
Polymethyl methacrylate	− 56.5	270

[a] Complete monomer conversion and heat capacity of 2 kJ/kg K assumed (and no dilution).

11.4.3 Production of LDPE

The formation of LDPE proceeds through a radical mechanism (see Scheme 11.2) without a catalyst. Therefore, a relatively high temperature and very high pressure are required, typically 470–570 K and 1500–3000 bar. At the high pressures used (above the critical pressure of ethene), ethene behaves like a liquid, and is a solvent for the polymer. LDPE is produced in either a stirred autoclave (CSTR) or a tubular reactor (PFR).

11.4.3.1 Reactors and process

The stirred autoclave (Figure 11.7) consists of a cylindrical vessel with a length-to-diameter ratio of 4:1 to 18:1. Reactors for the larger plants have volumes of about 1 m³, resulting in a residence time of 30–60 s. The reactor wall thickness, typically

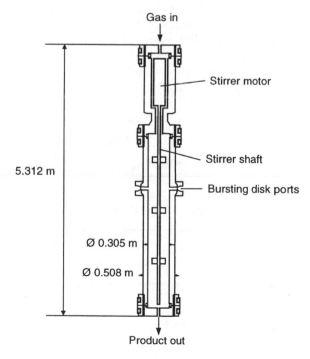

Figure 11.7   General arrangement of 0.25 m³ autoclave. After [8,9].

0.1 m, which is required to contain the high pressure of up to 2100 bar, severely limits the removal of heat.

In fact, the reactor can be considered to operate adiabatically; the fresh cold ethene entering the reactor removes the heat of reaction. Accordingly, the conversion must be limited (to about 20%) in order to limit the temperature increase. Another reason to limit the conversion is the increased viscosity with higher polymer content, which makes stirring difficult and can lead to local hot spots. Most modern reactors have two or more zones with increasing temperatures, the first zone typically operating at 490 K and the final at 570 K. Initiators can be injected at several points.

Bursting disks are mounted directly into the reactor walls to provide unrestricted passage of the reactor contents in the event of a pressure rise due to decomposition of ethene [8].

The tubular reactor is essentially a long double-pipe heat exchanger. Industrial reactors generally have a length of between 200 and 1000 m, with inside diameters of 2.5–7 cm. These reactors are built by joining straight lengths of 10–20 m in series, connected by U-bends (see Figure 11.8). The tubes are surrounded with jackets in which a heat transfer fluid circulates. In the first part of the reactor ethene is heated to a temperature between 370 and 470 K. The heat of polymerization raises the temperature to 520–570 K. In the final part of the reactor the temperature of the reaction mixture decreases. Optionally, initiator (oxygen or a peroxide) can be added at several points along the reactor. This increases the conversion but also increases the reactor length. Typically, the ethene conversion ranges from 15 to 22% with single initiator injection, and up to 35% with multiple injection.

In order to achieve the high velocity required for effective heat transfer to the cooling jacket, the reactor pressure is controlled by a cycle valve that opens periodically to reduce the pressure from about 3000 to 2000 bar. An additional advantage of this type of pressure control is that the periodically increased flow velocity through the reactor reduces contamination of the walls with polymer. Pressure cycling also has disadvantages, because the tubes have to endure both large pressure changes and temperature gradients through the walls. Especially fatigue plays an important role.

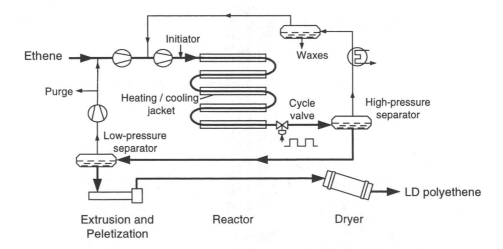

Figure 11.8   Production of LDPE in a tubular reactor (shown schematically).

QUESTION:

> *What problems could result from contamination of the reactor walls with polymer material?*

Due to the low ethene conversion, a recycle system is required, which is similar for both the autoclave and the tubular reactor. The high-pressure separator separates most of the polymer from the unreacted monomer. The separator overhead stream, containing ethene and some low molecular weight polymer (waxes) is cooled, after which the waxes are removed. The resulting ethene stream is recycled to the reactor.

The polymer stream leaving the high-pressure separator is fed to a low-pressure separator (0.02 bar), where most of the remaining ethene monomer is removed. This ethene is also recycled to the reactor. Some ethene is purged at this point to prevent accumulation of feed impurities. The molten polymer from the low-pressure separator is fed to an extruder, which forces the product through a die with multiple holes. Polymer strands are cut under water into pellets of about 3 mm diameter and then dried in a centrifugal drier.

The physical properties of the polymer can be controlled by the temperature, the type of initiator and its concentration. In addition, the polymer properties can be altered by adding comonomers, such as propene, isobutene, and acrylic acid, to the feed.

QUESTIONS:

> *What would be the source of ethene?*
> *What kind of impurities would be present in the ethene feed?*
> *Which would be the most important contributors to the polyethene production costs?*

11.4.3.2 Ethene compressors – high-pressure technology

High pressure was first applied at the end of the 19th century for the liquefaction of the so-called permanent gases such as oxygen and nitrogen. The use of high pressures in the chemical industry started at about the same time, and was triggered by the discovery of the dyestuff mauveine and the development of the dye industry. Some dye intermediates were produced in batch autoclaves at pressures of 50–80 bar. In the beginning of the 20st century, ammonia and methanol (see Chapter 6) were two of the first chemicals produced in continuous high-pressure plants. Nowadays, ammonia synthesis takes place at pressures ranging from 150 to 350 bar, although an early process operated at 1000 bar.

Ammonia and methanol plants currently use centrifugal compressors. These have several advantages over the earlier used reciprocating compressors, e.g. low investment and maintenance costs, high reliability, low space requirement, and high efficiency at high flow rates. However, for the production of LDPE the maximum attainable discharge pressure of centrifugal compressors (ca. 1000 bar at the lowest flow rates) is not nearly enough (see Figure 11.9).

Higher pressures can be achieved with reciprocating compressors than with centrifugal compressors. However, the early reciprocating compressors had too low throughputs for a high-pressure polyethene process to be economical. Improved techniques in high-pressure engineering enabled the increase in throughput of reciprocat-

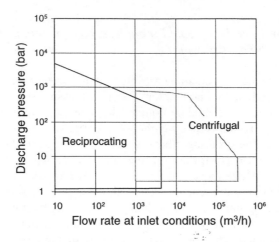

Figure 11.9 Operating ranges of reciprocating and centrifugal compressors. Adapted from [10].

ing compressors from about 5 t/h at 1250 bar in the early 1950s to 120 t/h at 3500 bar nowadays.

In LDPE plants a primary, centrifugal, compressor is used to raise the pressure of ethene to about 300 bar and a secondary, reciprocating, compressor, often referred to as a hypercompressor, is used to increase the pressure to 1500–3000 bar. The development of reciprocating compressors for these high operating pressures and high throughputs was not an easy task. Some problems that had to be solved are [11]:

- ethene leakage past the plunger;
- breakage of the plunger due to misalignment;
- fatigue failure of high pressure components;
- formation of low-molecular-weight polymer in the compressor, leading to obstruction of the gas ducts inside the compressor cylinder;
- decomposition of ethene into carbon, hydrogen, and methane due to overheating, for instance in dead zones where the gas is stagnant.

Ethene leakage not only means a loss of ethene but also represents an explosion danger. Plunger misalignment is the most common cause of failure of secondary compressors for LDPE. Plunger breakage may lead to serious fires due to large escapes of ethene. Fatigue failures in frame or cylinder parts of the compressor have led to serious damage [11]. In cases where fragments of material fell inside the cylinder, thus blocking the delivery duct, excessive overpressure led to explosion of the compressor. Obstruction of gas ducts by low-molecular-weight polymer and decomposition of ethene can have a similar effect.

11.4.3.3 Safety of high-pressure polymerization reactors

High-pressure polyethene reactors operate on the edge of stability, sometimes leading to sudden runaways. A runaway reaction is initiated by local hot spots developing in the reactor. In these hotter reactor parts the temperature becomes higher than in the surrounding fluid, and decomposition of ethene to carbon, methane, and hydrogen

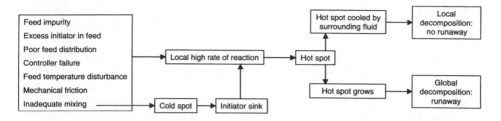

Figure 11.10 Possible causes of runaway reactions (autoclave process). After [12].

occurs. Figure 11.10 shows some of the incidents leading to hot spots and decomposition reactions in an autoclave reactor.

A runaway reaction occurs less frequently in tubular processes than in autoclave processes due to a smaller reactor content and larger cooling surface area. However, fouling of the reactor walls by polymer deposits in a tubular reactor is an additional possible cause of runaways. Since the reactor operates at its stability limit, only a small decrease in heat transfer, and hence, a small rise in (local) temperature, could trigger decomposition reactions.

The decomposition reactions are highly exothermic. Especially in the adiabatic autoclave reactors, once initiated these decompositions cause a runaway in a matter of seconds [12]. As an example, Figure 11.11 shows the response of an autoclave reactor to a disturbance in the amount of ethyne in the feed (simulation results [12]). This is not an academic exercise; small amounts of ethyne are present in the ethene feed; ethyne can decompose into free radicals and induce runaway reactions. The figure explains why the specification with respect to the maximum ethyne content is so severe (see also Table 11.3).

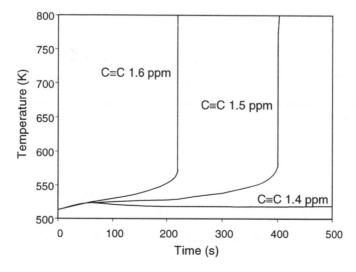

Figure 11.11 Dynamic response of autoclave temperature to different ethyne levels in the feed. Adapted with permission from [12].

QUESTION:

Why is it to be expected that ethene is contaminated with ethyne?

Similar behavior resulted from a disturbance of over 5 K in the reactor feed temperature (setpoint: 393 K), and a disturbance in the initiator concentration (from 7.3 to 8.0 ppm) [12].

In both autoclave reactors and tubular reactors the danger of a runaway reaction is usually held in check by adjusting the initiator concentration in the ethene feed. Even very small temperature changes (as measured at various points along the reactor) should be counteracted immediately. Therefore, a high temperature alarm for a polymerization reactor is set at only a few degrees above the normal operating temperature, whereas in other reactor types a margin of 10 to 20 K is common [13].

Decomposition reactions not only lead to a temperature excursion but also to a pressure peak. To protect the autoclave from excessive overpressure, without having to resort to a large vent area, it is fitted with pressure-relief valves (or bursting disks) that are actuated at an overpressure as small as practical. The hot decomposition products released (at sonic velocity) would be subject to severe reaction with air (decarbonization, cracking), and therefore are rapidly quenched before venting to the atmosphere [9].

In tubular reactors, the problem of excessive overpressure is not as serious as in autoclave reactors, because these reactors contain less ethene due to the smaller volume, and hence a possible pressure rise is slower. The first measure is fitting of one or more so-called dump valves [9] to the reactor. These dump valves are controlled externally. In addition, bursting disks, which are set to burst at a pressure much exceeding the normal operating pressure, are usually provided at several points along the reactor [9].

11.4.4 Production of HDPE and LLDPE

Although LDPE is a very versatile polymer, the technology for its production has now become obsolete due to the very high investment and operation costs (see Figure 11.13, Section 11.4.5) and the introduction of low-pressure processes for the production of linear low-density polyethene (LLDPE). LLDPE can be used in many of the traditional LDPE outlets. Low-pressure processes can also produce HDPE.

11.4.4.1 Introduction

Many of the common catalytic polymerization processes are suitable for the production of both HDPE and LLDPE. Such processes include bulk polymerization (polymer dissolves in monomer), solution polymerization (polymer and monomer dissolve in a hydrocarbon solvent), slurry polymerization (catalyst-polymer particles suspended in a hydrocarbon diluent), and fluidized-bed polymerization (catalyst-polymer particles fluidized in gaseous monomer).

LLDPE can also be produced in high-pressure processes, such as described in Section 11.4.3, in the presence of a Ziegler-type catalyst. Although this route is not seen as suitable for a new investment, several manufacturers have modified existing LDPE plants for the production of LLDPE.

This section focuses on the fluidized-bed polymerization process (also referred to as gas-phase polymerization) because this is the most flexible process; it is capable of producing both HDPE and LLDPE, and can use a wide range of solid and supported catalysts. It also is the most economical process (see Figure 11.13), because it does not require a diluent or solvent, and hence no equipment or energy for the recovery and recycle of diluent or solvent. Many fluidized-bed polymerization plants have been built as dual-purpose plants ('swing plants') with the ability to produce HDPE or LLDPE according to demand [8].

11.4.4.2 Fluidized-bed polymerization process

Figure 11.12 shows a flow sheet of a fluidized-bed process. A wide variety of heterogeneous catalysts can be used (e.g. modified chromium catalyst supported on silica and heterogeneous Ziegler catalysts). Recently, processes for the production of LLDPE have started to employ metallocene catalysts (see Figure 11.5), which enables the production of ethene copolymers of high compositional uniformity.

Ethene, hydrogen, and comonomer (for the production of LLDPE, mainly 1-butene and 1-hexene) are fed to the reactor through a grid, which provides an even distribution of the gas. The recirculating ethene gas fluidizes the catalyst (in the shape of spherical particles of ca. 50 μm) on which the polymer grows, and enables efficient mass and heat transfer. The characteristic shape of the reactor (ca. 25 m high, length-to-diameter ratio ca. 7 [14]), with a cylindrical reaction section and a larger diameter section where the gas velocity is reduced, allows entrained particles to fall back into the bed. Remaining polymer particles are separated from gaseous monomer in a cyclone and returned to the reactor.

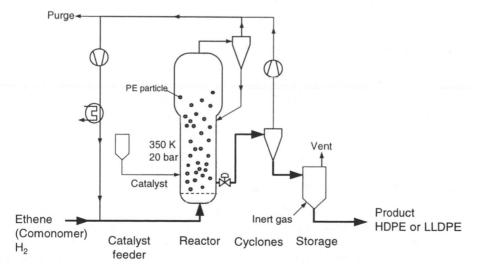

Figure 11.12   Union Carbide Unipol process for the production of HDPE and LLDPE.

QUESTIONS:

> *What is the purpose of adding hydrogen?*
> *Why is a purge stream required?*

The residence time of the PE-catalyst particles in the reactor ranges from 3 to 5 h, during which they reach an average size of about 500 μm [14]. The powder is taken off at regular intervals near the bottom of the reactor through valves. The polymer contains an appreciable amount of ethene, which is separated for the largest part in a cyclone operating at near atmospheric pressure and recycled to the reactor. The remaining ethene is purged from the polymer with nitrogen and stored.

Because of the high activity of current polymerization catalysts, over 100 kg of polymer is produced per gram of metal (Cr, Ti, etc., depending on the catalyst used). Therefore, the catalyst can remain in the product and no residual catalyst removal unit is required.

Temperature control is very important to keep the polymer particles below their melting point (398 and 373 K for HDPE and LLDPE, respectively [14]), and to prevent a thermal runaway reaction. When using cooling water from a cooling tower, the available temperature difference for cooling the gas should be at least 40–50 K. From this value it can be calculated that the single-pass conversion of ethene is limited to about 2% (somewhat higher if higher alkene comonomers are used, as in LLDPE production).

QUESTIONS:

> *What would be the consequences of melting of the polymer particles?*
> *How would the temperature be regulated?*
> *Verify the above conversion result assuming a constant heat capacity of ethene of 40 J/mol/K and a heat of reaction of 100 kJ/mol ethene. Also calculate the recirculation ratio (mol reactor out/mol feed ethene in). Neglect the purge stream in both calculations.*

In the case that a thermal runaway is detected, CO_2 can be injected into the reactor in order to poison the catalyst and stop the reaction.

Although the fluidized-bed polymerization process is often called a gas-phase process this is not actually the case. The catalyst is located either within or on the surface of existing polymer particles, and an appreciable amount of monomer is dissolved in the polymer. Hence, the actual polymerization takes place within the polymer particle, to which a continuous supply of monomer flows from the gas phase [8].

11.4.5 Economics of Polyethene Production Processes

Figure 11.13 compares the costs for some LDPE, HDPE and LLDPE processes. For more technical information on the various processes consult references [4,5,8].

Although the values given in Figure 11.13 are not of recent date, they are a good indication of the similarities and differences in the operating and investment costs of

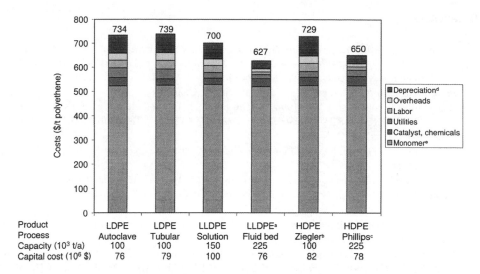

Figure 11.13 Production costs for polyethene in \$/t (1990, US Gulf Coast). Data from [8]. [a]Cost build-up for HDPE production in fluidized bed is similar (total costs: 632 \$/t). [b]Early slurry-phase process using Ziegler catalyst in CSTR. [c]Slurry-phase process using Phillips catalyst in loop reactor. [d]At 10%/year. [e]Comonomer costs for LLDPE production assumed to be the same as ethene (not true).

the various processes. Also, some differences occur as a result of the economy of scale due to the varying available sizes (basis: largest stream size available for licensing).

Clearly, the monomer costs (mainly ethene but also comonomers in the case of LLDPE) contribute most to the total cost (approximately 70%), followed by capital and catalyst and chemicals costs. The monomer and catalyst and chemicals costs per metric ton of polymer produced are similar for all processes. On the other hand, large differences are observed in capital costs. For instance, LDPE processes are not competitive with LLDPE processes (even at the same capacity) due to the large investment costs involved in high-pressure equipment. Furthermore, they have large energy requirements for compression. In spite of the high costs of the LDPE processes, they are not entirely obsolete, because for some applications LDPE cannot be replaced by LLDPE, yet. However, metallocene-based LLDPE may duplicate LDPE's properties in the future.

Another example of the contribution of the capital cost factor is the enormous difference between the solution process and the fluidized-bed process for LLDPE production. Even at much lower capacity the capital costs are much higher for the solution process. It must be noted, however, that the costs for the solution process are based on the DuPont process (see [8]), which at that time included a catalyst residue removal stage. Therefore, this process is probably not the most economical solution process available. A similar observation can be made when comparing the two slurry processes for HDPE production. The data for the Ziegler process are based on an early process, which also included a facility for the removal of catalyst residue. Since the late 1960s it has been possible to eliminate this step [8], resulting in appreciably lower costs.

References

1 Brydson JA (1995) *Plastics Materials*, 4th ed., Buttersworth-Heinemann, Oxford, Chapter 1.
2 Campbell IM (1994) *Introduction to Synthetic Polymers*, Oxford University Press, Oxford.
3 Van Santen RA, Van Leeuwen PWNM, Moulijn JA and Averill BA (eds.) (1999) *Catalysis: an Integrated Approach*, 2nd ed., Elsevier, Amsterdam, Chapter 6.
4 Pebsworth LW and Kissin YV (1996) 'Olefin polymers (polyethylene)' in: Kroschwitz JI and Howe-Grant M (eds.) *Kirk-Othmer Encyclopedia of Chemical Technology*, vol. 17, 4th ed., Wiley, New York, pp. 702–784.
5 Wells GM (1991) *Handbook of Petrochemicals and Processes*, Gower Publishing Company Ltd., Hampshire.
6 Sundaram KM, Shreehan MM and Olszewski EF (1994) 'Ethylene' in: Kroschwitz JI and Howe-Grant M (eds.) *Kirk-Othmer Encyclopedia of Chemical Technology*, vol. 9, 4th ed., Wiley, New York, pp. 877–915.
7 Nauman E (1994) 'Polymerization reactor design' in: McGreavy C (ed.) *Polymer Reactor Engineering*, Blackie, London, pp. 125–147.
8 Whiteley KS, Heggs TG, Koch H, Mawer RL and Immel W (1992) 'Polyolefins' in: Gerhartz W, et al. (eds.) *Ullmann's Encyclopedia of Industrial Chemistry*, vol. A21, 5th ed., VCH, Weinheim, pp. 487–530.
9 Bett KE (1995) 'High pressure technology' in: Kroschwitz JI and Howe-Grant M (eds.) *Kirk-Othmer Encyclopedia of Chemical Technology*, vol. 13, 4th ed., Wiley, New York, pp. 169–236.
10 Sinnott RK (1996) *Coulson and Richardson's Chemical Engineering*, vol. 6, 2nd ed., Butterworth-Heinemann, Oxford, p. 432.
11 Traversari A and Beni P (1974) 'Design of a safe secondary compressor' in: *Safety in Polyethylene Plants*, CEP Tech Manual, pp. 8–11.
12 Zhang SX, Read NK and Ray HW (1996) 'Runaway phenomena in low-density polyethylene autoclave reactors' *AIChE Journal* 42(10) 2911–2925.
13 Prugh RW (1988) 'Safety' in: Kroschwitz JI, et al. (eds.) *Encyclopedia of Polymer Science and Engineering*, vol. 14, Wiley, New York, pp. 805–827.
14 Brockmeier NF (1987) 'Gas-phase polymerization' in: Kroschwitz JI, et al. (eds.) *Encyclopedia of Polymer Science and Engineering*, vol. 7, Wiley, New York, pp. 480–488.

Chapter 12

Biotechnology ☆

12.1 INTRODUCTION

Biotechnology is both an old and a novel discipline. It is old in the sense that it has been applied in traditional processes like the production of beer and wine since before 6000 BC. However, it is fair to state that it was more an art than a scientific discipline then. The last decades this has changed. Biotechnology encompasses an array of sub-disciplines such as microbiology, biochemistry, cell biology, and genetics. It has become possible to describe life processes at cellular and molecular level.

Just like 'normal' chemical engineering, bioengineering includes kinetics, transport phenomena, reactor design, and unit operations: it is expected that chemical engineering will contribute enormously to the field. Often it is stated that biotechnology has a large potential for the future and that this will remain to be so. However, already today the life sciences affect over 30% of the global economic turnover, mainly in the sectors of environment, healthcare, food, energy, and agriculture and forestry. It is expected that the economic impact will grow as biotechnology provides new ways of influencing raw material processing. Biotechnology has enabled breakthroughs in the manufacture of new pharmaceutical drugs and the development of gene therapies for treatment of previously incurable diseases. In the production of chemicals it does not yet play a large role except in solving environmental problems. The most important example is water treatment. In water purification for civil purposes biotechnology largely contributes to the quality of life.

It is interesting to observe that the character of the biotechnological industrial sector is changing from a problem area oriented industry to a more generic one, based on enabling technologies such as genetic modification by recombinant DNA, and cell fusion techniques. The chemical industry has undergone the same development. In the past the process industry was organized along product lines, e.g. sugar industry, oil refining etc. Later it was discovered that generic sub-disciplines can be distinguished, viz. unit operations, chemical reactors, shaping techniques, etc. Subsequently, underlying disciplines such as transport phenomena, interface chemistry, and chemical reaction engineering were emphasized. More recently, integration has become more central, as reflected by the acknowledgement of areas as chemical reaction engineering and process integration. The same holds for biotechnology, where application areas have developed rather independently from each other, e.g. beer production, water purification.

☆ With acknowledgment to Prof. Dr. ir. J.J. Heijnen for his valuable comments.

Biotechnological industries will be based largely on renewable and recyclable materials and so will be able to adapt to the needs of a society in which energy is becoming increasingly expensive and scarce. Biotechnology can play a major role in the reduction of the greenhouse effect and, in more general terms, in the realization of a sustainable society. However, although biotechnology is generally considered clean technology, a critical note is in place; waste produced during the production and destruction of biocatalysts presents a considerable environmental problem.

Typical products of biotechnological processes are:

- cell biomass: yeast, single-cell protein. (Section 12.3);
- metabolic products of the cells (Sections 12.4 and 12.5):
  – anaerobic: alcohols, organic acids, hydrogen, carbon dioxide. (Sections 12.4.1, 12.4.2 and 12.5);
  – aerobic: citrate, glutamate, lactate, antibiotics, hydrocarbons, polysaccharides, (Section 12.5);
- products of reactions catalyzed by enzymes (Section 12.6); virtually all types of chemical reactions can be catalyzed by enzymes.

In general – and biotechnology is no exception – the cost price of a product decreases when the production rate increases due to development of the market with time. Figure 12.1 shows the historic development of the cost price of penicillin. It is expected that such a trend will be seen for many products that are still very expensive today.

Figure 12.2 shows an overview of some selected products and their prices. As one would expect, a distinct relation exists between cost price and production capacity. The prices differ enormously (note the logarithmic scale). For comparison crude oil and a number of petrochemical products (ethene, propene, and benzene) are included. The differences reflect the fact that crude oil can be mined, whereas most substances in the figure are products that are synthesized through expensive processes. Therefore, it is not surprising that crude oil is among the cheapest materials.

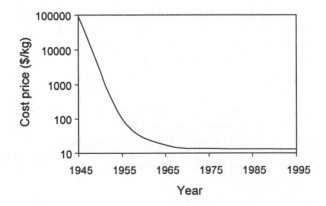

Figure 12.1   Historical development of the cost price of penicillin.

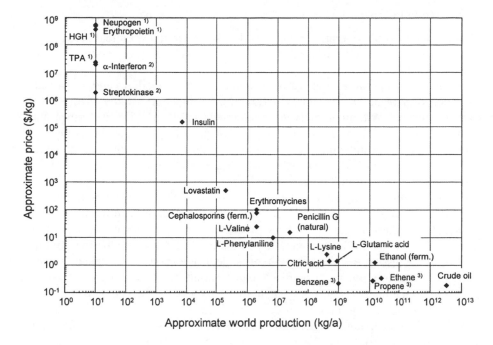

Figure 12.2 Overview of biotechnology products and their production capacities and prices; Most realistic figures for 2000; [1]1992 data; [2]estimated production costs for 1995, market price may be up to 100 times higher; [3]bulk petrochemicals. Data from [1].

12.2 CONVERSION PROCESSES

12.2.1 Introduction

In order to grow or to produce a metabolic product, a microorganism needs a source of carbon (the substrate), energy, nitrogen, minerals, trace elements, and frequently, vitamins. The substrate can consist of pure C-containing species such as polysaccharides, hydrocarbons, alcohols, and carbon dioxide or it can be a more complex material like molasses, cellulose waste liquors, pharmaceutical media, etc. Sources of nitrogen are, for instance, ammonia, urea, and amino acids. Minerals are organic salts such as phosphates, sulfates, and chlorides, while the most important trace elements are K, Na, Mg, Ca, Fe, Co, and Zn. In addition, aerobic organisms require oxygen for growth.

Similar to the statement that a catalyst particle is a reactor, the microorganism can be viewed as a reactor. Bio-transformations can be carried out by whole cells (yeast, plant or animal cells) or by part of the cell, in particular isolated enzymes; enzymes are biocatalysts. Although up to now their characteristics are mainly dictated by the microorganism itself, they can be modified by scientists. This is termed genetic engineering (recombinant DNA technique), and, in principle, offers unlimited opportunities to create new combinations of genes that do not exist under natural conditions. The technique also applies to plants and animals but then is referred to as pharming.

12.2.2 Mode of Operation

Bioreactors are often operated in batch or fed-batch (in the chemical industry usually referred to as semi-batch) mode with respect to the substrate, but continuous systems are also used.

In pure batch operation, the reactor is loaded with the medium and inoculated with the microorganism. During fermentation the components of the medium (carbon source, nutrients, vitamins, etc.) are consumed, while the biomass grows and/or a product is formed. One of the main disadvantages of batch reactors is the fact that loading, sterilizing, unloading, and cleaning of the reactor results in non-productive time. In this respect, the situation is completely analogous to fine chemistry and a continuous reactor would be more desirable.

During continuous operation, the medium is fed to the reactor continuously, while converted medium, loaded with biomass and product is continuously removed. Although continuous operation offers several advantages over batch operation, it is not used frequently in the fermentation industry. Often, the production volumes are too small to justify the construction of a dedicated continuous plant, while only sufficiently stable strains of organisms can be used due to the higher risk of mutations and contaminations [2,3]. Some biotechnological products that are or have been produced using a continuous process are high fructose corn syrups, single-cell protein (SCP), beer, and yogurt, all relatively large volume products [4]. Some of the continuous processes have been stopped, however, for economic reasons (SCP) or a combination of reasons (beer) [5,6]. In the case of beer, the fact that it was not possible to carry out downstream processing continuously was one of the reasons.

Although the continuous process has not conquered the biotechnological market, it should not be concluded that nothing has changed. Indeed, the classical batch system has been replaced to a large extend, not with continuous but with fed-batch operation. Fed-batch operation is a combination of batch and continuous operation. In fed-batch operation, after the start-up of the batch fermentation, substrate and possibly other components are fed continuously or step-wise. Fed-batch operation offers the possibility of better process control. For example, in the production of bakers' yeast the sugar substrate must be added gradually to assure the continuous growth of biomass and to prevent the formation of alcohol, which will occur at too high a sugar concentration (see Section 12.3).

The fed-batch system has relatively recently been developed further. For example, in repeated-fed-batch systems periodic withdrawal of 10–60% of the medium volume is applied. Another more sophisticated fed-batch strategy is the cell-retention system: the cells are retained in the reactor, while liquid that contains products or compounds that are toxic to the cells is continuously removed [4,6]. This is achieved by recycling part of the reactor contents through a membrane, which separates the cells from the liquid. Many analogous modes of operation can be thought of. In fact, they are fully analogous to those described in previous chapters on chemicals production.

QUESTIONS:

> *Why is the productivity of a continuous reactor higher than that of a (semi-)batch reactor? What are advantages and disadvantages of the operation modes discussed?*
> *Compare semi-batch processing in fine chemistry and biotechnology. Are there fundamental differences?*
> *Can aerobic fermentation be carried out batch-wise?*

12.2.3 Type of Reactor

As in the chemical industry, several basic reactor types and variations are used in practice. Fermentation reaction systems are nearly always multi-phase systems comprising a gas phase containing O_2 and/or N_2, one or more liquid phases, and a solid phase, including the microorganisms. A distinction can be made, based on the way of contacting the microorganisms with the substrate and air (in the case of aerobic reactors), between reactors where the microorganism are mobile or at a fixed position. In aerobic processes, the most important design factors are the contacting area between the microorganisms and the surrounding liquid and the rate of oxygen diffusion.

In chemical process technology the first class is referred to as slurry reactors, or, for certain types, fluidized-bed and ebullated-bed reactors. In biotechnology often the term *submerged reactors* is used. The second class includes the fixed-bed reactors. In biotechnology, these are referred to as *surface reactors*. In surface reactors, either the culture adheres to a solid surface that is continuously supplied with air and substrate, or the culture floats as a mycelium (a kind of network of threads) on the substrate.

Historically, surface reactors were used earlier than submerged reactors. Although they are very convenient in practice, at present surface reactors are not used frequently. However, the increased use of immobilized cells and, to lesser extent, enzymes has revived interest in surface reactors, mainly in wastewater treatment (see Sections 12.5 and 12.6).

From a reaction-engineering point of view, reactors for biotechnological and chemical processes are very similar. However, the specific properties of the medium and the quite different industrial tradition justify a separate discussion of bioreactors.

12.2.3.1 Submerged reactors (slurry reactors)

Figure 12.3 shows the most common representatives of three classes of submerged reactors (in batch mode). The same types have been treated in Chapters 8 to 10. The major difference between the types shown is the way mixing is accomplished. For aerobic processes, good mixing is very important in order to guarantee sufficient oxygen transfer from the gas phase to the liquid phase and to avoid anaerobic regions in the reactor.

The energy input for mixing is provided as follows:

- mechanical stirring: mechanically stirred tank reactor;
- forced convection of the liquid: plunging-jet reactor;
- operation using pressurized air: bubble-column and air-lift reactor.

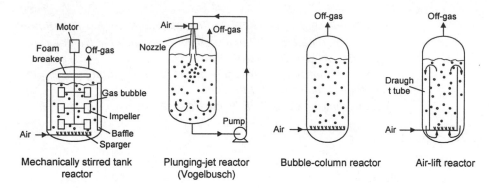

Mechanically stirred tank reactor | Plunging-jet reactor (Vogelbusch) | Bubble-column reactor | Air-lift reactor

Figure 12.3   Submerged reactors.

QUESTIONS:

>   *Compare these reactors with those treated in previous chapters.*
>   *What is specific for biotechnology?*

Mechanically stirred tank reactor

The mechanically stirred tank reactor is still used almost universally in the fermentation industry, although it is not necessarily the best solution. Figure 12.3 shows the standard mechanically stirred fermenter. The reactor consists of a cylindrical vessel with a height-to-diameter ratio of 1–3, often equipped with baffles to avoid the formation of vortices. The reactor is usually only filled for 2/3 in order to leave room for foam formation, which is one of the most important problems in many fermentation processes. A foam breaker can be incorporated at the waste gas outlet, while in addition antifoaming agents can be used. In the case of aerobic fermentation, air or oxygen is sparged into the reactor at the bottom. Mixing is achieved by a mechanical impeller equipped with one or more stirrers. An external jacket or an internal coil provides the surface area required for cooling or heating of the reactor contents.

In some cases, the mechanically stirred tank is not a suitable reactor. In scale-up the problems are mainly concerned with heat and mass transfer, in particular the transfer of oxygen. In some cases, the latter would require such large power input that the use of a simple mechanical stirrer is not practical or even impossible. These problems become still more severe with highly viscous or non-Newtonian reaction mixtures. Moreover, the biomass might suffer from too high shear. Therefore, in the recent past reactors have been developed that are better adapted for the specific circumstances of a certain fermentation (see below).

Plunging-jet reactor

In the chemical industry these reactors are commonly referred to as loop reactors and occasionally as Busch reactors. In biotechnology often the colorful name plunging-jet reactor is used. In these reactors, the gas is dispersed by a liquid stream in downward direction, the jet, that is created in a nozzle or a slot placed above the liquid surface. In the constriction of this nozzle, the liquid velocity is increased to typically 8–12 m/s. The jet entrains a mantle of gas, impinges on the surface, breaks through it, and penetrates into the liquid volume. The jet exists as long as it is surrounded by the gas mantle. The breakdown of this mantle leads to the formation of a swarm of small bubbles, which move downward and sideward in the liquid.

Plunging-jet reactors are used primarily for purposes of improving heat and mass transfer. A heat exchanger can easily be incorporated in the external loop, which enables the independent control of mass and heat transfer. Other advantages compared to the conventional mechanically stirred tank reactor are the lower power requirement per amount of oxygen transferred, the more reliable scale-up, and the possibility of the use of larger reactors.

Although the most important field of application for plunging-jet reactors is still in wastewater treatment, they are also used on a large scale for other applications. Examples are the production of yeast in so-called 'Vogelbusch' reactors, which have been scaled up to 2000 m^3, and the production of single-cell protein (SCP) [7].

QUESTION:

In the Vogelbusch reactor liquid is recycled by an external pump. Does the pump have to fulfill specific requirements (and if so, which)?

Bubble-column and air-lift reactor

The principle of both the bubble-column reactor and the air-lift reactor is that mixing takes place solely by the dispersion of pressurized air into the reactor. The bubble column is the simplest reactor type, and has long been used in the chemical industry, because of its low investment and operating costs, and simple mechanical construction. The bubble-column reactor is characterized by a large height-to-diameter ratio. Sparging air at the bottom of the reactor in most cases results in sufficient mixing.

Air-lift reactors are similar to bubble columns but have additional provisions for control of the bulk liquid flow. The circulation of the liquid is due to the difference in the densities (gas contents) of the gas–liquid mixture in the aerated section and the downcomer regions; air is sparged at the bottom of the reactor, and in its upward movement drags the liquid along. At the top of the column, most bubbles are separated, resulting in a larger apparent density of the mixture in the downcomer, which will flow downward. Some advantages of air-lift reactors over mechanically stirred reactors are the simple mechanical construction, the easier scale-up, better bulk mixing, and the possibility of using larger reactors. On the other hand, the energy and investment costs are higher.

Bubble columns are applied on a large scale in the production of, e.g. beer and vinegar. Air-lift fermenters are applied, in, for example, the production of SCP by Hoechst AG (40 m^3, ≈1000 t/year) and ICI (3000 m^3!, ≈100,000 t/year) [7,8]. Air-lift reactors, with internal separation and recirculation of the biomass (b air-lift suspension reactors), are also applied in aerobic wastewater tre [9,10].

QUESTIONS:

> *Compare the four reactors in Figure 12.3 with respect to pressure drop and mass transfer (see also Table 12.1). What would be a typical biomass production rate per unit volume of reactor?*

12.2.3.2 Surface reactors

Figure 12.4 shows three types of surface reactors, also referred to as biofilm reactors.

Tray reactor

The tray reactor is a classical surface reactor. The substrate flows from the top to the bottom via overflow of the liquid from one tray to another. The application of this type of reactor is limited to cell cultures that can form a coating that is sufficiently stable to be reused. An example of the use of the tray reactor is the production of citric acid.

Trickle-flow reactor

Trickle-flow reactors are frequently used in the oil refining and chemical industries. They are also increasingly used in biotechnology, particularly in wastewater treatment, but also in, for example, the production of vinegar. In trickle-flow reactors, the microorganisms are attached to the packing as a 'biological film'. The nutrient solution is evenly distributed through a feed device and flows downward. In contrast with customary operation in, for instance, hydrodesulfurization of a heavy gas oil fraction (see Chapter 3), the required air is often fed from the bottom, leading to countercurrent operation. The flow of air is brought about by the fact that it is warmed by the heat of fermentation and rises due to natural convection. The disadvantage of countercurrent operation in fixed-bed reactors is that the gas velocity must be kept low, because else flooding can occur (see Chapter 9).

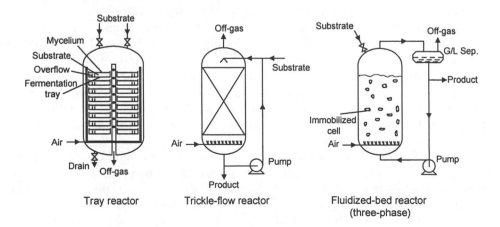

Figure 12.4   Surface reactors.

QUESTION:

> *Why is counter-current operation common in biotechnology, while it is not often applied in bulk chemicals production?*

Fluidized-bed reactor

The use of fluidized-bed reactors in biotechnology has increased considerably in recent years. They are mainly used with cells that are immobilized on solid particles, e.g. in wastewater treatment. Fluidized-bed reactors in biotechnology are operated with one fluid phase (liquid, anaerobic) or two fluid phases (air and liquid, aerobic). In order to increase the fluid velocity (in the case of relatively heavy particles) a liquid recirculation loop can be added. In addition to the general advantages of fluidized-bed reactors, such as the superior mass and heat transfer characteristics and good mixing, in biotechnological applications the low shear stresses in the reactor make the reactor suitable for cells that are sensitive to shear stresses (animal and plant cells). Another advantage is that plugging of the reactor does not occur as easily as in a fixed-bed reactor.

QUESTIONS:

> *Why is the fluidized-reactor considered a surface reactor and not a submerged reactor?*
> *In biotechnology the kinetics are often of the so-called Michaelis–Menten type with respect to reactant concentration. This is similar to Langmuir-Hinshelwood kinetics in catalysis. Furthermore, in biotechnology often products are inhibiting. Rank the reactors described in order of decreasing suitability for this type of kinetics.*

12.2.3.3 Oxygen supply in fermenters

Most fermentations are carried out in submerged reactors, as described in the section on submerged reactors above. The main problem of oxygen supply to fermentation processes is its low solubility (approximately $7–8$ g m_l^{-3}, but of course also depending on the type of substrate, the temperature, the oxygen partial pressure, etc.). Fast-growing microorganisms will consume oxygen at a rate of between 2 and 6 g m_l^{-3} s^{-1} [8]. This explains why even in batch processes, oxygen has to be fed continuously. In most aerobic fermentation processes, oxygen transfer from the gas phase to the liquid is the limiting step. Oxygen transfer is enhanced by increasing the gas–liquid interfacial area, a_l and/or the mass-transfer coefficient k_l, which are usually combined in one parameter (see Table 12.1). The main operation variables controlling this variable are the intensity of mixing (power input) and the gas velocity.

QUESTIONS:

> *Estimate the value of the gas-liquid mass transfer parameter, $k_l a_l$ in case of fast-growing microorganisms in a stirred-tank reactor and a trickle-flow reactor. What can you conclude?*

Biofilm reactors are used for applications that use slow-growing organisms or for diluted feed (substrate) streams [10]. A typical example is the treatment of large

Table 12.1 Important parameters in submerged and surface reactors

Reactor type	$k_l a_l$ (s^{-1})	$k_s a_s$ (s^{-1})	Solids hold-up (m_s^3 m_r^{-3})	Specific biofilm area ($m_{biofilm}^2$ m_r^{-3})
Stirred tank	0.15–0.2	0.1–0.5	0.01–0.01	
Bubble column	0.05–0.15	≈0.25	0.01–0.01	
Plunging Jet	0.2–1.2	0.1–1	0.01–0.01	
Three-phase fluidized bed	0.05–0.3	0.1–0.5	0.1–0.5	≈2000
Trickle flow	0.01–0.8	0.06	0.55–0.6	≈200

volume wastewater streams with very low substrate concentration. Here, it is important to retain the microorganisms in the reactor for sufficient growth of biomass, and consequently, sufficient conversion, while the rate of oxygen transfer is less important than in typical 'production' fermentations with fast-growing organisms (maximum specific growth rate > 0.1 h^{-1}) or with concentrated feed streams [10]. This is also illustrated by the values of the mass-transfer parameters and the solids hold-up, as summarized in Table 12.1.

The most important design parameter of aerobic biofilm reactors is the biofilm area per unit of reactor volume. For high reactor capacities, oxygen transfer may become limiting as a result of insufficient specific biofilm surface area. A typical value for the oxygen flux in biofilm reactors is 0.4–0.5 g O$_2$/($m_{biofilm}^2$·h).

QUESTIONS:

Estimate (calculate using estimated values) the mass-transfer parameters and the oxygen transfer rate per unit volume of reactor for the trickle-flow reactor and the three-phase fluidized bed reactor. Assume an oxygen concentration of 7 g m$_l^{-3}$. Hints: (1) See also Section 9.3.2; (2) You also require the liquid hold-up, see Table 8.9. Are the values you calculated reasonable?

12.2.4 Sterilization

The presence of contaminating microorganisms in a bioprocess may have unfavorable consequences such as loss of productivity (medium has to support the growth of both the production organism and the contaminant), product contamination (e.g. single-cell protein, bakers' yeast), product degradation (antibiotic fermentations), etc. [11].

Sterilization practices for biotechnological media must achieve maximum kill of contaminating microorganisms, with minimum temperature damage to medium components. The most convenient method of sterilization is by heating to a sufficiently high temperature to kill living organisms, maintaining that temperature long enough to achieve sterility and then cooling to culture temperature. For materials liable to damage by heat sterilization, e.g. some nutrient media, alternative methods should be used, such as filtration, radiation, or treatment with a chemical sterilization agent (e.g. ethene oxide) [12].

Two alternative heat sterilization procedures are available, viz. combined sterilization and separate sterilization. Combined sterilization involves loading of the reactor with part or all of the growth medium and subsequent sterilization. In separate sterilization, the sterilized medium is charged aseptically to the already sterile reactor. For batch processes, both procedures can be used, while continuous processes require separate sterilization.

The advantage of separate sterilization is that the medium can be sterilized in a specifically-designed unit that provides sterile medium for several fermenters. The disadvantage is the risk of contamination during transfer from the sterilizing unit to the fermenter.

QUESTION:

> *What factors will determine when separate and when in-situ sterilization is used for batch processes?*

In the case of combined sterilization in a batch reactor, heating is carried out by passing steam through the bioreactor coils or jacket or by direct sparging of steam into the liquid medium. The latter results in very rapid heating, but in many media produces excessive foaming. Continuous sterilization of a medium is carried out by passing it through heat exchangers or a venturi steam injection device. Sterilization of empty vessels is commonly carried out by direct sparging with wet steam, which results in a much more rapid sterilization than dry saturated or superheated steam at a given temperature [12].

The rate at which organisms are killed increases rapidly with temperature. For example, in a particular continuous sterilization system with steam injection the holding time ranged from 2.44 min at 403 K to 2.7 s at 423 K [11]. Of course, the sterilization time also depends on the type of microorganism.

For the sterilization of air for aerobic processes, filtration is most commonly used [11]. Filters for the removal of microorganisms may be divided into two groups, so-called 'absolute filters', with pores smaller than the particles to be removed, and fibrous-type filters, with pores larger than the particles to be removed. The former type of filter (e.g. ceramic or plastic membrane) is claimed to be 100% efficient, explaining the name 'absolute filter'. Filters of the latter type are made of materials such as cotton, steel wool, etc. and, in theory, the removal of microorganisms cannot be complete. The advantages of fibrous filters are their robustness, low cost, and lower pressure drop compared to the absolute filters.

12.3 FERMENTATION TECHNOLOGY – CELL BIOMASS (BAKERS' YEAST PRODUCTION)

A typical industrial example of fermentation technology for biomass production is the production of bakers' yeast. Bakers' yeast is used in the production of bread. It provides the rising power of the dough, and therefore, the airiness of the bread during its baking process. Bakers' yeast is produced by the growth of microorganisms on a substrate consisting of sugars (e.g. glucose, molasses) under aerobic conditions, i.e. with excess oxygen. Under anaerobic conditions, i.e. with a shortage of oxygen, ethanol will be produced, which is not a desired reaction.

12.3.1 Process Layout

Bakers' yeast is produced in the fed-batch mode. This has the advantage of a higher efficiency, a better control of the dynamics of the overall process and a better quality of the Bakers' yeast compared to batch processing. As the fermentation reaction is exothermic and the process operates at 298–303 K, while the temperature of the cooling water is 283–288 K, cooling is limited, which is a common problem in bioprocesses.

Yeast plants are typically run between 5 and 7 days a week. In every production cycle, a series of reactors with increasing size is used (small-medium-large volume) as shown in Figure 12.5. Prior to the production, the reactors are cleaned and sterilized with steam in order to make the reactor free of germs.

Beet and cane molasses serve as the sugar providers. Molasses is a waste product from the sugar industry, and the cheapest source of fermentable sugars. Even so, the substrate costs are responsible for 60–70% of the cost of bakers' yeast. Beet and cane molasses differ in sugar composition, proteins, salts, and vitamins content. Therefore, an additional supply of nutrients (salts, vitamins) is necessary.

The raw molasses is diluted to facilitate pumping and fermentation, and it is treated with acid (to pH 4) by which precipitation of some organic material occurs. Then the molasses is centrifuged and sterilized ('heat shocked') at 410 K, usually by steam injection for a couple of seconds. The sterilized solution is stored in sterilized vessels. Ammonia is added to supply the necessary nitrogen for yeast growth and to adjust the pH in the fermenter.

The specific growth rate of the biomass may vary between 0.05 and 0.6 h^{-1}. In the beginning, the pH of the substrate solution is around 4 to ensure optimal growth of the biomass and to prevent contamination as a result of incomplete removal of proteins

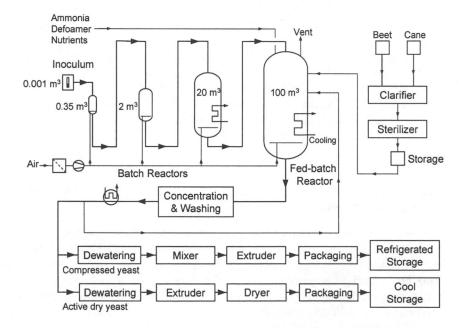

Figure 12.5   Process steps in the production of bakers' yeast.

from the substrate. At the end of the production cycle, the pH is increased to a value of 5 to prevent strong coloring of the yeast.

QUESTION:

Which sterilization procedure is used?

12.3.2 Cultivation Equipment

To start up simply using one large reactor is impossible. Due to the long residence time for this type of operation, the chances for contamination of the substrate solution are large. The process consists of a number of growth stages, the first of which is the laboratory scale culture (inoculum). The next stages (two are shown in Figure 12.5) are again pure batch with progressively increasing reactor volume, while in the final production stages several reactors (shown as one reactor) are used in the semi-batch (fed-batch) mode, with respect to the sugar substrates (cycle time ≈ 16 h). The air flow is continuous when the reactors are in operation.

In the old plants, all reactors were mechanically stirred. Nowadays the large reactors are more frequently bubble columns. The initial liquid volume in the semi-batch reactor is 20% and the final volume is 70% of the reactor volume.

QUESTION:

What effects would (local) insufficient mixing have on the production of Bakers' yeast?

12.3.3 Downstream Processing

At the end of the production cycle, the fermentation product contains about 5 wt% yeast. The biomass is separated from the remaining liquid by centrifuging in several stages. In the transfer of the biomass from one to another centrifuge, the biomass is washed with water. After the last separator the so-called yeast cream (called yeast cream because of its off-white color) is cooled to 277 K and stored in tanks. A small part of the yeast cream, after acid treatment, is used as starting material for the next fermentation cycle. The remainder is processed into either compressed or dried yeast.

Compressed yeast

The yeast cream from the yeast cream storage is filtered. The dewatered yeast is continuously cut from the filter surface. It is mixed with emulsifiers and the moisture content adjusted to 70 wt%. The yeast is then extruded in the shape of thick strands, cut, packaged, and stored at low temperature.

Active dry yeast

The quality of dried bakers' yeast is less than that of compressed yeast. However, it has a better stability, and hence can be used in (sub)tropical countries. Production and downstream processing for so-called active dry yeast is similar to that of compressed yeast. The yeast is extruded to fine strands (2–3 mm thick) directly after filtration. These strands are chopped to a length of about 7 mm and then dried. Dr commonly of the fluid-bed drier type. The active dry yeast, containing onl 8% moisture, can be stored at higher temperature than compressed yeast.

12.4 FERMENTATION TECHNOLOGY – METABOLIC PRODUCTS (BIOMASS AS RENEWABLE ENERGY SOURCE)

The progressive depletion of fossil fuel resources has led to the consideration of other materials as sources of energy. A possible source is biomass, i.e. any material derived from photosynthesis, which can be converted by biotechnology into more useful and valuable fuels. Generally, biomass is converted into ethanol or methane. Thermal cracking of biomass to obtain gasoline-like products, gasification of biomass to produce synthesis gas, and combustion of dried biomass for power generation, are also feasible, at least technically, but beyond the scope of this chapter.

12.4.1 Ethanol

As discussed in Section 12.3 yeast cells grow on sugars in the presence of oxygen, while they produce ethanol under anaerobic conditions:

$$\underset{\text{glucose}}{C_6H_{12}O_6} \overset{\text{yeast}}{\to} \underset{\text{ethanol}}{2C_2H_5OH} + 2CO_2 \tag{1}$$

Figure 12.6 shows a schematic of the production of ethanol by yeast. The yeast has been cultivated in advance in aerated fermenters as discussed for the production of Bakers' yeast (Section 12.3).

Yeast from the last fermenter is added to a mixture of cane molasses and water (to yield a sugar content of 17 wt%) in large, unaerated fermenters. HCl or H_2SO_4 is added to obtain a pH of 4.5. Heat is removed by external heat exchangers in order to keep the temperature at 303 K. The residence time in the fermenters is about 40 h, and after this time the 'beer' contains 8–12 vol% ethanol [13]. The yeast cells are removed by filtration in a rotary-drum vacuum filter, and either recycled or used in animal feed after further handling.

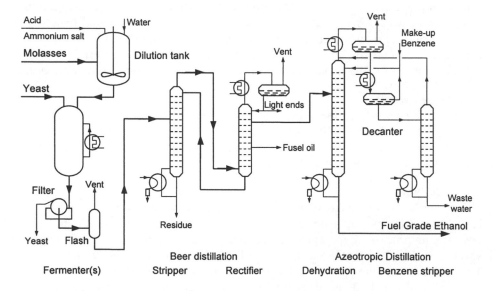

Figure 12.6   Production of fuel-grade ethanol from biomass.

In order to obtain high yields on an industrial scale in an economical way, yeast strains must be selected that are resistant to ethanol. As far as the yeast is concerned alcohol is not only a waste product, but it is also harmful to the yeast. Normally, microorganisms are killed when the alcohol concentration exceeds 12–15 vol%. New types of yeast have been developed that withstand much higher alcohol levels, so that downstream processing is facilitated. The separation of ethanol and water is very energy intensive. Furthermore, the complete separation of ethanol and water, which is required for ethanol as fuel, cannot be attained by normal distillation. Ethanol and water form a homogeneous minimum boiling azeotrope at a temperature of 351 K, where the mixture contains 95 wt% ethanol. Therefore, one needs to resort to azeotropic distillation (see Section 8.3), or more advanced separation methods, e.g. based on membrane technology.

The beer is first fed to a flash drum in order to remove dissolved CO_2 and then distilled in a series of four distillation columns. The bottom product from the first distillation column (non-fermentable molasses solids) is an additive for animal feed. The overhead vapor contains a mixture of ethanol (appr. 50 vol%), water, and other volatile components (e.g. acetaldehyde) and is fed to the base of a rectifying column. In the rectifying column light ends and so-called fusel oil (a mixture of higher alcohols) are removed overhead and as a side stream, respectively. The azeotrope (95 wt% ethanol) is removed as a liquid side stream somewhat below the top of the column and fed to the azeotropic distillation unit.

The overhead vapor of the dehydration column is a ternary minimum boiling azeotrope of ethanol, water, and benzene (the azeotropic agent), while anhydrous motor fuel grade ethanol is produced at the bottom. The azeotrope is separated by distillation and benzene is recycled to the dehydration column. With respect to heating and cooling facilities, Figure 12.6 only shows the overhead condensers and the reboilers of the columns involved. However, in this separation sequence heat integration and energy recovery play a vital role in reducing energy requirements.

QUESTIONS:

> *In fact, the higher alcohols are later blended with ethanol and serve to add fuel value. What could be the reason that they are first removed?*
>
> *Where in Figure 12.6 do you see possibilities for heat integration? What are your thoughts about benzene as the azeotropic agent?*

12.4.2 Biogas

Biomass can be converted into methane-containing gas, called biogas. This is done in anaerobic purification of wastewater (see Section 12.5). It is also done on a small scale in agriculture based on conversion of animal waste. Fermentation of the waste, in the absence of air, produces biogas containing about 60% methane. The solid residue left after fermentation is a good fertilizer. The biogas can be used as fuel and for the production of electricity locally.

HYDROCARB Process

Several processes based on renewable energy carriers have been suggested as part of the solution to the CO_2 problem. Since in the growth of biomass (plants, trees) CO_2 is consumed, the use of this biomass as a source of fuel would result in zero net CO_2 emissions. An example is the HYDROCARB process [14]. This process is based on the simultaneous hydrogenolysis of biomass and fossil fuels according to the general reaction:

$$CH_xO_y + (2 + y - x/2)H_2 \rightarrow CH_4 + yH_2O \qquad \Delta H^0_{298} < 0 \qquad (2)$$

Hydrogen is produced by catalytic decomposition of methane at 1273–1373 K [14]:

$$CH_4 \rightarrow C + 2H_2 \qquad \Delta H^0_{298} = 75 \text{ kJ/mol} \qquad (3)$$

Another part of the methane reacts with the steam formed in reaction (2) to produce synthesis gas (steam reforming, see Chapter 5). The syngas is used in the synthesis of methanol (see Chapter 6), and, after separation, the remaining hydrogen can be used in the hydrogenolysis processes, while carbon can be applied as a fuel or stored permanently or for future use. Figure B.12.1 shows a block diagram of the HYDROCARB process.

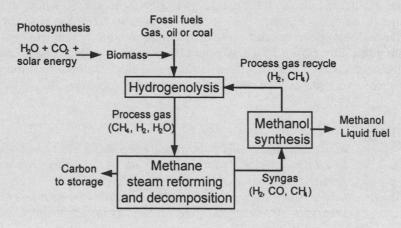

Figure B.12.1   HYDROCARB process.

12.5 ENVIRONMENTAL APPLICATION – WASTEWATER TREATMENT

12.5.1 Introduction

Wastewater treatment is the largest application of biotechnology. Wastewater is the combination of water-carried wastes arising from domestic and industrial use, together with ground water, surface water, and storm water. The decomposition of organic materials in wastewater can produce foul smelling gases and lead to a reduction in the dissolved oxygen content thus killing aquatic life. Furthermore, wastewater can

Table 12.2 Major wastewater contaminants [15]

Contaminant	Comments
Suspended solids	Lead to sludge deposits
Biodegradable organics	Mainly proteins, carbohydrates and fats, leading to reduced dissolved oxygen concentration
Pathogens	Can cause diseases in humans and animals
Nutrients (N and P)	Can cause eutrophication of lakes and reservoirs leading to algae blooms
Priority pollutants	May be carcinogenic, mutagenic, teratogenic or highly toxic (e.g. benzene, chloro-hydrocarbons)
Refractory organics	Include surfactants, phenolics and pesticides, and are often not removed by conventional wastewater treatment processes
Heavy metals	May arise from industrial processes
Dissolved inorganics	Mainly calcium, sodium and sulfate arising from domestic use

contain microorganisms, heavy metals and other toxins that may be detrimental to both plant and animal life. Clearly, removal of these potentially hazardous components from the wastewater is essential. Table 12.2 summarizes the major contaminants of concern. It is clear that wastewater is a very complex mixture.

A major problem in wastewater treatment is the varying composition and flow rate with time. In particular the composition of industrial wastewaters can vary considerably, depending on the nature of the chemical process. The flow rate also is an important factor. Daily as well as seasonal variations in flow rate occur.

12.5.2 Process Layout

Wastewater treatment proceeds in three general steps, viz. so-called primary, secondary, and tertiary treatment processes; they are also referred to as pretreatment, biological treatment, and advanced treatment (see Figure 12.7).

The aim of the pretreatment process is to remove suspended solids (e.g. sand and possibly fats). This is done by means of physical processes such as sedimentation and flotation. Industrial wastes sometimes have pH values far from the pH range required

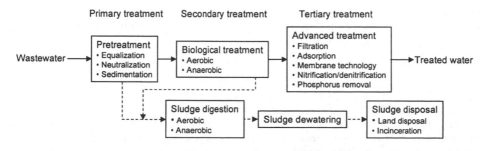

Figure 12.7 Outline of wastewater treatment.

for optimal performance during the subsequent biological processes. Then, neutralization is required as pretreatment step. Apart from reducing the load on subsequent processes, the primary treatment process also has the effect of minimizing variations in the wastewater flow rate (equalization).

In secondary treatment processes, the concentration of organics (suspended and soluble) is reduced. Secondary treatment processes are biological processes and can be divided into aerobic and anaerobic processes. The microorganisms responsible for the degradation of the organic compounds are already present in the wastewater.

Tertiary treatment processes aim to further improve the quality of the wastewater. Conventional biological treatment produces an effluent containing typically about 30 g/m^3 suspended solids and 20 g/m^3 BOD[1]. In tertiary treatment processes these values are both reduced to 2 g/m^3. In addition, the total nitrogen and phosphorous concentrations are reduced as well as the amounts of heavy metals and pathogenic microorganisms. Various processes are used, including grass plots (gently sloping grass fields, on which water purification is achieved by the action of soil microorganisms, by the uptake of nutrients by grass, and by the filtering action of the soil), filtration, adsorption, and membrane technology.

In addition to the treatment of the wastewater, processing and disposal of the resulting sludge is required. This sludge, which is formed during primary and secondary treatment in large amounts, is not very concentrated (about 98% water) and has a bad odor. Stabilization of the sludge takes place by either aerobic or anaerobic digestion similarly to the secondary treatment discussed in Section 12.5.3 (activated sludge process). Reduction of the water content is achieved by various methods, such as sedimentation, air drying, filtering, and centrifuging. The sludge can then be disposed off on land or be incinerated. The latter is increasingly used.

12.5.3 Biological Treatment Processes

Biological treatment of wastewater streams can be either aerobic or anaerobic. The choice of process depends on the concentration of biodegradable organics and the volume of wastewater to be treated. Aerobic treatment is generally applied to mildly polluted (relatively low concentration of biodegradable organics) and cooler wastewaters, whereas anaerobic treatment is often employed as a pretreatment step for highly polluted and warmer wastewaters.

QUESTION:

    Explain the logic of the above remark concerning aerobic and anaerobic treatment.

---

[1] BOD = biochemical oxygen demand: the amount of oxygen required for the biochemical oxidation per unit volume of water at a given temperature and for a given time. It is used as a measure of the degree of organic pollution of water. The more organic matter the water contains the more oxygen is used by the microorganisms [10]. (COD = chemical oxygen demand = amount of oxygen required for chemical oxidation).

12.5.3.1 Aerobic treatment processes

In aerobic treatment of wastewater basically the following reaction takes place:

$$\text{organics} + O_2 + N + P \xrightarrow{\text{Cells}} \text{new cells} + CO_2 + H_2O + \text{soluble microbial products} \quad (4)$$

Roughly one half of the organics removed is oxidized to CO_2 and H_2O, the other half is synthesized to biomass. Besides a source of carbon and oxygen, the micro-organisms also require nitrogen (as ammonia or nitrate) and phosphorus (as phosphate), sulfur, and trace quantities of many other substances for growth. Domestic sewage normally contains a balanced supply of nutrients. For industrial effluents this is not the case. Addition of nutrient-rich domestic wastewater is then often applied.

Activated sludge process

The so-called activated sludge process, developed by Arden and Lockett (1914) [15], is still widely used today. Figure 12.8 shows a schematic of this process.

The wastewater is introduced into the reactor, where bacteria convert the organic compounds according to Eqn. (4). The mixture of cells and treated water is then passed into a settling tank. Part of the settled sludge is recycled to the reactor while the remainder is further treated and disposed off.

In the conventional process aeration takes place uniformly over the length of the reactor tank. This is not optimal. The oxygen requirement at the inlet of the aeration tank (where the substrate concentration is largest) is larger than at the outlet and, as a consequence, an oxygen deficiency might occur at the inlet, while an oxygen surplus can occur at the outlet, which is unnecessarily expensive. Furthermore, peaks in the wastewater supply and poisonous substances could disturb the purification process.

Solutions to these imperfections are staged introduction of either the air or the wastewater. In the first case, 'tapered aeration', aeration is decreased along the reactor tank by reducing the number of aerators per section. In the second case, wastewater is introduced along the length of the tank instead of only at one point.

QUESTIONS:

> *What are advantages and disadvantages of both solutions. Why is part of the sludge recycled?*

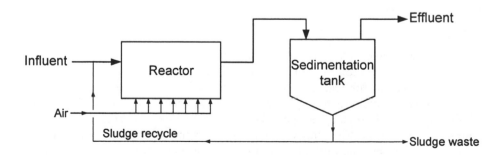

Figure 12.8 Schematic of activated sludge process. After [11].

In the activated sludge processes described above the microorganisms are suspended in the wastewater (submerged reactor). An alternative is the use of so-called fixed-film processes (surface reactor). In the latter case, the natural ability of organisms to grow on surfaces can be exploited to keep the biomass in the reactor and avoid separation from the treated water. Two examples of such surface reactors have already been discussed in Section 12.2, viz. the countercurrent trickle-flow reactor (also called 'trickling filter') and the three-phase fluidized-bed reactor. The most important design parameter is the biofilm area per unit volume of reactor (see Section 12.2).

In the case of wastewater treatment, the trickle-flow reactor is filled with stone or (recently) plastic media on which the microorganisms grow as a film (typically 200 $m^2_{biofilm}/m^3_{reactor}$). These microorganisms utilize the soluble organic material in the wastewater, but little or no degradation of suspended organics is achieved. Hence, the reactor effluent is separated in a sedimentation tank, where these suspended organics are removed.

In fluidized-bed reactors for wastewater treatment, usually sand is used as support for the microorganisms to grow on, because sand is a cheap and convenient material (typically 2000 $m^2_{biofilm}/m^3_{reactor}$).

More advanced particle-based biofilm reactors with large specific surface areas (up to 3000 $m^2_{biofilm}/m^3_{reactor}$) ensuring that conversions are not strongly limited by the biofilm liquid mass-transfer rate, have recently been developed. These reactors are capable of processing a higher load of wastewater and are more efficient than the trickle-bed and fluidized-bed reactors [10].

QUESTION:

Why can the biofilm area per unit of reactor volume be so much larger in the fluidized-bed reactor than in the trickle-flow reactor?

Nitrogen and phosphorus removal

Biological treatment of wastewater presently aims at the removal not only of organic substances, but also of nitrogen and phosphorus compounds, because these compounds cause eutrophication of surface waters. Nitrogen removal is based on two different processes, nitrification:

$$NH_4^+ + 2O_2 \rightarrow NO_3^- + H_2O + 2H^+ \tag{5}$$

and denitrification in the presence of readily biodegradable organic compounds such as methanol or acetic acid:

$$6NO_3^- + 5CH_3OH \rightarrow 3N_2 + 6OH^- + 5CO_2 + 7H_2O \tag{6}$$

Since the removal of nitrogen requires the presence of organic compounds, it seems logical to carry out the removal of carbon and nitrogen containing substances from wastewater streams more or less simultaneously. However, for denitrification an oxygen free environment is required, whereas the degradation of organic compounds and the nitrification require oxygen. Hence, a separate denitrification reactor is required at least. Several reactor configurations for the removal of nitrogen can be used, see e.g [16]. For the biological removal of phosphorus, an additional reactor is required.

Rotating biological contactor

A special type of surface reactor is the rotating biological contactor (see Figure B.12.2). This reactor comprises a series of discs (2–3 m diameter) arranged in compartments and mounted on a rotating horizontal shaft, which is typically positioned above the liquid level so that the discs are only partially submerged. The biofilm growing on the discs is exposed alternatively to the atmosphere, where oxygen is absorbed, and to the liquid phase where soluble organic material is utilized.

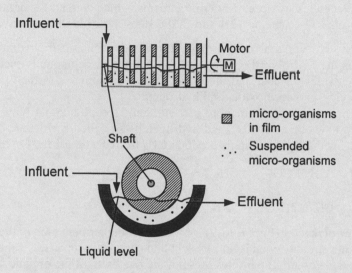

Figure B.12.2   Rotating biological disc contactor; top: reactor; bottom: one disc.

Aerobic treatment of gases and solid materials [17,18]

Polluted gases [17] and solid materials [18] can also be treated successfully by biotechnological methods. An example of gas treatment is the biological oxidation of hydrogen sulfide in gas streams (e.g. natural gas, Claus plant tail gas, see Section 3.6.1) [17]. The following reactions proceed in two separate reactors:

Absorption and hydrolysis of H_2S : $H_2S + OH^- \rightarrow HS^- + H_2O$ (7)

Biological sulfur formation : $HS^- + \frac{1}{2}O_2 \rightarrow S + OH^-$ (8)

This process is a viable alternative for liquid redox processes used in small-scale operation (see Section 3.6.1). Its main advantage is that, unlike in liquid redox processes, the elemental sulfur is not formed in the absorber but in the bioreactor. This solves the main problem of liquid redox processes, namely severe clogging of the absorber by sulfur.

12.5.3.2 Anaerobic treatment processes

Anaerobic treatment of wastewater has important advantages over aerobic treatment, such as the lower sludge production, the formation of a gas with high energy content ('biogas'), and the lower processing costs (no energy required for aeration). However, anaerobic processes require much more careful control of specific process parameters and the microorganisms grow quite slowly. Furthermore, the degree of removal of biodegradable substance is generally lower than in aerobic processes. Therefore, anaerobic treatment is usually applied as a pretreatment for wastewaters with a high concentration of biodegradable components.

The anaerobic decomposition of organic substances in the wastewater is accomplished by a complex mixture of microorganisms, which convert the organic material to methane and CO_2. Scheme 12.1 shows the steps involved in this conversion.

1. In the first step, high molecular weight compounds such as carbohydrates, proteins, and lipids are hydrolyzed to yield soluble monomer compounds such as sugars, amino acids, fatty acids, and alcohols. These compounds are then fermented to organic acids, carbon dioxide, and hydrogen.
2. Organic acids with three or more carbon atoms (propionic and butyric acid) are converted to acetic acid, CO_2, and hydrogen by so-called acetogenic bacteria.
3. The last, 'methanogenic', steps produce CH_4 from acetic acid, CO_2, and H_2:

$$CH_3COOH \rightleftarrows CH_4 + CO_2 \qquad \Delta H^0_{298} = -34 \text{ kJ/mol} \qquad (9)$$

$$CO_2 + 4H_2 \rightleftarrows CH_4 + 2H_2O \qquad \Delta H^0_{298} = -164 \text{ kJ/mol} \qquad (10)$$

The choice of reactor (Figure 12.9) for the anaerobic fermentation of the organics in the wastewater stream depends on physical and chemical properties of the substances present, i.e. the amount of particles present, the reactivity of the organic components, etc. Table 12.3 summarizes characteristics and applications of the main types of reactors used for different wastewater streams.

In the stirred-tank reactor mixing of the contents can take place by mechanical agitation (as shown), a water recirculation pump, or by injection of oxygen-free gas. The residence time of the sludge is increased by the presence of a settling tank from where settled particles (sludge and unconverted substrate in suspension) are recycled. This is necessary because the methane producing microorganisms grow slowly (minimum residence time of sludge ca. 20 days).

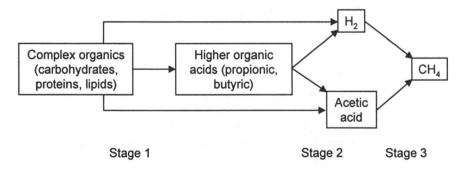

Scheme 12.1 Anaerobic decomposition of organic substances. After [19]

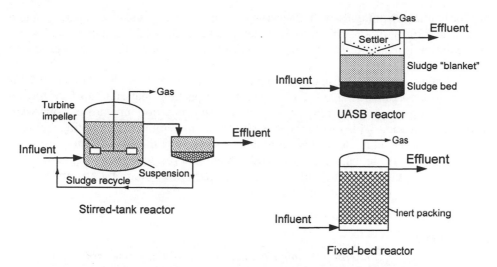

Figure 12.9 Reactors for anaerobic wastewater treatment. After [20].

The upflow anaerobic sludge blanket (UASB) reactor usually consists of three zones, viz. a dense sludge bed, a sludge 'blanket' (with less concentrated sludge), and a separation zone where entrained sludge particles are separated from the liquid and gas. For this type of reactor, the sludge must have a relatively large density in order to prevent it from being washed out of the reactor with the wastewater flowing upward (although with low velocity). A critical factor is the formation of 'granules' (biomass aggregates). Advantages of the UASB are its simple construction, the high obtainable throughput and the lack of mixing costs (mixing occurs by the gas

Table 12.3 Anaerobic reactors for different types of wastewater [20,21]

Water type	Reactor	Examples
Water with 0.5–25 kg/m^3 easily digestible lower carbohydrates, large particle content allowed	Stirred-tank reactor, WHSV = 5–10 kg COD[a]/ (m$^3_{reactor}$ · day), Conversion: 80–95%	Manure, distilleries, chemical industry, pulp and paper industry, preserved food plants
Water with 0.5–25 kg/m^3 easily digestible higher carbohydrates, small particle and protein content	UASB reactor, WHSV = 10–15 kg COD[a]/ (m$^3_{reactor}$ · day), Conversion: ca. 95%	Food industry, starch industry, breweries
Water with 25–200 kg/m^3 slowly digestible carbohydrates, small particle content	Fixed-bed reactor, WHSV = 10–20 kg COD[a]/ (m$^3_{reactor}$ · day), Conversion: 90–98%	Fermentation industry, dairy industry

[a] COD = chemical oxygen demand.

produced). Disadvantages are the necessity of an efficient gas/sludge separator and a good water distributor. Furthermore, sometimes (at small loads) gas or treated water recirculation is required. The reason is that at small loads not enough gas is produced for sufficient mixing.

The fixed-bed reactor is filled with an inert packing material, such as gravel, rock, or plastic. Microorganisms attach themselves to the packing material and thus form a biofilm that remains in the reactor. The wastewater may flow upward or downward through the reactor. This reactor type is similar to an aerobic trickling-filter reactor. These reactors are suitable for high throughput, especially for decomposition of soluble easily convertible substrates. Evidently, fixed-bed reactors are suited mainly for the treatment of dissolved organic matter, since particles would plug the channels in the bed and thus decrease its efficiency.

QUESTIONS:

> *Is anaerobic water treatment a multi-phase process? What phases are present?*
>
> *Instead of the combination of a stirred tank and a settling tank a stirred tank reactor without a settling tank could be used. Compare both configurations with respects to (reactor and total) volume and operation parameters (energy requirements for stirring/pumping). Discuss mixing in the UASB reactor.*

12.6 ENZYME TECHNOLOGY – BIOCATALYSTS FOR TRANSFORMATIONS

Enzymes in principle are homogeneous catalysts, although a colloidal solution is a better description. They are complex chains of amino acids with a molecular weight ranging from 20,000 up to 200,000. Enzymes are present in living cells where they act as catalysts. Although enzymes are only formed in living cells, they can also function outside the cell. The use of enzymes can be of two kinds, viz. as biocatalysts in a reaction system, or as the final product, e.g. for food additives. Here, we will focus on the former application. Two examples of the use of immobilized enzymes will also be discussed in this section.

12.6.1 General Aspects

Enzymes can operate as part of a living cell, but they can also be extracted from living cells while keeping their catalytic activity. The immobilization of enzymes on a fixed support is a relatively recent development that shows promise for the future. Immobilization was developed in the 1960s because the amount of enzymes present in biomass was very small, resulting in diffusion and stability problems. However, current recombinant DNA techniques enable the cheap production of enzymes, which makes them suitable for once-only use. This has the advantage that the enzymes can be used directly, while immobilization is not required. Enzymes compete with conventional catalysts, mainly in fine chemicals production (see Chapter 10). Table 12.4 shows a comparison of the three technologies.

An important advantage of enzymes over most conventional catalysts is their stereoselectivity (see also Section 10.1). In general, chemical processes lack the ability to discriminate between the L- and D-enantiomers of asymmetric (chiral) molecules and, as a consequence, produce a racemic mixture. It should be noted that chemocatalysts can also possess stereoselectivity. A commercial example of a chemical process route in fine chemistry with a stereoselective homogeneous catalyst is in the production of L-dopa (see Section 10.1) [22]. Stereoselectivity allows the production of one stereo-isomer exclusively. This is often important, especially in pharmaceuticals production, because generally one of the optical isomers is inactive, or worse, has a damaging effect. Understandably, the separation of two optical isomers is very difficult using common separation methods such as distillation, crystallization, etc., because all physical properties are identical for both enantiomers. Furthermore, even if the undesired stereo-isomer has no averse effects, the productivity of a process decreases by its formation. It is obvious that it is preferred to exclusively produce the desired enantiomer.

Analogously to homogeneous catalysis much research is carried out on the subject of immobilizing enzymes. Supports such as polymers are used, or membrane reactors are applied. The advantage of immobilized enzymes over their homogeneous analogues is the possibility to recover the enzymes from a reaction mixture and to re-use them repeatedly, like heterogeneous catalysts in the chemical industry. In addition, it has been discovered that in immobilized form enzymes are thermally more stable, while they can also be used in non-aqueous environments. There will also be a reduction in the amount of waste produced, which may lead to lower effluent treatment costs [23]. Finally, the product is not contaminated with the enzyme, which is particularly important in food and pharmaceutical applications [24]. Of course, the additional costs for development time and immobilization must be balanced against these advantages. For the production of enzymes and immobilization techniques see, for instance, [24,25].

The choice between the use of whole cells and enzymes, and between soluble and immobilized form, depends on many factors, such as the nature of the conversion, stability and reuse, technical improvements, and eventually, cost. In principle, enzymes contain orders of magnitude more active sites per unit mass than whole

Table 12.4 Comparison of technologies for chemicals production. Adapted from [7]

Parameter	Classical fermentation	Enzymes	Chemo catalysts
Catalyst	Living cells	Enzymes	Metals, acids, …
Catalyst concentration (kg/m^3)	10–200	50–500	50–1000
Specific reactions	Sometimes	Often	Often
Reaction conditions	Moderate	Moderate	Moderate–extreme
Sterility	Yes	Yes	No
Yield (%)	10–95	70–99	70–99
Important cost item	Cooling water	Enzyme	Varies
Problems	Regulation of microorganisms	Stability	Selectivity, stability

cells. On the other hand, the activity per site usually is considerably lower. Technical improvements can result directly from immobilization (e.g. increased product purity and/or yield, reduced waste production) but also indirectly. Immobilization of cells or enzymes enables the use of continuous rather than batch operation, thus simplifying control and reducing labor costs due to the possibility of increased automation of the process. Of course, this is only advantageous for large-scale processes, whereas most bioproducts are only produced on a small scale.

Immobilized enzymes are mainly used in the production of fine chemicals and pharmaceuticals, because currently they cannot compete economically with conventional catalysts in the bulk chemical industry. Present commercial applications of immobilized enzymes are the production of L-amino acids, organic acids, and fructose syrup.

12.6.2 Production of L-Amino Acids

The demand for L-amino acids for food and medical applications is growing fast. Both chemical and microbial processes can be used for their production. However, the chemical routes lack stereoselectivity, thus leading to lower productivity. In Japan the immobilized enzyme aminoacylase has been used for the production of L-amino acids, of which methionine is the most important, since 1969 [7]. Aminoacylase catalyzes the stereospecific de-acylation of acyl-amino acids

$$
\begin{array}{lll}
\text{L-R-CH-COOH} & & \text{L-R-CH-COOH} \\
\quad\text{NHCOR'} & \xrightarrow{\text{aminoacylase}} & \quad\text{NH}_2 \\
\qquad\qquad + \text{H}_2\text{O} & & \qquad\qquad\qquad\qquad + \text{R'COOH} \quad (11) \\
\text{D-R-CH-COOH} & & \text{D-R-CH-COOH} \\
\quad\text{NHCOR'} & & \quad\text{NHCOR'}
\end{array}
$$

Figure 12.10 shows a flow scheme of the Tanabe Seiyaku process that utilizes immobilized aminoacylase for the continuous production of L-amino acids.

The de-acylation reaction is carried out in a reactor packed with aminoacylase adsorbed on an ion-exchange material. The desired L-amino acid can be separated from the unconverted acetyl-D-amino acid by crystallization, as a result of their different solubilities. The acetyl-D-amino acid is racemized and the resulting racemic mixture is returned to the enzyme reactor together with feed acetyl-DL-amino acid. Racemization of acetyl amino acids can be accomplished, for example, by heating at 373 K with acetic anhydride in an acetic acid solution [26].

Aminoacylase can also be applied in solution, i.e. as a homogeneous catalyst. Figure 12.11 shows a comparison of the relative costs involved in the production of amino acids with homogeneous aminoacylase in a batch process and immobilized aminoacylase in a continuous process.

Clearly, the main differences can be attributed to catalyst (enzyme or enzyme + support) and operation costs. The latter include labor, fuel, etc. The labor costs, in particular, are greatly reduced in the continuous process. The reduction in catalyst

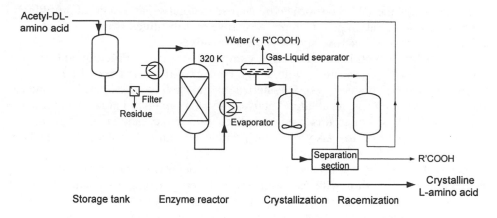

Figure 12.10   Production of L-amino acids using aminoacylase. Adapted from [26].

costs when using the continuous process is a result of the higher stability and the reusability of immobilized aminoacylase.

QUESTION:

> *Analyze the batch and continuous process in terms of the factors mentioned in Section 12.6.1. Is Figure 12.11 a fair comparison?*

12.6.3 Production of Artificial Sweeteners

Probably the most successful application of immobilized enzymes is the production of fructose syrups using immobilized glucose isomerase for the conversion of glucose. Another artificial sweetener, D-mannitol can also be produced using this enzyme.

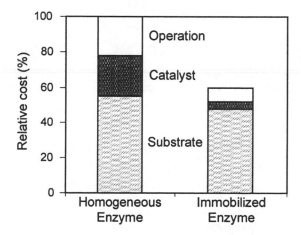

Figure 12.11   Relative production costs of L-amino acids using homogeneous (batch process) and immobilized (continuous process) aminoacylase. Adapted from [23].

At the end of the 1970s a shortage of sucrose (natural sugar) occurred, resulting in high prices. It was attempted to replace sucrose by substitutes. Table 12.5 gives the relative sweetness of sucrose and possible alternatives.

Glucose has a much lower sweetness, compared to sucrose. Fructose adds most to sweetness, and therefore sugars with a high content of fructose are desired.

Glucose can be obtained by depolymerization of starch by acid hydrolysis, and can then be isomerized using an alkaline catalyst into a 1:1 mixture of glucose and fructose (see Eqn. (12)). Although this product is highly desirable, neither the HCl induced depolymerization nor the alkaline isomerization are attractive steps due the large amounts of byproducts formed.

A novel process became feasible in which the depolymerization is catalyzed by enzymes. In combination with the discovery of the enzyme glucose isomerase, which catalyzes the interconversion of glucose into 1:1 mixture of glucose and fructose, an attractive process could be developed. This corn-derived syrup is referred to as high-fructose corn syrup, abbreviated as HFCS.

Initially (in the mid-1960s), the isomerization of glucose into fructose was carried out in batch reactors with soluble glucose isomerase. In the late 1960s the cost-reducing advantages of using immobilized glucose isomerase were demonstrated, and this led to the rapid development and commercialization of immobilized glucose isomerase technology [27]. The production of HFCS is the largest commercial application of immobilized enzymes. An alternative to immobilization of the enzyme is the immobilization of the complete cell. This technique has also been applied for glucose isomerase [27].

The interconversion of glucose into fructose is an equilibrium reaction:

The thermodynamic equilibrium concentrations vary depending on the reaction conditions, but generally an approximately 1:1 mixture of glucose and fructose is obtained at temperatures between 300 and 350 K [27] (see Figure 12.12).

Table 12.5 Relative sweetness of sucrose and alternative sugars

Sugar	Relative sweetness
Saccharose (sucrose)	100
Glucose syrups	40–60
Fructose	114

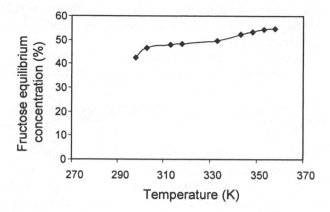

Figure 12.12 Equilibrium concentration of fructose in a glucose/fructose mixture as a function of temperature. Based on data from [27].

QUESTION:

In practice, temperatures of 325–335 K are applied. Why, despite the better equilibrium, are higher temperatures not used?

Figure 12.13 shows a simplified flow scheme of the continuous production of HFCS starting from glucose. Glucose isomerase requires the presence of metal ion such as Co^{2+}, Mn^{2+}, Mg^{2+}, or Cr^{2+} for its catalytic activity. Usually, a magnesium salt is added to the glucose substrate. In additions, salts are added for adjustment of the pH.

Treatment of the product with carbon and by ion exchange is required for the removal of the added salts and impurities that produce undesired coloring. Organic compounds responsible for color in sugar syrups are removed by adsorption on carbon in most sugar plants. Two ion exchangers are used for removal of the remaining salts and impurities, the first is a strong acid cation-exchange resin in the hydrogen form,

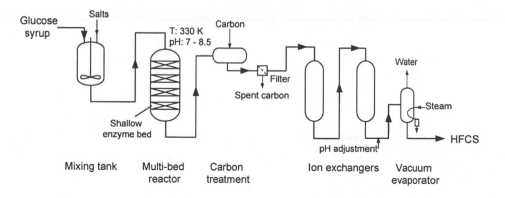

Figure 12.13 Continuous production of HFCS (High-fructose corn syrup).

the second a weak base anion-exchange resin in the free base form [27]. The product pH after ion exchange is adjusted for maximum syrup stability, and the syrup is concentrated by vacuum evaporation.

The pressure drop over the reactor is a point of concern. Particle sizes vary from 150 to 3000 μm. The use of shallow beds as shown in Figure 12.13 reduces the pressure

---

Production of mannitol

D-Mannitol is a valuable sweetener, because it is non-hygroscopic and has no teeth decaying effects. Mannitol can be produced in the 'Combi-Process' (see Scheme B.12.1) by combination of the enzymatic interconversion of glucose and fructose with the simultaneous selective hydrogenation of fructose into mannitol.

In this case, the equilibrium of the isomerization reaction is shifted towards D-fructose by the selective hydrogenation of D-fructose to D-mannitol. For the conversion of glucose into fructose a commercial immobilized enzyme (glucose isomerase) is applied, while the selective hydrogenation is carried out using a heterogeneous metal catalyst, based on copper.

Optimal mannitol yield requires:

- selective hydrogenation of fructose with respect to glucose;
- high selectivity towards mannitol in the hydrogenation of fructose;
- relatively fast glucose to fructose interconversion to maintain fructose concentration at maximum.

Mannitol yields over 60 % are obtained when starting with glucose (see Figure B.12.3). When a mixture of glucose and fructose is used, essentially the same yield is obtained. This shows that the isomerization reaction is fast and the hydrogenation reaction is relatively slow.

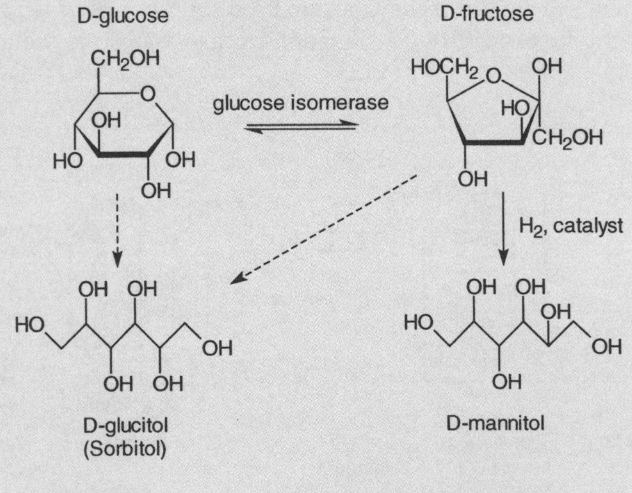

Scheme B.12.1 Combi-process: Simultaneous enzymatic isomerization and metal-catalyzed hydrogenation of D-glucose/D-fructose mixtures [28].

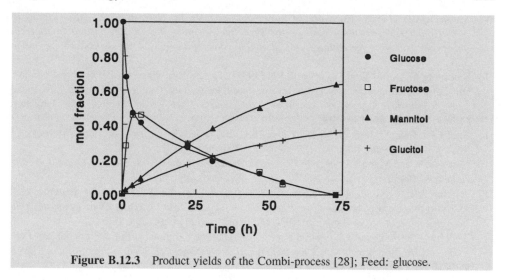

Figure B.12.3 Product yields of the Combi-process [28]; Feed: glucose.

drop across the reactor. In addition, some enzyme systems would be compressed in deep-bed operation. Systems of enzymes supported on porous ceramic have also been developed. Such systems reduce the pressure drop [27].

References

1 Bruggink A (2000) personal communication.
2 Aiba S, Humphrey AE and Millis NF (1973) *Biochemical Engineering*, 2nd ed., Academic Press, New York.
3 Righelato RC (1976) 'Selection of strains of *Penicillium chrysogenum* with reduced penicillin yields in continuous cultures' *J. Appl. Chem. Biotechnol.* 26 153.
4 Heijnen JJ, Terwisscha van Scheltinga AH and Straathof AJ (1993) 'Fundamental bottlenecks in the application of continuous bioprocesses' *J. Biotechnol.* 22 3–20.
5 Hough JS, Keevil CW, Maric V, Philliskirk G and Young TW (1975) *Continuous Culture in Brewing*, pp. 226–237.
6 Lee YL and Chang HN (1990) 'High cell density culture of a recombinant *E. coli* producing penicillin acylase in a membrane recycle fermenter' *Biotechnol. Bioeng.* 36 330–337.
7 Liers S (1997) 'Bioreactoren' in: *Procestechnieken en Engineering* 37540.
8 Winkler MA (1983) in: Wiseman A (ed.) *Principles of Biotechnology*, Surrey University Press, London, Chapter 4.
9 Van Benthum WAJ, Van den Hoogen JHA, Van der Lans RGJM, Van Loosdrecht, MCM and Heijnen JJ (1999) 'The biofilm airlift suspension extension reactor. Part I: design and two-phase hydrodynamics' *Chem. Eng. Sci.* 54 1909–1924.
10 Nicolella C, Van Loosdrecht MCM and Heijnen JJ (2000) 'Particle-based biofilm reactor technology' *TIBTECH* 18 312–320.
11 Stanbury P and Whitaker A (1984) *Principles of Fermentation Technology*, Pergamon Press Ltd., Oxford, Chapter 5.
12 Winkler MA (1983) 'Application of the principles of fermentation engineering to biotechnology' in: Wiseman A (ed.) *Principles of Biotechnology*, Surrey University Press, New York, pp. 94–143.

13 Moon Jr GD (1985) 'Recovery of biological products by distillation' in: Moo-Young M (ed.) *Comprehensive Biotechnology, the Principles, Applications & Regulations of Biotechnology in Industry, Agriculture and Medicine*, vol. 2, Pergamon Press, Oxford, pp. 557–566.

14 Steinberg M, Dong, Y and Borgwardt RH (1994) 'The coprocessing of fossil fuels and biomass for CO_2 emission reduction in the transportation sector' in: Paul J and Pradier C-M (eds.) *Carbon Dioxide Chemistry: Environmental Issues*, The Royal Society of Chemistry, Cambridge, UK, pp. 189–199.

15 Snape JB, Dunn IJ, Ingham J and Prenosil JE (1995) *Dynamics of Environmental Bioprocesses, Modelling and Simulation*, VCH, Weinheim, Chapter 2.

16 Simmler W and Mann T (1992) 'Water' in: Gerhartz W, et al. (eds.) *Ullmann's Encyclopedia of Industrial Chemistry*, vol. B8, 5th ed., VCH, Weinheim, pp. 36–37.

17 Janssen AJH, Dijkman H and Janssen G (2000) 'Novel biological processes for the removal of H_2S and SO_2 from gas streams' in: Lens PNL and Hulshoff Pol L (eds.) *Environmental Technologies to Treat Sulfur Pollution*, IWA Publishing, London, pp. 265–280.

18 Tichý R (2000) 'Treatment of solid materials containing inorganic sulfur compounds' in: Lens PNL and Hulshoff Pol L (eds.) *Environmental Technologies to Treat Sulfur Pollution*, IWA Publishing, London, pp. 329–354.

19 Erickson LE (1988) in: Erickson LE and Yee-Chak Fung D (eds.) *Handbook of Anaerobic Fermentations*, Marcel Dekker, New York, p. 325.

20 Hosten L and Van Vaerenbergh E (1993) 'Biologische eenheidsbewerkingen' in: *Procestechnieken en Engineering* 34170.

21 Barford JP (1988) in: Erickson LE and Yee-Chak Fung D (eds.) *Handbook of Anaerobic Fermentations*, Marcel Dekker, New York, p. 803.

22 Parshall GW and Nugent WA (1988) 'Making pharmaceuticals via homogeneous catalysis, Part 1' *CHEMTECH* 3 184–190.

23 Lilly MD (1977) in: Bohak Z and Sharon N (eds.) *Biotechnological Applications of Proteins and Enzymes*, Academic Press, New York, Chapter 8.

24 Messing RA (1975) *Immobilized Enzymes for Industrial Reactors*, Academic Press, New York.

25 Bohak Z and Sharon N (1977) *Biotechnological Applications of Proteins and Enzymes*, Academic Press, New York.

26 Chibata I (1974) 'Optical resolution of DL-amino acids' in: Kaneko T, Izumi Y, Chibata I and Itoh T (eds.) *Synthetic Production and Utilization of Amino Acids*, Wiley, New York, Chapter 2.

27 Antrim RL, Colilla W and Schnyder BJ (1979) 'Glucose isomerase production of high-fructose syrups' in: Wingard Jr LB, Katchalski-Katzir E and Goldstein L (eds.) *Applied Biochemistry and Bioengineering*, vol. 2, Academic Press, New York, pp. 97–155.

28 Makkee M (1984) *Combined Action of Enzyme and Metal Catalyst, Applied to the Preparation of D-Mannitol*, Thesis, Technical University of Delft.

Literature

Averill BA, Laane, NWM, Straathof AJJ and Tramper J (1999) 'Biocatalysis' in: van Santen RA, van Leeuwen PWNM, Moulijn JA and Averill BA (eds.) *Catalysis: an Integrated Approach to Homogeneous, Heterogeneous and Industrial Catalysis*, 2nd ed., Elsevier, Amsterdam.

Flickinger MC and Drew SW (eds.) (1999) *Encyclopedia of Bioprocess Technology: Fermentation, Biocatalysis, and Bioseparation*, Wiley, New York.

Ho CS and Oldshue JY (eds.) (1987) *Biotechnology Processes, Scale-up and Mixing*, AIChE, New York.

Präve P, Faust U, Sittig W and Sukatsch DA (eds.) (1987) *Basic Biotechnology, a Student*'s *Guide*', VCH, Weinheim.

Wingard LB, Katchalski-Katzir E and Goldstein L (eds.) (1979) *Applied Biochemistry and Bio-engineering*, vol. 2, Academic Press, New York.

Zaborsky OR (1973) *Immobilized Enzymes*, CRC Press, Cleveland, OH.

Chapter 13

Process Development

The strategy for developing new technologies or products is an important concern in research and development (R&D) departments for the chemical industry. The development of a particular process is successful if the desired product is produced in time at the planned rates, the projected manufacturing cost, and the desired quality standards. Included in the term cost are obvious items such as raw materials cost, but also safety and environmental compatibility. Timing is also a critical factor: A technically perfect plant that comes on stream too late, and therefore has to operate in a changed market might turn out to be worthless. So, cost and planning have to be carefully monitored throughout the route from laboratory to process plant.

13.1 DEPENDENCE OF STRATEGY ON PRODUCT TYPE AND RAW MATERIALS

Referring to Figure 2.1 the position of a product in the tree-structure determines the type of R&D that is conducted. R&D for base chemicals and intermediates (bulk chemicals, also referred to as commodities) is basically different from that for consumer products (specialties), as shown in Table 13.1. Fine chemicals will be somewhere in between.

The development of new technologies for the production of base chemicals and intermediates may also be dictated by the raw materials situation. For instance, acetaldehyde production (see Section 8.2) was based on coal derived ethyne until the 1960s, but when the oil era started it became much cheaper and safer to use petroleum derived ethene as the feedstock. In the future, as the crude oil reserves decrease, C_1 chemistry, based on natural gas will probably gain importance for the production of many nowadays oil-derived base and intermediate chemicals.

Bulk chemicals have a wide range of possible uses, and therefore have a long life. The cost of process development is justified by the increased profit that will result from the lower costs of the new process. Examples are the introduction of synthetic methanol after World War I and the commercialization of the methanol carbonylation route to acetic acid in the 1960s, which has quickly replaced the old acetaldehyde oxidation process (see box and Section 8.3).

On the other hand, consumer products will often be replaced by better ones eventually. Thus, process development for bulk chemicals usually aims at *process improvement*, while *product improvement* is the common strategy for consumer products.

At present, the technologies for the production of the base chemicals and intermediates are generally well-established. Therefore, development activities will

Replacement of acetaldehyde oxidation process by methanol carbonylation route

According to [1] Celanese will close its 165,000 t/a acetic acid plant in Cangrejera, Mexico and start up a new 500,000 t/a acetic aid plant in Singapore. Earlier, the company had closed a plant in Frankfurt, Germany with a capacity of 180,000 t/a and one in Celaya, Mexico with a capacity of 65,000 t/a. The technology of the new plant is methanol carbonylation, whereas the plants closed or to be closed use(d) acetaldehyde oxidation technology.

usually result in minor process improvements (e.g. energy savings in methanol and ammonia production, environmental measures, improved catalyst) which, however, can have a large impact on total profits due to the large volumes involved. Still, it is possible that novel concepts are introduced, e.g. the use of a multifunctional reactor for the production of MTBE (see Section 6.2.5) and the ongoing efforts to replace the chlorohydrin process for the production of propene oxide by the direct oxidation route (see Section 9.5.2). A more general drive is towards sustainability and process intensification [2].

In the specialty chemicals (consumer products) industry the development of modi-

Table 13.1 Differences in process development for bulk and specialty chemicals

Feature	Bulk chemicals	Specialty chemicals
Product Life Cycle	Long (>30 years)	Short (<10 years)
Focus of R&D	Driven by cost and environment; Process improvement (modified or new technology)	Driven by yield; Product improvement
Competing processes	One route usually the best	Competitors may use variety of routes
Route selection		
Number of feasible routes	Few	Many, possibly hundreds
Number of feasible feedstocks	Few	Many
Number of reaction steps	Usually one or two	Several (from 3 to 20)
Effort as a proportion of total R&D	Low	High
Impact of poor route selection	High	Moderate
Patent protection	On processes or technologies	On chemistry, often to block competitors' use of similar routes
Technology	Usually continuous	Batch and continuous
Scale-up	Pilot plants and simulation tools commonly used	Use of rules of thumb prevail

fied or new products and their accompanying new processes is much more motivated by market demands concerning product quality and by environmental and safety considerations. Examples of product modifications are the enormously changed quality of car tyres (rubber), paints, and detergents in the past 10–20 years. Examples of new products developed to meet current and future demands are advanced materials such as composites (fiber reinforced resins, e.g. for use in aviation and aerospace) and engineering plastics (high-grade plastics for constructive uses, e.g. in cars). Another example is the increasing interest in chiral drugs (see also Chapters 10 and 12). Catalysts and biocatalysts also belong to the class of new and modified product development.

For bulk chemicals, which can be produced from only a limited number of feedstocks by a limited number of reactions, the majority of the R&D activities concern process design and scale-up. In contrast, in R&D for specialty chemicals a lot of effort is put into the selection of the chemical route. This is the result of the large number of potentially feasible routes; pharmaceuticals and agrochemicals can be manufactured by hundreds of different routes.

In the following sections we focus on research and development of continuous bulk chemicals processes. For a focus on fine chemistry see [3].

13.2 THE COURSE OF PROCESS DEVELOPMENT

Process development is a continuous interaction between experimental programs and design and economic studies. The starting point is a chemical reaction discovered in the laboratory, often accompanied by a catalyst, and the outcome is the production plant. A large part of the development activities is concerned with scaling up from laboratory to full-scale industrial plant, with the intermediate miniplant and pilot-plant stages.

Until about the Second World War, scale-up basically was an empirical process. The common way to deal with scale-up was enlargement in small steps as schematically depicted in Figure 13.1 for a chemical reactor (drawn on scale). This procedure is still quite common in fine and specialty chemicals production, where most processes are batch operations.

Figure 13.1 also shows the approximate production rates and the time intervals associated with the design/construction and operation. Depending on the nature of the process these time intervals may differ greatly from those given. Large safety margins were incorporated, which often resulted in plants with a much larger capacity than what they were designed for. Clearly, scale-up in small steps is very expensive and insecure as long as no predictive models are used.

The time scale in Figure 13.1 is extremely long according to modern standards. Hardly any process development will be approved by management if the horizon for the start-up is 8 years or more. There is a need for a drastic reduction of the time for process development. Fortunately, the discipline chemical engineering has matured and promising developments are taking place. Progress has been and is still being made in software and hardware improvements. Many process steps at present may be captured in a mathematical model with sufficient predictive value. So, the scale-up factor may be enhanced, making it possible to extrapolate directly from miniplant to production scale, thus saving time and money. Therefore, nowadays the usual practice

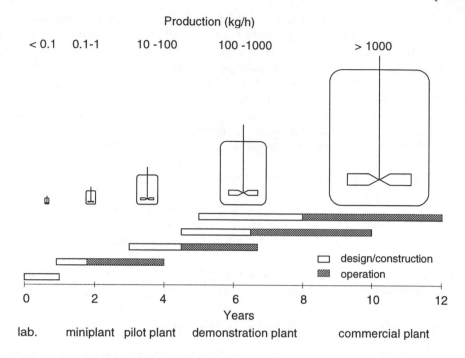

Figure 13.1 Conventional scale-up procedure (production applies to bulk chemicals).

in scale-up is to design an industrial process based on chemical engineering principles, and laboratory experiments on a very small scale. Next, analysis shows which parts of the process need to be examined in further detail on a small scale.

Figure 13.2 shows the cyclic nature of the development of a chemical process. Although the steps outlined in Figure 13.2 are typical, they are not mandatory. For instance, in fine chemicals manufacture the sequence of steps often is laboratory-miniplant-commercial production.

Figure 13.2 assumes a successful completion of the exploratory phase, i.e. the reaction provides satisfactory yields and selectivities in the laboratory reactor. The laboratory (and literature) data are the basis for the process concept. Once this has been produced, the individual steps are further developed and tested on laboratory scale (Is a certain separation possible at all?). Once all the individual steps have been successfully tested, a reliable flow sheet of the entire process can be drawn. A mini-plant is then designed, based on scale-down of the full-scale plant, in order to evaluate the performance of the entire process (e.g. build-up of impurities in recycle streams). Some equipment might have to be tested on a larger scale, the pilot stage, in order to be able to scale up without too large a risk. An additional step might be to design and build a complete pilot plant before the full-scale production plant comes on stream. An evaluation follows each step in the process development. It is then decided whether to continue with the development, stop, or start again at an earlier stage. This decision is based on technical, cost, and market considerations.

It should be stressed that modern process development is done in a parallel rather than a serial way. This calls for close co-operation between the various disciplines

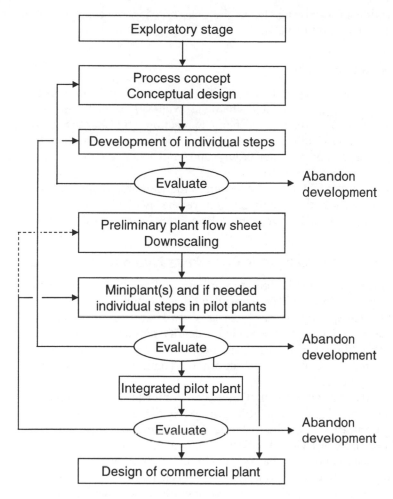

Figure 13.2 Steps in the development of a process according to [4].

involved. One of the advantages is that the time required for development can be greatly reduced.

Also in experimentation the world is changing. Combinatorial methods in catalyst development are applied (also referred to as High Throughput Screening (HTS)); also in other parts of laboratory work a robotic approach is worthwhile. The latter includes experimental programs in miniplants and pilot plants. It is expected that the experimental procedures will change remarkably in the next decade.

13.3 DEVELOPMENT OF INDIVIDUAL STEPS

This section discusses the chemical and chemical engineering steps (shown in Figure 13.2) in the development of a new process. Of course, not only the technical and economical feasibility of a process is important for the successful development of laboratory results to a process plant. Other factors determining the success are the market situation, patenting and licensing situation, etc. This section does not focus upon retrofitting for existing chemical processes, when the changes are only marginal from a scientific point of view.

13.3.1 Exploratory Phase

Process development typically starts with the discovery and subsequent research of a promising new product, or new chemical synthesis route to an existing product. Research in this phase focuses on chemical reactions and catalysis. It includes obtaining information on what reactions take place, the thermodynamics and kinetics of the reaction(s), dependence of selectivity and conversion on the process parameters, catalyst and catalyst deactivation rate, etc.

During process development, research in the laboratory continues. For instance, in the early stages kinetic data might not yet be available, but as the development continues these data will have to be collected for proper reactor and process design.

13.3.2 From Process Concept to Preliminary Flow Sheet

After optimization of the laboratory synthesis and assembly of other important information, a process concept is developed and the economic potential is determined. The availability and quality of raw materials must also be determined. In this phase conceptual process designs are made to identify promising options (see [5]). At this point several different versions of the process may be drawn up (and even different synthesis routes might have come out of the laboratory), from which it is very hard to choose the best one.

Many of the alternatives can of course be eliminated for trivial reasons (industrial, safety, economic, legal, etc.). Still, a large number remains and it is impossible to calculate costs for all of them. Finding the optimum choice requires knowledge, experience, and creativity. Tools that help making the decision are continuously improving and include data banks and simulation programs.

It is important that an economic evaluation is part of the design process, because only then can promising alternatives be identified from the start and hopeless ones discarded as soon as possible. At the early stages, a comparison of the raw material costs with the product value can already give an indication; if even at 100% conversion the product value does not exceed the raw material costs, the process is certainly not viable at that time.

13.3.2.1 Reactor system

The more information becomes available about the process, the more detailed the conceptual design can become. The reactor system (reactor, including fresh feed, product, and gas and liquid recycle streams) must be specified early on because it influences the product distribution and the separation section. The effect of reactor type on product distribution is illustrated by the coal gasification process (see Chapter 5). In this process, the product distribution in a moving-bed, fluidized-bed, and entrained-flow reactor are very different (e.g. much more hydrogen and carbon monoxide are formed in an entrained-flow reactor, while the content of methane and other hydrocarbons in the product gas from a moving-bed reactor is very large).

Example: ammonia synthesis

In ammonia synthesis (see Section 6.1), a solid-catalyzed gas-phase reaction, low temperatures are required to achieve acceptable conversions, while the rate of reaction increases with temperature (this is a dilemma all reversible exothermic reactions have in common). Hence, the reaction is carried out in an adiabatic fixed-bed reactor with intermediate cooling, thus compromising between high temperature (small reactor volumes) and high equilibrium conversion. The exploratory laboratory research reveals that high conversions are not possible, even at high pressures. Therefore, a gas recycle loop has to be incorporated. A purge is needed to prevent impurities build-up. Because of the damaging effect of CO and CO_2 in the synthesis gas on the catalyst activity, purification of the feed is necessary. Figure B.13.1 shows a simplified conceptual flow sheet, which incorporates these results. The source of synthesis gas (coal, natural gas, etc.) has not been specified. The choice of feedstock and process depends on (local) availability, cost, etc., but nowadays steam reforming of natural gas followed by partial oxidation in air, is the predominant technology. Also, decisions must be made concerning feed purification steps (shift reactions, CO removal by methanation or some type of physical separation process). Further research and development activities determine the ammonia reactor design in more detail (i.e. whether intermediate cooling will be accomplished by cold gas injection or by inter-stage heat exchangers, etc.). The reactor size can be determined once the kinetic data become available.

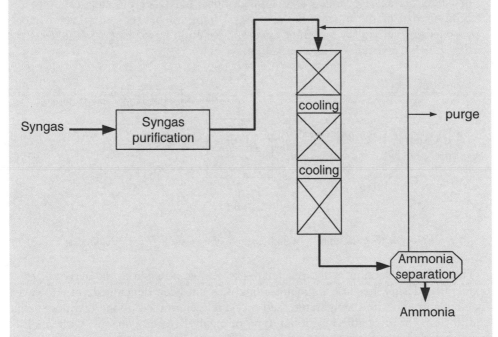

Figure B.13.1 Simplified initial conceptual flow sheet of ammonia synthesis.

Example: Fluid Catalytic Cracking (FCC)

A process where the size of the catalyst particles plays an important role is
catalytic cracking (Section 3.4.1). The earliest commercial catalytic cracking
process was based on fixed-bed reactors, developed and patented by E. Houdry.
Oil companies were very interested in the process, but because of the high
licensing costs, they decided to develop their own processes. In these processes
fine catalyst particles were used instead of the larger particles used in a fixed bed,
and thus, they evaded the patent. The advantage of the usage of small particles is
the high catalyst effectiveness factor. Another reason for using fine particles was
that transport between reactor and regenerator would be possible. The primary
advantage of catalyst recirculation is the fact that the heat needed for the
endothermic cracking reactions can be supplied by the combustion of coke in
the regenerator, the catalyst being the heat carrier. In the Houdry process this
'heat-balanced' operation is not possible.

The choice of reactors with fine catalyst particles is limited. If the reactants are
in the liquid phase a slurry reactor can be applied. However, because of the high
temperatures required for the endothermic cracking reactions, the hydrocarbons
are present in the gas phase, so this option can be discarded immediately. With
gaseous reactants a fluidized-bed or entrained-flow reactor can be used. Figure
B.13.2 shows a simplified conceptual flow sheet for the FCC process.

The use of fine catalyst particles, either in a fluidized-bed or entrained-flow
reactor, implies a separation between gas and catalyst. The hydrocarbon
products are also separated to obtain gasoline and other products. The most
difficult part of the process is the catalyst transport between the reactor and
regenerator and vice versa. Much development effort has to be put in the design
of a reliable and safe catalyst recycling system.

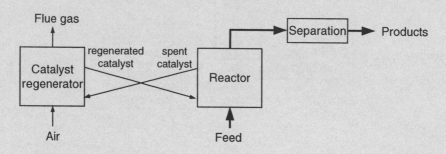

Figure B.13.2   Simplified initial conceptual flow sheet of the FCC process.

The definite selection of the reactor is not always possible at the early stages of
conceptual design, because a kinetic model has not been developed yet. However,
factors such as desired temperature and pressure (determined by the thermodynamic
equilibrium for reversible reactions), type of catalyst (heterogeneous, with small or
large particles, homogeneous, none), and the phases of the reactants and products, will

narrow the choice. It has to be noticed that sometimes the particle size can only be decided upon after elucidation of the kinetics.

13.3.2.2 Separation system

When the reaction product contains multiple components, it can be difficult to decide which is the best separation system. An example of the determination of a distillation sequence is given next.

QUESTION:

> *Check the sequence in the flow sheets for FCC (Section 3.4.1), steam cracking (Section 4.4), methanol production (Section 5.2), and acetic acid production (Section 8.3).*

13.3.2.3 Testing of the individual steps

When reactor design and separation processes have been decided upon, a conceptual flow sheet for the industrial plant can be drawn. The individual unit operations (reactor, distillation, etc.) are designed based on existing information (approximate size of columns is determined, etc.). The process design at this early stage is, of course, preliminary: not all data are available yet, and some pieces of equipment either are not included in the flow sheet or are not sized accurately. Individual process steps have to be examined in more detail, and the results fed back to the conceptual flow sheet stage in order to produce a reliable preliminary process flow sheet. For instance, the first conceptual flow sheet is often based on the assumption that the (liquid) reaction products can be separated by distillation relatively easily, because no vapor-liquid equilibria have been determined yet. When the individual steps are developed and tested, the occurrence of azeotropes, etc. might be observed, resulting in more expensive azeotropic distillation or some other separation technique. Temperature limitations may also be discovered, e.g. in the distillation of the products from styrene manufacture, distillation under reduced pressure is necessary in order to prevent styrene from polymerization.

The laboratory experiments are usually carried out with a mixture prepared from pure materials. Therefore, accumulation of impurities or by-products in recycle streams is not accounted for in the tests. Furthermore, laboratory experiments often are of short duration, and usually initial catalyst activities are measured. Catalysts may appear very active in the laboratory, but they could loose their activity and/or selectivity within days or even hours (e.g. as a result of coke formation by undesired side reactions). This at least calls for special measures concerning catalyst replacement and/or regeneration, but might even lead to the abandonment of the entire process development. Miniplants and pilot plants can reveal such problems.

13.3.3 Pilot Plants/Miniplants

A pilot plant may be defined as an experimental system, at least part of which displays operation that is representative (either identical or transposable) of the part which will correspond to it in the industrial unit [6]. Pilot plants range in size from a laboratory unit (miniplant) to an almost commercial unit (demonstration plant) but are usually intermediate in size. The development of a new process or product usually requires miniplant or pilot plant work or both. The reasons are various. It might be important to generate larger quantities of a product in order to develop a market. It might be advisable to confirm the feasibility of the process because serious doubts might

Example: cyclohexanone production

In the production of cyclohexanone from cyclohexane, a series of four distillation columns is required to separate the product mixture, consisting of five components. This means that 14 possible sequences can be distinguished. The question: what is the best sequence? In this case it is relatively easy to come up with the right answer, based on common heuristic rules. These rules, which apply to other separation processes also, are [5]:

1. Remove corrosive or hazardous components as soon as possible.
2. Remove reactive components or monomers as soon as possible.
3. Remove products as distillates.
4. Remove recycles as distillates, particularly if they are recycled to a fixed-bed reactor.
5. Most plentiful first.
6. Lightest first.
7. High-recovery separation last.
8. Difficult separation last.

QUESTION:
          Explain the logic of these rules.

The product mixture from the cyclohexane oxidation reactor has the following composition (the products are arranged in order of increasing boiling point):

Cyclohexane: 94.0% to be recycled to reactor
Light products: 0.5%
Cyclohexanone: 3.0%
Cyclohexanol: 1.5%
Heavy products: 1.0%

All components possess about the same properties as far as corrosiveness, hazard ratings, and reactivity are concerned. Based on rule 3 we want to remove cyclohexanone and cyclohexane as distillates. Cyclohexane is both the most plentiful and the lightest component, so it is removed first. The separation of cyclohexanone and cyclohexanol is most difficult and requires the highest recovery, so this separation is saved till last. This only leaves the choice of whether to remove the heavy or the light components first. The heavies are more plentiful so they are removed first. The distillation sequence thus is as shown in Figure B.13.3.

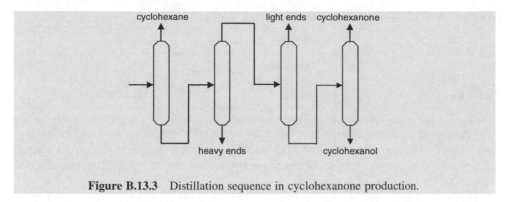

Figure B.13.3   Distillation sequence in cyclohexanone production.

exist on individual steps. It is often necessary to check design calculations and generate design data for a commercial plant. In particular with novel processes scale-up problems can often only be solved by experimental demonstration. Often it is rewarding to gain operational process know-how (e.g. start-up procedures). In practice, often long-term effects are encountered. Notorious examples are catalyst deactivation and the build-up of impurities. In particular, when the process is really novel it is advisable to demonstrate it.

Pilot plants are very costly. Typically, the initial investment is at least 10% of the investment costs of the industrial plant. Therefore, there is a large incentive to use miniplants combined with mathematical simulation of the process.

13.3.3.1 Miniplants

If the objective of a pilot plant is to evaluate the feasibility of a new process technology or generate design data for a new process, then the miniplant strategy may be most appropriate. The miniplant serves to test the entire process, including all recycles, on a small scale. Its design is based on scaling down of the projected commercial plant. The miniplant technique has the following characteristics:

- the miniplant includes all recycling paths and it can consequently be extrapolated with a high degree of reliability;
- the components used (columns, pumps, etc.) are often the same as those used in the laboratory. In addition, they usually contain standardized equipment that can be reused in other miniplants, resulting in reduced investment costs and high flexibility;
- the miniplant is operated continuously for weeks or months, and therefore is automated to a large degree;
- the miniplant is used in combination with the mathematical simulation of the industrial process.

A miniplant typically produces 0.1–1 kg product per hour. In fact, it can also be regarded as a small pilot plant.

QUESTION:

What are the advantages of a miniplant compared to a large pilot plant?

Table 13.2 Typical production rates and scale-up factors

	Production rate (kg/h)	Scale-up factor
Industrial plant	1000–10000	–
Pilot plant	10–100	10–1000
Miniplant	0.1–1	1000–100000

The use of a miniplant together with mathematical simulation often makes it possible to skip the pilot-plant stage, though retaining the same scale-up safety. As a result it is possible to speed up process development, which can have a noticeable effect on the cost effectiveness of a new process.

Miniplants are particularly suitable to test long term catalyst stability under practical conditions, i.e. with real feeds and recycle streams. The incorporation of recycle streams is important to detect the effects of trace impurities or accumulated components, that are not observed in laboratory scale experiments (see also Section 13.4.3).

The miniplant technique has certain limitations, however, in that the scale-up of individual process steps such as extraction, crystallization, and fluidized-bed operation from miniplant to full-scale plant often is too risky in technical terms. Table 13.2 shows typical production rates and scale-up factors (size of industrial plant/size of mini or pilot plant).

For some important process steps, Table 13.3 shows typical maximum scale-up values above which reliable scale-up is no longer possible. As can be seen from Table 13.3, fluidized-bed reactors are governed by a much smaller scale-up factor than other reactor types. The scale-up problems are mainly caused by the change of flow characteristics when going from a small unit to an industrial reactor. Calculations show that the conversion in fluidized beds may vary from that in plug flow to well below

Table 13.3 Typical maximum scale-up values of some important process steps [7]

Process step	Maximum scale-up value
Reactors	
Multi-tubular and adiabatic fixed-bed reactor	>10000[a]
Homogeneous tube and stirred tank	>10000
Bubble column	<1000
Gas-solid fluidized bed	50–100
Separation processes	
Distillation and rectification	1000–50000
Absorption	1000–50000
Extraction	500–1000
Drying	20–50
Crystallization	20–30

[a] Scale-up factors of over 50000 have already been achieved.

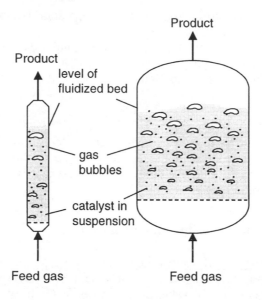

Figure 13.3 Small and large scale fluidized-bed reactors.

mixed flow, and it is very difficult to reliably estimate the conversion for a new situation. The problem is that in a laboratory scale reactor the gas bubbles are of the same order of magnitude as the column diameter (ca. 0.1 m). This bubble diameter is only slightly dependent on the scale of the fluidized bed, which means that in an industrial reactor the bubbles are much smaller than the reactor diameter. Figure 13.3 illustrates this. The result of the different bubble-to-column diameter ratios is that the flow patterns on the small and large scale differ, and therefore also mixing, mass transfer, and heat transfer.

The disadvantage of the limited scale-up factor can be circumvented by isolating the critical process steps and working on them at an intermediate stage, the pilot stage. If enough representative feed can be supplied to the critical step for sufficiently long time, then the entire plant can again be extrapolated to full scale without too much risk.

QUESTION:

> *Why are the hydrodynamics for a laboratory size fluidized bed often very different from those in the commercial one?*

13.3.3.2 Pilot plants

A pilot plant should be designed as a scaled-down version of the industrial plant, not as a larger copy of the existing miniplant. Pilot plants vary greatly in size and complexity. In the case of new technology and for market testing an integrated pilot plant is required. When the feasibility of a process cannot be proven on miniplant scale, a demonstration plant may be built. This often is the case with solid handling processes, e.g. coal gasification and coal liquefaction. Its minimum size is determined by the minimum size of a particular unit operation or piece of equipment. In coal liquefaction for instance, the process chemistry can be proven in miniplant operation. However, the uncertainties in the scale-up of the liquid/solid separation equipment requires pilot plants of much larger scale.

Critical steps may be examined in individual pilot units. These pilot units need not be a complete but smaller replica of the envisioned commercial unit, as long as they yield the required data for scale-up. Some examples will be given in Section 13.4.

13.3.3.3 Mock-ups

Mock-ups are useful tools in scale-up. Mock-ups or 'cold flow models' are usually used to simulate those aspects of the process related to fluid flow. They do not necessarily physically resemble the part of the process that is studied. Compared to pilot units they are relatively cheap.

Example: FCC catalyst testing in a miniplant

An example of the use of a mock-up at the miniplant stage is found in FCC catalyst testing. Figure B.13.4 shows a schematic of a novel miniplant for the testing of FCC catalysts [8]. Table B.13.1 shows scale-up data for an industrial FCC riser reactor, the miniplant, and the miniplant mock-up.

Although the miniplant (microriser) and industrial unit (see Section 3.4.1) have quite different appearances, the miniplant still resembles the industrial unit, e.g. similar length, residence time, oil/catalyst contact time, contacting pattern, etc. Industrial riser reactors have a length of 20–30 m and a diameter of 1–2 m. In the miniplant the same length is used but due to spatial restrictions the tube has to be folded. It is important to determine whether the folding of the reactor influences the contact between gas and catalyst, and thus conversion, product distribution, etc. It is conceivable, for instance, that the catalyst particles and the gas are separated in the bends due to centrifugal forces.

The mock-up, a glass model of the folded pilot riser reactor, is ideal to examine this phenomenon. As the problem is essentially of a hydrodynamic nature, it is important to keep the Reynolds number constant (hydrodynamic similarity, see also Section 13.4 for the use of similarity in scale-up). This is achieved by using a glass reactor with the same diameter as the pilot plant, and nitrogen at room temperature as feed. This way, the kinematic viscosity (η/ρ) is the same as that of the pilot plant feed at reaction temperature. In the glass mock-up, which can be shorter than the miniplant reactor, observation of the behavior of the catalyst particles is possible. In addition, this behavior can be quantified with techniques such as residence time distribution measurements of colored catalyst particles. The results obtained from both the miniplant (chemical behavior) and the mock-up (hydrodynamics) can be incorporated into a reactor model to predict the catalyst performance in an industrial riser.

QUESTIONS:

During catalyst development research for FCC usually fixed-bed and occasionally fluidized-bed reactors are used. Compare the three reactor types. Which type would you choose? Explain.

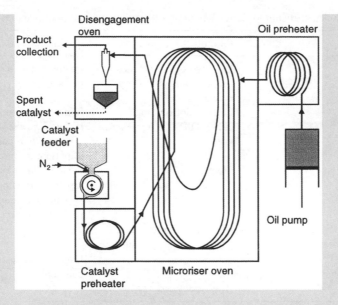

Figure B.13.4 Miniplant for testing FCC catalysts [8].

Table B.13.1 Data for industrial FCC unit, miniplant for FCC catalyst testing and mock-up

	Industrial reactor	Miniplant	Mock-up
Feed	FCC feed + steam	FCC feed + nitrogen	Nitrogen
Temperature (K)	790	790	298
Reactor length (m)	20–30	20	4–20
Reactor diameter (m)	1–2	0.004	0.004
Residence time (s)	2–5	2	0.4–2
Reynolds$(\rho u D/\eta)^a$	$\approx 5 \times 10^5 - 10^6$	≈ 3000	≈ 3000
Catalyst feed rate (kg/min)	$10^4 - 3 \times 10^4$	≈ 0.001	≈ 0.001

$^a\rho$ = fluid density (kg/m^3); u = fluid velocity (m/s); D = reactor diameter (m); η = dynamic viscosity (kg/(m s).

13.3.3.4 General remarks on pilot plants/miniplants

As has become clear not all unit operations need to be pilot planted (see Table 13.4). For instance, for distillation columns the required information such as vapor-liquid equilibrium data, etc., is either available from literature or can easily be determined in the laboratory. The methodology for design of distillation equipment is well known and straightforward. Reactors, on the other hand, often require pilot plant testing.

13.4 SCALE-UP

This section is not intended to treat the theoretical aspects of scale-up, but rather to give some examples of the application of scale-up methods to processes that already have been treated in this book. The reactor is the core of the process plant. It also is the part of the plant where most scale-up problems are encountered. Therefore, this

Table 13.4 Rules of thumb to decide whcih unit operations require pilot plant testing [9]

Operation	Pilot plant required?	Comments
Distillation	Usually not. Sometimes needed to determine tray efficiency	Foaming may become a problem
Fluid flow	Usually not for single phase. Often for two-phase flow	Some 1-phase polymer systems are also difficult to predict. CFD[a] can be an important tool
Reactors	Frequently	Scale-up from lab often justified for homogeneous systems and single-phase packed-bed reactors
Evaporators, reboilers, coolers, condensers, heat exchangers	Usually not unless there is a possibility of fouling	
Dryers, solids, handling, crystallization	Almost always	Usually done using vendor equipment[b]
Extraction	Almost always	

section will concentrate on reactor scale-up. Scale-up makes use of laboratory and/or pilot plant data, complemented by mock-up studies and mathematical modeling, to determine the size and dimensions of the industrial unit. Two reactor categories will be dealt with in this section, namely (1) homogeneous reactors with a single fluid phase and (2) fixed-bed reactors with one or more fluid phases. Some attention will also be paid to a typical scale-up problem being the build-up of impurities in recycle streams.

13.4.1 Reactors with a Single Fluid Phase

Reactors involving a single fluid phase present the least difficulty in scaling up. Homogeneous reactions may be carried out using one of the following reactor systems:

- (semi)-batch reactor (fine chemistry, biotechnology);
- continuous tubular reactor (steam cracking);
- continuous stirred-tank reactor/cascade of stirred-tank reactors (homogeneous catalysis).

13.4.1.1 Batch reactor

Batch and semi-batch reactors are frequently used for the production of fine chemicals and specialties such as cosmetic products, floor polishes, etc. Bioreactors are also often batch reactors. The concerns involved with the scale-up of batch reactors have been dealt with in Chapters 10 and 12. As discussed there, the greatest problems evolve when highly exothermic reactions are involved. The solution to this problem is to separate reaction volume and heat exchange area, e.g. by applying external heat exchangers. In this respect it is convenient to work at the boiling point of the reaction mixture, allowing heat removal with a reflux system. Then, the batch reactor may be scaled up from laboratory size to industrial plant simply based on reaction time.

Nowadays often a special type of laboratory reactor is applied to determine relations to describe heat production and transfer. The reactor used is a calorimeter in which parameters such as the thermal resistances and the heat evolved from reactions can be measured accurately. From the results, reactor specific constants can be derived, based on which scale-up calculations for reactors of similar geometry can be made.

13.4.1.2 Continuous tubular reactor

An example of a process employing a continuous tubular reactor involving one fluid phase is steam cracking of naphtha (see Chapter 4). The most important concern in the scale-up of a steam-cracking unit is heat transfer and, associated with this, the temperature profile. On a small scale often an electric furnace is used, whereas on an industrial scale the furnace is always heated by burners [10]. To be able to translate the laboratory results to the industrial unit, a kinetic model of the reaction system is necessary because the interaction of temperature and chemical reaction has to be predicted. The problem is that many different molecules are involved in the reaction and that naphtha does not always have the same composition. An accurate kinetic model can be derived from experimentation on bench/miniplant scale and pilot plant scale together with the introduction of the 'lumping principle', i.e. various molecules are grouped together to form e.g. the n-alkanes, the iso-alkanes, etc. lumps. The model treats these lumps as if they were one species, where in fact the lumps consist of several species (n-alkanes: C_4, C_5, etc.). The model will then be able to predict the effect of the operating parameters on performance of different naphthas, based on their composition (i.e. % n-alkanes, % iso-alkanes, etc.), without extensive additional experimentation.

This 'lumping' approach is also commonly used in modeling of, for instance, the kinetics of catalytic cracking and catalytic reforming. The problem is to determine to what extent lumping is tolerated (varying from considering every molecule separately to putting all molecules in one lump).

Nowadays, due to the increasingly faster computers, lumping models may contain hundreds of different lumps. A good example is naphtha cracking where models are used based on real elementary reaction steps. In fact extended de-lumping has taken place. Another example is combustion of methane. In that case essentially all reaction steps can be taken into consideration in a model. This leads to several thousands of equations of the concentrations of components and the associated reaction rate constants. It is not surprising that the best examples are non-catalyzed reactions. Also for FCC and hydrotreating de-lumping is practiced by applying sophisticated models, but this is much more difficult than for non-catalyzed reactions. It is expected that molecular modeling will give this field a push.

13.4.1.3 Continuous stirred-tank reactor

The scale-up of this type of reactor usually involves the transposition from a batch laboratory reactor to a continuous stirred-tank industrial reactor (or cascade of reactors). If the kinetics are known from small-scale experimentation, this transposition is

Example: lumping models for catalytic cracking

Figure B.13.5 shows two lumping models that are used in catalytic cracking.The three-lump model considers a gas oil feed (F), a gasoline (G) and a light gas + coke lump (C). In the ten-lump model the gas oil lump is divided in a heavy and a light gas oil (subscripts h and l), which each contain four lumps, i.e. alkanes (P), naphthenes (N), aromatic substituent groups (A) and aromatic rings (C_A). The two remaining lumps again are the gasoline (G) and the light gas + coke lump (C). The arrows represent the conversion of one lump into the other with the corresponding rate constants.The three-lump model is useful to model data obtained in the laboratory, but has no predictive value for other feedstocks. One of the reasons for this is the oversimplification. Another, more fundamental reason is that the definition of conversions and selectivities is not straightforward. For instance, the feed F already contains gasoline components, which are partly converted to gas and coke. The ten-lump model is more complex and does have more predictive value. However, determination of the model parameters in this case is more elaborate.

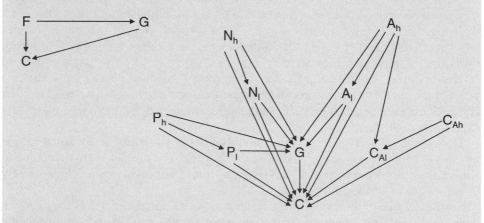

Figure B.13.5   Three-lump model [11] and ten-lump model [12] for catalytic cracking; explanation of symbols is given in the text

relatively easy because scaling up is possible on the basis of the kinetic model. However, a pilot plant reactor may still be built. The motivation then is not a scale-up problem but the need to answer some questions, for instance about the influence of impurities on the catalyst activity, mechanical stability of the catalyst particles, and catalyst removal, particularly in homogeneously catalyzed systems.

13.4.2 Fixed-bed Catalytic Reactors

Many applications of heterogeneous catalysis can be found in oil refining and the petrochemical industry. Examples that have been treated in this textbook are steam reforming, catalytic reforming, ammonia synthesis, methanol synthesis, and hydro-treating. Major issues in the design and scale-up of these processes are:

- temperature control;
- pressure drop;
- catalyst deactivation.

13.4.2.1 Temperature control

A distinction is made between endothermic and exothermic reactions. In very endothermic reactions such as steam reforming, the temperature decrease may be so severe that the reactor would have to be excessively long to compensate for the low reaction rate. In this case the reaction mixture has to be heated rapidly to high temperature in order to achieve satisfactory production. In steam reforming heating is accomplished by mounting tubes containing the catalyst in a furnace in which the heat flux is very high. In order to achieve uniform heating the maximum tube diameter must not exceed 0.1 m.

Exothermic reactions such as ammonia production and hydrodesulfurization are usually carried out in multi-bed reactors with intermediate cooling. Cooling is either done in external heat exchangers, or by injection of cold feed gas.

13.4.2.2 Pressure drop

The pressure drop in a catalyst bed must be limited, particularly when recycle streams are involved (ammonia, methanol, ethene oxide). In order to decrease the pressure drop the bed height may be reduced or larger catalyst particles may be applied. Both solutions have their drawbacks. Another solution is to apply axial flow, or a structured reactor.

13.4.2.3 Catalyst deactivation

An important issue in the choice of reactor type for hydrocarbon conversions is the deactivation of the catalyst by coke deposition. To maintain the desired conversion level, regeneration by burning off the coke is necessary. This may be achieved by installing a number of parallel reactors, or by using moving-bed or fluidized-bed operation. Examples are catalytic reforming and catalytic cracking.

13.4.2.4 Show-tube concept and downscaling

In the scale-up of fixed-bed reactors the 'show-tube' concept (see Figure 13.4) is very useful. The 'show-tube' concept is based on similarity. It is highly successful, since most, if not all parameters are kept the same.

The show-tube concept is shown for a large diameter fixed-bed reactor (used in e.g. hydrotreating), for a multi-tubular system (used in e.g. the Lurgi methanol process), and for monolith reactors.

In the scale-up of a large diameter fixed-bed reactor a characteristic slice of the bed is studied; the reactor is scaled up in diameter and not in height. In the case of a multi-

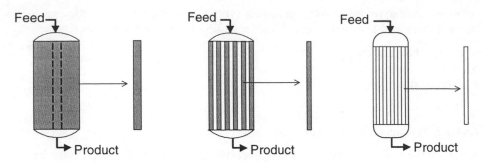

Figure 13.4 Show-tube concept applied to large diameter fixed-bed reactor (left), show-tube reactor (middle), and monolith reactor.

tubular reactor or a monolith reactor, the show tube can be one full-sized tube or one monolith channel, respectively, with the same dimensions as the ones used on the commercial scale. Scale-up is then simply linear, adding more tubes or channels in a bundle. Some points of concern in the use of the 'show-tube' concept are:

- fluid distribution over bed or multiple tubes is not addressed;
- heat distribution over bed or tubes is not addressed;
- inconveniently long pilot reactors are needed.

The problem of fluid distribution is most severe in trickle-flow reactors (gas/liquid/catalyst), because situations may prevail in which liquid preferentially flows through a certain part of the bed, while the gas flows through another part (see Figure 13.5). In fact, often not only the non-ideality of the bed itself is the cause, but the inlet device

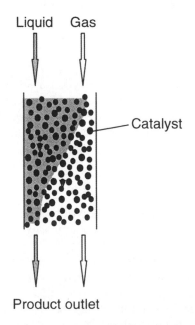

Figure 13.5 Non-ideal trickle-flow reactor.

Table 13.5 Data for commercial, pilot and microflow trickle-flow reactors; LHSV $= 2$ m_{liq}^3/m_{cat}^3·h, $d_p = 1.5$ mm [14]

	Commercial	Pilot plant	Microflow
Catalyst volume (m^3)	100	0.01	8×10^{-6}
Diameter (m)	2.5	0.04	0.01
Length (m)	20	8	0.1
Liquid velocity (cm/s)	1.1	0.4	0.006
Reynolds number, Re_p	55	22	0.3

also plays a crucial role. This holds even more for monolithic reactors: if the distribution at the inlet is right it will be right everywhere in the reactor, while the opposite, i.e. maldistribution at the inlet, will also persist throughout the monolith. In this case a mock-up is very useful to study the liquid-phase distribution [6]. Recently, also Computational Fluid Dynamics (CFD) has become a great help in designing inlets and other internals.

Presently, the design criteria for fixed beds are well-known. When a packed-bed reactor is used in catalyst testing, downscaling is the objective. The issue now is to find the smallest scale reactor that could provide accurate data, e.g. in determining the kinetics. The trend is to scale down to microflow level [13,14]. The 'show-tube' principle is no longer used in this case. Table 13.5 shows some typical data of for trickle-flow reactors of varying size.

The liquid hourly space velocity (LHSV) is kept constant (and thus the average residence time) but due to the smaller reactor length, the liquid velocity has to be lower in going to a smaller scale. This results in different fluid dynamics in the large and the small reactor.

The particle diameter, d_p, is the main factor determining the minimum dimensions of the small reactor. A particle size of 1.5 mm is commonly used in hydrotreating processes. With this particle size, reactors with bed volumes of the order of 10^{-4} m^3 (single fluid) and 10^{-3} m^3 (trickle-flow) would be required to insure an even flow distribution. These volumes are based on relations determining the minimum D/d_p and L/d_p ratios in order to minimize wall effects and deviations from plug flow, respectively. One way to still be able to use a microflow reactor is to dilute the catalyst bed with smaller inert particles. Then the hydrodynamics are determined by the smaller particles and the kinetics by the 1.5 mm diameter catalyst particles. The catalyst bed dilution technique has proven to be a convenient tool to allow downscaling of laboratory reactors to microreactor size, both for single-phase and two-phase fluid flow, while generating data that are relevant for commercial operation. The small size has many advantages such lower costs, low materials consumption, enhanced safety, etc.

QUESTIONS:

Why is the length not kept constant in the three reactors in Table 13.5? What is the consequence of the choice not to vary the LHSV?

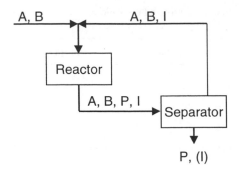

Figure 13.6   Accumulation of an impurity in a recycle stream.

13.4.3 Catalyst Stability and Accumulation of Impurities

Catalyst stability is a very important aspect of every catalytic process. It is often frustrating that it is close to impossible to obtain the experimental data on deactivation. In particular, a problem exists when deactivation is caused by the accumulation of impurities whose presence was not detected on the laboratory scale.

Impurities virtually become lost in small-scale experimentation, e.g. due to settling as a result of long standing feedstocks [15]. Impurities may not be found in laboratory work because their concentrations are too small to be detectable. In the laboratory usually the reaction is carried out batch-wise without recycle streams. Furthermore, usually short runs are carried out which do not allow enough time for the impurities to build up for their effect to be noticeable. However, in commercial operations with long operating periods, these impurities may well reach concentrations that are significant and sometimes very damaging. Examples are deactivation of catalysts due to very small quantities of impurities in feed streams or the build-up of impurities in recycle streams, and lowering of product purity due to enrichment of a solvent with by-products. Figure 13.6 shows an example of the accumulation of impurities in a recycle stream.

Two gaseous reactants A and B react to produce liquid product P. In the separator the gas and liquid phase are separated. The gas phase is recycled to the reactor. However, some gaseous by-product I, which is only slightly soluble in P, is formed during the reaction. In a once-through laboratory process this often remains unnoticed. The danger is that the impurity unexpectedly accumulates in the system. Impurities can be detected in miniplants or integrated pilot plants, which include all the recycle streams. Once they have been detected, the appropriate measures can be taken, e.g. installing pretreating units, purging some of the recycle stream, etc.

13.5 SAFETY AND LOSS PREVENTION

13.5.1 Introduction

Over the last three decades the chemical process industry has grown very rapidly, with corresponding increases in the quantities of hazardous materials in process, storage, or transport. Plants are becoming larger and are often situated in or close to densely

Table 13.6 Types of chemical plant accidents [16]

Type of accident	Fire	Explosion	Toxic release
Probability of occurrence	High	Intermediate	Low
Potential for fatalities	Low	Intermediate	High
Potential for economic loss	Intermediate	High	Low

populated areas. Accidents in chemical plants can result in large human and/or economic losses and environmental pollution. Increasing attention must therefore be paid to the control of these hazards, and in particular to the use of appropriate technologies to identify the hazards of a chemical process plant, and to eliminate them before an accident occurs; this is *loss prevention*.

Major hazards in chemical plants are generally associated with the possibility of explosion, fire, or the release and dispersion of toxic substances, with some plants combining some or all of these hazards. As shown in Table 13.6, fires are the most common chemical hazards, while explosions, and in particular unconfined vapor cloud explosions, lead to the largest economic losses. The potential for fatalities is highest in the case of toxic release (the most notorious being the Bhopal disaster, which killed over 2500 civilians at the time (the present death toll has exceeded 8000) and seriously injured an estimated 200,000 more [16] (see Section 13.5.2).

A number of the most severe accidents are described throughout this section, but it should be noticed that lesser incidents with smaller consequences occur every day across the globe. Major disasters often lead to major changes in legislation. Unfortunately, minor incidents are seldom documented, often inadequately investigated, and the lessons that could be learnt are quickly forgotten. For instance, prior to the Bhopal disaster, minor incidents had already occurred at the plant site.

Similar accidents as in Bhopal, although improbable, are certainly possible. An analysis that compared chemical incidents in the U.S. in the early to mid-1980s to the Bhopal incident revealed that of the 29 incidents considered, 17 incidents released sufficient volumes of chemicals with such toxicity that the potential consequences could have been more severe than in Bhopal (depending on weather conditions and plant location) [17].

Figure 13.7 shows the main causes of chemical plant accidents. Most of the losses as a result of accidents can be attributed to human error. For instance, mechanical failures could all be due to human error as a result of improper design, maintenance, inspection, or management. Piping system failures represent the largest part of the incidents (for instance, the Flixborough disaster in 1974 (see Section 13.5.2)).

The analysis of accidents has shown that they are a result of a chain of events, often starting with a relatively trivial incident which (either because it goes unnoticed, or because of inappropriate response) triggers a chain reaction that can rapidly lead to a catastrophe.

13.5.2 Safety Issues

Most accidents in chemical plants result in spills of flammable, explosive, and/or toxic materials, e.g. from holes and cracks in tanks and pipes, and from leaks in flanges, pumps, and valves.

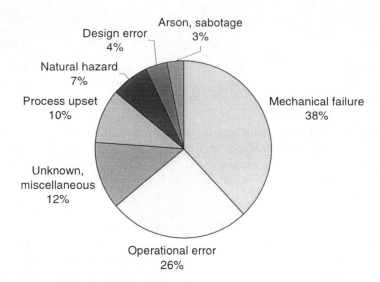

Figure 13.7   Causes of chemical plant accidents. Based on data from [18].

13.5.2.1 Flammability – fires and explosions

The hazard caused by a flammable material depends on a number of factors, i.e. its flashpoint, its autoignition temperature, its flammability limits, and the energy released in combustion (see Tables 13.7 and 13.8).

   The main difference between fires and explosions is the rate of energy release. Fires release energy slowly, while explosions release energy very rapidly (microseconds typically). Whether the presence of a flammable material will result in a fire or explosion depends on several factors. The essential elements for combustion are fuel (e.g. gasoline, wood, propane), oxidizer (e.g. oxygen, chlorine, hydrogen peroxide), and ignition source (e.g. spark, flame, static electricity, heat). When one of these is lacking (or is present in too low quantity) a fire will not occur.

Table 13.7   Flammability characteristics of liquids and gases

Characteristic	Description
Flashpoint	Lowest temperature at which a liquid will ignite from an open flame
Autoignition temperature	Temperature at which a material will ignite spontaneously in air, without any external source of ignition (flame, spark, etc.)
Flammability limits	Lowest and highest concentrations of a substance in air, at normal pressure and temperature, at which a flame will propagate through the mixture
Lower flammability limit (LFL)	Below LFL mixture is too lean to burn (not enough fuel)
Upper flammability limit (UFL)	Above UFL mixture is too rich to burn (not enough oxygen)

Table 13.8 Toxicity and flammability characteristics of liquids and gases [16,19,20]

Compound	TLV[a] (ppm)	Flash point (K)	LFL (vol% in air)	UFL (vol% in air)	Autoignition temperature (K)	Heat of combustion (MJ/kg)
Acetone	750	253	2.5	13	738	28.6
Ethyne	2500[b]	Gas	2.5	100	578	48.2
Benzene	10[c]	262	1.3	7.9	771	40.2
Butane	800	213	1.6	8.4	678	45.8
Cyclohexane	300	255	1.3	8	518	43.5
Ethanol	1000	286	3.3	19	636	26.8
Ethene	2700[b]	Gas	2.7	36.0	763	47.3
Ethene oxide	1[c]	244	3.0	100	700	27.7
Hydrogen	4000[b]	Gas	4.0	75	773	120.0
Methane	5000[b]	85	5.0	15	811	50.2
Toluene	100 (skin)	278	1.2	7.1	809	31.3

[a] TLV = threshold limit value (see Section 13.5.2.2).
[b] Simple asphyxiant; value shown is 10% of LFL.
[c] Suspected carcinogen; exposures should be carefully controlled to levels as low as reasonably achievable below TLV.

With the kind of information presented in Table 13.8, process design can eliminate or reduce the existence of flammable mixtures in the process during start-up, steady-state operation, and shut-down.

Example: Basel, Switzerland

In November 1986, the Sandoz chemical factory in Basel, Switzerland had a warehouse fire [21]. While the firemen were extinguishing the flames they sprayed water over drums of chemicals that were exploding due to the heat. The water and chemical mixture was washed into the Rhine, dumping approximately 30 t of pesticides and other toxic chemicals into the river. As a result the river life died up to 150 km downstream. Things could have been worse though. A nearby building contained sodium, a metal that reacts violently with water. If the fire hoses had been sprayed on the stored sodium, the explosion could have destroyed a group of storage tanks holding the nerve gas phosgene. An investigation of the spill revealed that there were no adequate catch basins at the site for collecting the run-off firewater with the chemicals [21].

After the fire had been put out, the German government checked the water as it passed through Germany. They discovered twelve major pollution incidents (not related to Sandoz) within 1 month.

QUESTION:

Into which category does this accident fit (Figure 13.7)?

Example: Flixborough, UK

The accident at the Nypro Ltd. caprolactam plant outside of Flixborough, UK (June, 1974) is a typical example of a vapor cloud explosion (VCE). An intermediate in the production of caprolactam is cyclohexanol, which is produced by partial oxidation of cyclohexane (see Figure B.13.6). The conversion must be kept low (usually below 10%) in order to avoid complete oxidation of the feed and product [22].

A sudden failure of a temporary cyclohexane line bypassing a reactor that had been removed for repair (see Figure B.13.6) led to the vaporization of an estimated 30 metric tons of cyclohexane. The vapor cloud dispersed throughout the plant and was ignited by an unknown source. The entire plant site was leveled and 28 people were killed.

The failure of the by-pass pipe was attributed to inadequate support of the pipe and a poor design (consisting only of a full-scale sketch in chalk on the workshop floor [23]) that failed to account for the movement of the pipe due to the pressure inside it. When the pressure rose a little above the normal level, the bellows at each end of the pipe began to rotate and failed.

QUESTION:

What could have been done to prevent this incident, or, at least, limit its consequences?

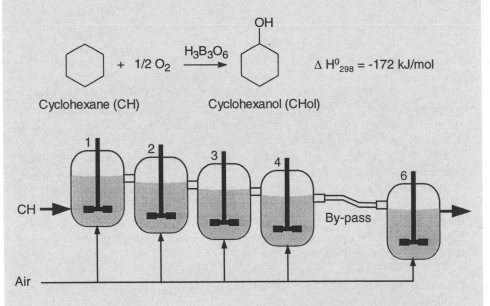

Cyclohexane (CH)   Cyclohexanol (CHol)

$$\Delta H^0{}_{298} = -172 \text{ kJ/mol}$$

CH →   By-pass

Air

Figure B.13.6   Oxidation reaction and arrangement of reactors and temporary pipe.

Many accidents have occurred as a result of plant modifications which had unforeseen side effects. Modifications may be required for start-up, during maintenance, may be temporary modifications when a piece of equipment fails, or process modifications as a result of changed feedstock, etc. A well-known example of an accident involving a temporary modification is the Flixborough disaster (see box).

13.5.2.2 Toxicity – toxic releases

Nearly all of the materials used in the production of chemicals are poisonous, to some

---

Example: Bhopal, India

The most notorious example of the hazards involved when dealing with highly toxic materials is the Bhopal, India accident (December 3, 1984). The Bhopal plant (partially owned by Union Carbide and partially owned locally) produced pesticides. An intermediate compound in this process is methyl isocyanate (MIC), which is extremely dangerous. It is reactive, toxic, volatile, and flammable. The TLV for MIC is 0.02 ppm.

The disaster started with the contamination of a large MIC storage tank with a large amount of water (how is still unclear), leading to an exothermic chemical reaction. Insufficient cooling of the storage tank led to a runaway reaction with an accompanying temperature increase past the boiling point of MIC (312 K). The MIC vapors traveled through a pressure-relief system and into a scrubber and flare system installed to consume the MIC in case of a release. Unfortunately, the scrubbing and flare systems were not operating. An estimated 40 metric tons of MIC vapor was released, the toxic cloud spread to the adjacent town (vapors stayed close to the ground due to the density of about twice that of air), killing over 2500 civilians and injuring an estimated 200,000 more.

The exact cause for the contamination of the MIC is not known. However, several design measures could have prevented this disaster [16].

- A well-executed safety review/HAZOP study (see Section 13.5.3 and references) would possibly have identified the problem.
- Adequate refrigeration of the storage tank could have prevented the runaway reaction.
- The scrubber and flare system should have been fully operational to prevent the release.
- The inventory of MIC should have been minimized, e.g. by redesigning the process to produce and consume MIC in a highly localized area (inventory <10 kg [16]); although storage of MIC was convenient, it was not necessary.
- An alternative reaction route, not involving MIC, could have been adopted (see Scheme B.13.1).

The Bhopal event and other major chemical plant incidents have changed the chemical processing industry profoundly; the result has been new legislation with better enforcement, enhancement in process safety, development of inherently safer plants, harsher court judgements, management willing to invest in safety related equipment and training, etc.

Methyl isocyanate (MIC) route

$$CH_3NH_2 \ + \ COCl_2 \longrightarrow CH_3N=C=O \ + \ 2\,HCl$$

Methyl amine   Phosgene   Methyl isocyanate (MIC)

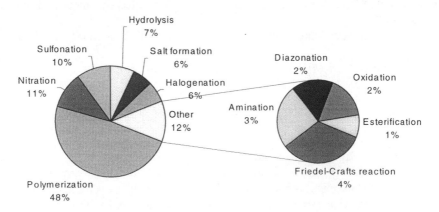

$$CH_3N=C=O \ +$$

1-Naphthol   Carbaryl

Alternative route

$$+ \ COCl_2 \longrightarrow$$

$$+ \ HCl$$

1-Naphthol chloroformate

$$+ \ CH_3NH_2 \longrightarrow$$

$$+ \ HCl$$

Scheme B.13.1   Alternative routes for the production of the insecticide carbaryl.

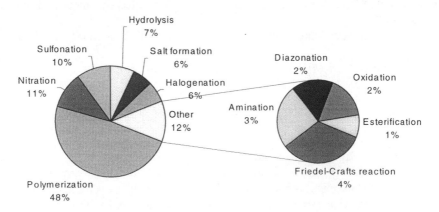

Hydrolysis
7%

Sulfonation
10%

Salt formation
6%

Diazonation
2%

Oxidation
2%

Nitration
11%

Halogenation
6%

Amination
3%

Esterification
1%

Other
12%

Friedel-Crafts reaction
4%

Polymerization
48%

Figure 13.8   Contribution of processes to chemical plant accidents involving runaway reactions (% of incidents). Based on data from [18].

extent. The potential hazard will depend on the inherent toxicity of the material and on the frequency and duration of exposure. Toxic effects on humans can be acute or (short-term effects) or chronic (long-term effects). The inherent toxicity of a material is measured by tests on animals. The acute effects of materials are expressed by the LD_{50} value, the lethal dose at which 50% of the test animals are killed. Table 13.9 gives LD_{50} values of a selection of materials.

The LD_{50} value only gives a crude indication on possible chronic effects. The most commonly used guide for controlling long-term exposure of workers to contaminated air is the 'Threshold Limit Value' (TLV). The TLV is defined as the concentration to which it is believed the average worker could be exposed to, for 8 h a day, day by day, 5 days a week, without suffering harm. Table 13.8 shows TLV values for some materials. Chronic effects are much less easy to identify since years may pass before these effects become visible. Furthermore, the cause then might be difficult to trace, e.g. the effect of asbestos, and solvents in the paints industries.

Some points to consider in the minimization of the risks involved when using hazardous substances are substitution (using less hazardous materials), containment (avoid leaks), ventilation, disposal provisions (vent stacks, vent scrubbers), and good operational practice (written instructions, training of personnel, protective clothing, monitoring of the plant environment to check exposure levels, etc.).

13.5.3 Reactivity hazards

It is known that every year there are a significant number of exothermic runaway reactions, particularly in batch and semi-batch processes, that can halt production and result in serious injuries, or even death, to plant operators. At the top of the critical list are reactions such as polymerization, nitration, sulfonation, and hydrolysis (see Figure 13.8).

QUESTIONS:

What do all of the processes in Figure 13.8 have in common?
The number of accidents in batch processes is much larger than in continuous processes (see Section 10.5). Give at least three reasons.

The consequences of an exothermic runaway reaction can be as severe as those from the ignition and explosion of a fuel/air mixture. In addition, the release of toxic materials can occur. Chemical hazards principally arise from:

- thermal instability of reactants and/or products;
- rapid exothermic reactions which can raise the temperature to the decomposition temperature or cause violent boiling of the reactants;
- rapid gas evolution which can pressurize and possibly rupture the plant.

These chemical hazards must be considered in assessing whether a particular chemical process can be operated safely on the industrial scale. Runaway reactions frequently result from poor understanding of the chemistry involved, leading to a badly designed plant. Knowledge of the reaction enthalpies associated with the desired

Example: Seveso, Italy

The Seveso (Italy) accident happened in 1976 at a chemical plant manufacturing pesticides and herbicides. An exothermic runaway reaction occurred in a reactor used for the production of 2,4,5-trichlorophenol (TCP) from 1,2,4,5 tetrachlorobenzene (TCB) by hydrolysis with NaOH:

How could this runaway reaction have happened? The plant was being shut down for the weekend, leaving the reactor half full of unreacted material at elevated temperature. The runaway reaction in the unstirred mixture was triggered by the slight, unexpected, heat input from the hot, dry wall of the reactor to the upper layer of the reaction mixture. The temperature of this upper layer reached a level at which slow and weak exothermic reactions started. After 7 h (!) these reactions resulted in the start of other, more rapid, exothermic reactions [24]. The rupture disk in the safety valve burst as a result of excessive pressure, and an aerosol cloud containing 2,3,7,8-tetrachlorodibenzo-*p*-dioxin (TCDD) was released into the air. TCDD was a by-product of the uncontrolled exothermic reaction:

The reactor had no automatic cooling system and only maintenance personnel were there at the time of the accident, so there was no one to start cooling manually and suppress the reaction. Fortunately, a worker noticed the cloud and stopped the release after only 20 min. TCDD is commonly known as dioxin, which is widely believed to be one of the most toxic man-made chemicals. Although no immediate fatalities were reported, between a few 100 g and a few kilograms of dioxin (lethal to man even in microgram doses (see Table 13.9)) were widely dispersed, which resulted in an immediate contamination of some 30 square kilometers of land and vegetation. More than 600 people had to be evacuated and 2000 were treated for dioxin poisoning.The best-known consequence of the Seveso disaster and previous serious chemical accidents (e.g. Flixborough) was the impulse that it gave to the creation of the European Community's Seveso Directive in 1982, a new system of industrial regulation for ensuring the safety of hazardous operations [25,26].

Table 13.9 LD_{50} values of some materials (oral, rat) [20]

Compound	LD_{50} value (mg/kg)	Category (EEC guideline)[a]
Dioxine (TCDD)	0.001	Very toxic
Potassium cyanide	10	Very toxic
Tetraethyl lead	35	Toxic
Lead	100	Toxic
DDT	150	Toxic
Aspirin	1000	Harmful
Sodium chloride	4000	–

[a] <25, very toxic; 25–200 toxic; 200–2000 harmful.

reaction and possible undesired reactions, and information on the thermal stability, i.e. the temperature at which any decomposition reaction may occur and its magnitude are essential to evaluate the hazards. The consequences of possible process maloperation must also be considered, e.g. agitation or cooling water failure, overcharging or omitting one of the reactants, or charging the wrong reactant. Adequate control and safety back-up systems should be provided, as well as adequate operational procedures, including training.

The occurrence of runaway reactions is not limited to chemical reactors; they may also start in storage tanks (see Section 10.5 and the Bhopal disaster).

13.5.3.1 Catalysts

One should be aware of the hazards associated with the use of catalysts. Many metal-based catalysts are pyrophoric in their reduced state and must be kept away from flammable materials. When not in use, they must be blanketed with an inert gas, such as nitrogen.

In, for example, FCC or hydrocracking operations, when unloading any coked catalyst, the possibility exists for iron sulfide fires. Iron sulfide will ignite spontaneously when exposed to air, and therefore must be wetted with water to prevent it from igniting vapors. Coked catalyst may be either cooled below 320 K before being dumped from the reactor, or dumped into containers that have been purged and inerted with nitrogen and then cooled before further handling.

Catalysts might form dangerous compounds when exposed to gases containing carbon monoxide. For instance, catalysts containing metallic nickel might form $Ni(CO)_4$ at temperatures below 430 K. $Ni(CO)_4$ is an extremely toxic, almost odorless gas which is stable at low temperatures. It might be formed, for instance, in methanation reactors (see Chapter 5) upon cooling down, unless the system has been thoroughly purged with nitrogen.

13.5.4 Design Approaches to Safety

Many of the chemicals used in industry can present safety, health, and environmental problems, particularly if their inherent hazards or those arising from specific operations have not been identified, evaluated, and a basis for safe operation of the process developed and implemented [27]. Furthermore, possible (probable) deviations from the design operating conditions (e.g. temperature and pressure in a reactor) that could result in hazardous situations must be foreseen, and proper action taken to prevent these deviations from occurring, or to limit their effect. Processes that cause no or negligible danger under all foreseeable circumstances (all possible deviations from the design operating conditions) are inherently safe, and naturally are preferred if possible. However, most chemical processes are inherently unsafe to a greater or lesser extent. There are opportunities for hazard reduction at every level of the design process. The steps to be taken in the design of a safe chemical plant are summarized below.

13.5.4.1 Identification and assessment of the hazards

For the safe design of a chemical process it is important that hazardous chemical and physical properties of all reactants and products involved (including undesired by-products) are known. These properties are often not inherent characteristics of the substance, but depend on the plant situation (e.g. operating temperature and pressure, presence of other chemicals, etc.). Important properties from a safety viewpoint include vapor pressure (related to overpressure), flash points and flammability limits, and toxicity (see Tables 13.8 and 13.9). Furthermore, all reactions (desired and undesired) should be identified if possible, especially those that may result in overpressure or the production of dangerous chemicals (e.g. explosive or toxic). Data are required on reaction rates and energies for exothermic reactions and unstable chemicals, on temperature limits beyond which explosive decompositions or other undesirable behavior can occur, and on rates of gas or vapor generation (overpressure). The strength and corrosion rates of materials of construction should also be evaluated.

The operating conditions at which hazardous situations could occur (during normal operation and as a result of failures) should be identified. Normally, the design team prepares a HAZOP (hazard and operability) study, in which all of the possible paths to an accident are identified. Then, a fault tree is created and the probability of the occurrence of each potential accident computed. See [20] and [28] for a brief outline of a HAZOP study and references to more detailed accounts.

13.5.4.2 Control of the hazards

Control of material hazards can be achieved, for example, by the replacement of hazardous chemicals by less dangerous chemicals if possible, and by minimization of the amount of hazardous materials present. Similarly, replacing a hazardous process route by an alternative route, or minimization of the amount of hazardous operations leads to increased safety.

One method of preventing fires and explosions is inerting, i.e. the addition of an inert gas to reduce the oxygen concentration so that the mixture is below its LFL. This method is used in, for example, the production of ethene oxide, where methane is used for inerting (see Section 9.5.2.1). Another method involves avoiding the build-up of static electricity and its release in a spark, e.g. by the installation of grounding devices, and the use of antistatic additives. In addition, explosion-proof equipment and instruments are often installed, e.g. explosion-proof housings that absorb the explosion shock and prevent the combustion from spreading beyond the enclosure. Another approach is good ventilation or construction in the open air to reduce the possibility of creating flammable mixtures that could ignite.

In the design of the plant, containment of flammable and toxic materials and using less severe operating conditions of pressure and temperature will result in a less hazardous process. Furthermore, inventories of hazardous chemicals should be minimized.

13.5.4.3 Control of the process

The safety of a process that is inherently unsafe depends on proper design of the process, provision of automatic control systems, alarms, interlocks, trips, etc. together with good operating practices and management.

13.5.4.4 Limitation of loss

If an accident occurs, the damage and injury caused should be minimized. Some measures are the installation of sprinkler systems, provision of fire-fighting equipment, and a safe plant lay-out (separate people from processes, etc.). In processes where pressures can build up rapidly (e.g. ethene polymerization – see Chapter 11), especially during an accident, it is crucial that pressure-relief devices are installed for venting to the atmosphere, or to scrubbers, flares and condensers. The devices include relief and safety valves, knock-out drums, rupture disks, etc.

13.6 PROCESS EVALUATION

13.6.1 Introduction

After each development stage the status of the process should be evaluated (see Section 13.2). The basis for this evaluation is the documentation of the knowledge of the process gained so far. The important questions to be answered are:

* Is the production process technically feasible in principle?
* Is it economically attractive?
* How big is the risk in economic and technological terms?

The technical feasibility of a process is proven by research in the laboratory, mini-plant, and pilot plant. If certain process steps present difficulties, these problems should be solved either by improving these steps or by selecting a different procedure. The technical evaluation of a process aims to steer the process research and development in the right direction. Factors increasing the technical risk of a process are:

* exceeding of limits (e.g. if the dimensions of the largest extraction column previously operated by a company are considerably exceeded);
* unfamiliarity of the company with the particular technology (e.g. continuous processes, high pressure, per-acids, fluidized beds);
* use of units that are difficult to scale up (e.g. solids processing);
* use of technically non-established equipment or unit operations (e.g. monolithic reactors in multiphase applications).

The technical risk of a process can be reduced in a number of ways. The first is to increase the expenditure on research and development of the weak points, e.g. pilot plant testing of the problematic unit, find well-established alternative equipment, or another way to perform the unit operation. A second option is to develop failure scenarios, i.e. determine what can be done if problems should occur (e.g. backup units, additional instrumentation). The decision basically is an economic question; the costs of increased research and development expenditure have to be weighed against the cost of eliminating the risk when the plant is started up or in operation. It should be noted that safety should have equal weight as economy, i.e. processes that are either uneconomical or unsafe should not be brought to commercialization.

A very important factor in the decision to pursue or abandon the development of a process is the economic perspective of a process. As the purpose of investing money in a chemical plant is to earn money, some means of comparing the economic performance or profitability is needed. To be able to assess the profitability of a process, estimates of the investment and operating costs are required. For small projects and for simple choices between alternative processes or equipment, usually a comparison of the investment and operating costs is sufficient. More sophisticated techniques are required if a choice has to be made between large, complex processes with different scope, time scale and type of product.

The economic evaluation of processes is based on criteria set by the company, such as payback period (PBP), net present value (NPV), etc. A detailed description of the economic criteria and their use can be found in references [29–31] and numerous other texts on the subject. These kinds of evaluations are usually done by a specialist group within the company, but chemical engineers and industrial chemists should have a basic knowledge in order to be able to communicate.

13.6.2 Capital Cost Estimation

Several methods are used to estimate the capital cost required for a chemical plant. These range from preliminary estimates (or study estimates) with an accuracy of 20–30% to detailed estimates with an accuracy of within 5%. The latter are usually only obtained in the final development stage of a process, when the engineering drawings have been completed and all equipment has been specified. Here, we will focus on the preliminary estimates, which are adequate for comparing processes in the early stages of development, and for left-or-right and go-or-no go decisions (see Figure 13.2).

Preliminary estimates usually start with the use of cost charts (see for instance [31–33]) for estimating the purchase cost of major equipment items, the other costs being estimated as factors of the equipment cost. Computing systems using equations based on cost charts are already available (e.g. ASPEN PLUS), and more advanced systems (adding more details of equipment design, materials, etc. and related costs such as site preparation, service facilities, etc.), leading to more accurate estimates (10–20%) are becoming available.

Step counting methods are also frequently used for order-of-magnitude estimates. See, for example, references [34–38]. These techniques are based on the presumption that the capital cost is determined by a number of major processing steps (e.g. a distillation column with its reboiler and condenser) in the overall process. The main weakness of step-counting methods is the ambiguity in defining a 'major processing

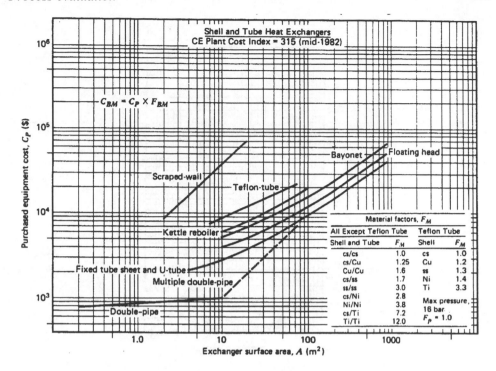

Figure 13.9 Purchased equipment cost chart for shell-and-tube and double-pipe heat-exchangers. Bare-module factors are derived from Figure 13.10 [31]. Used with permission of Process Publishing, Lee, NH.

step'. Clearly, these methods can only give order-of-magnitude estimates of the plant cost. However, due to their speed and simplicity, they are very useful in the conceptual stage of process design, when comparisons between alternative process routes are being made.

13.6.2.1 Cost charts

Cost charts (and equations based on them) represent purchase costs of equipment items as a function of key equipment sizes, e.g. the heat-transfer area for heat exchangers, as shown in Figure 13.9 [31]. In this figure, the logarithm of the purchased cost C_P is plotted as a function of the logarithm of the heat-exchange area A for a variety of shell-and-tube and double-pipe heat exchangers.

The equipment costs in Figure 13.9 are for carbon-steel constructions and operation at atmospheric pressure. Corrections are needed when more expensive materials (e.g. corrosion resistant or able to resist elevated pressure) are used. Material factors F_M are provided in Figure 13.9 for different combinations of shell and tube materials. Figure 13.10 shows the pressure factor F_P as a function of pressure.

The product of $F_M F_P$ is used in Figure 13.11 to determine the bare-module factor F_{BM}, which can then be used to obtain the purchased equipment cost of heat exchan-

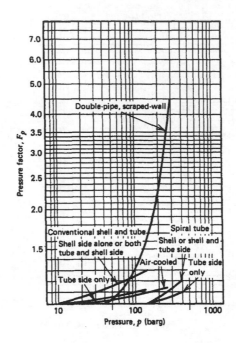

Figure 13.10   Pressure factors [31]. Used with permission of Process Publishing, Lee, NH.

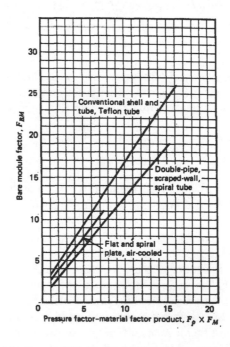

Figure 13.11   Bare-module factors [31]. Used with permission of Process Publishing, Lee, NH.

Table 13.10 Exponents for equipment costs as a function of capacity [29,31]

Equipment	Unit	Size range	Exponent
Agitator			
Propeller	kW	5–75	0.50
Turbine	kW	4–40	0.50
Blower, single stage, 1.4 bar	m^3/s	0.05–0.4	0.64
Centrifuge			
Horizontal basket	m (diam.)	0.5–1.0	1.3
Vertical basket	m (diam.)	0.5–1.0	1.0
Compressor			
Centrifugal	kW	20–500	0.80
Reciprocating	kW	200–3000	0.70
Conveyor, belt	m^2	5–20	0.75
Cyclone	m^3/s	0.0001–0.33	0.61
Dryer			
Drum	m^2	5–40	0.63
Pan	m^2	2–10	0.35
Vacuum shelf	m^2	10–100	0.53
Evaporator			
Falling film	m^2	3–6	0.55
Vertical tube	m^2	10–1000	0.53
Filter			
Plate and frame	m^2	5–50	0.60
Vacuum drum	m^2	1–10	0.60
Furnace (process)			
Box	kW	10^3–10^5	0.77
Cylindrical	kW	10^3–10^4	0.77
Reactor, jacketed	m^3	3–30	0.40
Refrigeration unit	kW	25–14000	0.72
Stack	m	6–50	1.00
Tank			
Horizontal	m^3	5–20	0.60
Vertical	m^3	1–40	0.52
Storage	m^3	1000–40000	0.80
Tower (process)	m^3	10–60	0.60

gers of materials of construction other than carbon steel and/or operating at higher pressure (the so-called bare-module cost, C_{BM}):

$$C_{BM} = C_P \cdot F_{BM} \tag{1}$$

It can be seen from Figure 13.9 that for shell-and-tube exchangers with $100 \leq A \leq 1000$ m^2 the following equation is valid (the slopes of the lines equal 0.7):

$$\frac{C_{P_1}}{C_{P_2}} = \left(\frac{A_1}{A_2}\right)^{0.7} \tag{2}$$

The fact that the exponent is less than unity expresses the *economy of scale*, i.e. one large heat exchanger is less costly than several smaller ones having the same overall heat-exchange area (and being of identical type and material of construction, and at the same moment in time!). Economy of scale applies to many types of equipment, as shown in Table 13.10, which is based on data from [29,31].

QUESTIONS:

Why have some equipment items exponents of 1 or even >1? What unit, not present in the table also has an exponent of about 1? Why are size limits given in the table? (Consider what happens for very small equipment and extremely large equipment).

Economy of scale also applies to entire plants (see Table 13.11), provided that the materials of construction are the same for each capacity figure. Of course, the technologies should also be the same.

Table 13.11   Exponents for plants as a function of capacity [39]

Compound	Process	Exponent
Acetaldehyde	Wacker oxidation of ethene	0.70
Ammonia	Steam reforming of natural gas	0.70
Methanol	Steam reforming of natural gas	0.71
Polyethene	Low pressure polymerization process	0.70
Sulfuric acid	Contact process	0.67

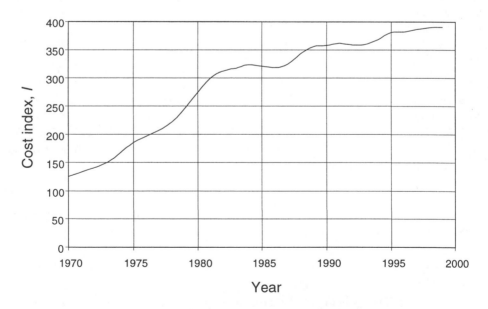

Figure 13.12   Chemical engineering plant cost index; $I = 315$ in mid-1982.

13.6.2.2 Cost indices

The purchased costs (C_P) of equipment have increased over the years as a result of inflation. This increase is represented by the chemical engineering plant cost index published bi-weekly by *Chemical Engineering Magazine* and shown graphically in Figure 13.12.

13.6.2.3 Total capital investment

The total capital investment, C_{TPI}, consists of the total permanent investment (or total fixed capital), C_{TPI}, and the working capital, C_{WC}. Purchase costs alone will not suffice to build a plant; the equipment has to be installed (including piping and instrumentation), there are costs for spare equipment items, storage vessels, initial catalyst load (if catalyst cannot be reused), site preparation, etc. These costs together are the total permanent investment C_{TPI}, which is usually estimated by using a factored-cost method. The C_{TPI} is the total cost of the plant ready for start-up. Additional investment, the working capital, is needed to start the plant up and operate it up to the point when income is earned. The working capital includes the initial catalyst load (if catalyst can be regenerated), inventories of raw materials and products, etc.

A quick estimate of the total capital investment can be obtained by using the overall factor method developed by Lang [40]. The purchase cost of each piece of equipment is estimated (as above), these costs are summed and used in the following equation:

$$C_{TPI} = 1.05 \cdot f_L \cdot \sum \left(C_{P_i} \cdot F_{BM_i} \right) \tag{3}$$

in which the factor 1.05 covers the delivery cost to the plant site and f_L is the Lang factor, which accounts for costs associated with the installation of the equipment and any additional investment costs. Values are given in Table 13.12.

More detailed factor estimates can be obtained by considering each of the items included in the Lang factor separately – see e.g. [29,32,33,41,42].

Many companies also use a factor to account for the different costs of construction at different plant locations around the world – see e.g. [32].

Calculation of bare-module costs of cylindrical process vessels, blowers, and compressors

Most process vessels are cylindrical, and positioned either vertically or horizontally. Vertical vessels include distillation, absorption, and stripping towers, and flash vessels. Horizontal vessels include decanters and settling tanks. Many reactors are also cylindrical vessels. Ulrich [31] gives cost charts for both vertical and horizontal vessels. Table 13.13 gives a semi-empirical correlation for the bare-module cost of a vertical vessel.

Table 13.12 Factors recommended by Lang for use in various plant types [40]

Plant type	Lang factor f_L
Solids processing	3.9
Solids-fluids processing	4.1
Fluids processing	4.8

Table 13.13   Bare-module cost (C_{BM}) [32]

Cylindrical process vessels (vertical)

$$C_{BM} = 1780 \cdot L^{0.87} \cdot D^{1.23} \cdot [2.86 + 1.694F_M(10.01 - 7.408\ln p + 1.395(\ln p)^2)] \tag{5}$$

with: L = height (m)
        D = diameter (m)
        p = pressure (barg)[a]

Additional for mist eliminators or sieve trays

$$C_{BM} = (193.04 + 22.72D + 60.38D^2) \cdot F_{BM} \cdot N_{act} \cdot f_q \tag{6}$$

with: F_{BM} = bare-module factor: [31]

	Trays	Mist eliminators
Carbon steel	1.2	–
Stainless steel	2.0	1.2
Fluorocarbon	–	2.0
Nickel-alloy	5.0	4.2

N_{act} = actual number of trays:	1	4	7	10	>20
f_q = quantity factor:	3.0	2.5	2.0	1.5	1.0

Blowers and compressors (excluding the drive, e.g. electric motor or steam turbine)

$$C_{BM} = 334 \cdot W^{0.95} \cdot F_{BM} \quad \text{with} \quad W = m \int_{p_1}^{p_2} \frac{dp}{\rho} = E \cdot W_s \quad \text{(kW)} \tag{7}$$

with: W = power imparted on the gas being transported (kW)[b]
        m = mass flow rate of gas (kg/s)
        p_1 = suction pressure (bar)
        p_2 = discharge pressure (bar)
        ρ = gas density (kg/m^3), depends on pressure and temperature
        E = efficiency (energy losses occur as a result of fluid friction)
        W_s = brake power of the compressor (kW)

[a] barg is the pressure of a system which one would see displayed on a normal pressure gauge; it is the pressure of the system, over and above atmospheric pressure.
[b] See, for example, [29,31,43] on how to calculate W.

Blowers are used for pressure increases form 0.03 to 5 bar, while compressors are used for higher discharge pressures. The cost of compressors may represent a large part of the equipment costs. Compressors are very expensive because they require large volumes and thick walls and involve large moving parts. When comparing Figure 13.13, showing the cost chart for blowers and gas compressors, with Figure 13.9 and Eqn. (5) for heat exchangers and vessels, respectively, it is immediately evident that for large power input compressors are very costly. Table 13.13 gives a semi-empirical correlation for the bare-module costs of blowers and compressors.

13.6.3 Operating Costs and Earnings

Apart from the total capital investment, the annual operating costs and annual earnings (pretax or after-tax) must be estimated in order to obtain an approximate measure of the profitability of the plant.

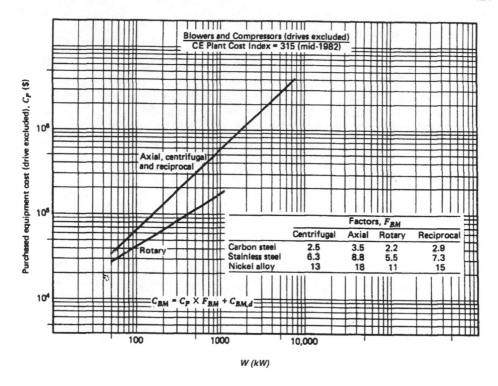

Figure 13.13 Cost chart for blowers/compressor; $C_{BM,d}$ = bare-module cost of drive [31]. Used with permission of Process Publishing, Lee, NH.

Table 13.14 Summary of operating costs (in $, etc). Adapted from [29]

Cost item	Typical values
Fixed costs (A)	
1. Maintenance	5–10% of C_{TPI}
2. Operating labor	From manning estimates and labor costs
3. Laboratory costs	20–23% of (2)
4. Supervision	20% of (2)
5. Plant overheads	50% of (2)
6. Depreciation	10% of C_{TPI}
7. Insurance	1% of C_{TPI}
8. Local taxes	2% of C_{TPI}
9. Royalties	1% of C_{TPI}
Variable costs (B)	
10. Raw materials	From flow sheets and unit cost
11. Miscellaneous materials	10% of (1)
12. Utilities	From flow sheets and unit costs
13. Shipping and packaging	Usually negligible
Indirect costs (C)	20–30% of (A + B)
Annual operating costs	A + B + C
Operating costs ($/kg)	(A + B + C)/annual production rate

Example: distillation column

For a certain separation a distillation tower has been designed, with a length, L, of 110 m, a diameter, D, of 4 m, and containing 300 sieve trays. Both tower and trays are made of carbon steel. The design pressure is 6 barg. Determine the bare-module cost of the tower plus trays for the year 1995.

Solution

- Tower

    Material factor: $F_M = 1$ (carbon steel)
    Bare-module cost of tower (Eqn. (5), CE = 315): $2.88 \cdot 10^6$ $

- Sieve Trays

    Material factor: $F_M = 1.2$ (carbon steel trays)
    Quantity factor: $f_q = 1$ (>20 trays)
    Bare-module cost of sieve trays (Eqn. (6), CE = 315): $0.45 \cdot 10^6$ $
    Total bare-module cost (1995): $(3.33 \times 10^6) \cdot (381/315) = 4.0 \cdot 10^6$ $

QUESTION:
> *What is the scaling factor?*

Operating costs are divided in fixed costs and variable costs. The fixed costs do not vary with production rate and have to be paid whatever the quantity produced. The

Compressor cost in methanol production from syngas

Consider a 1500 t/d methanol plant, in which the methanol reactor operates at 70 bar and syngas is delivered at 17 bar. In such a plant, consisting of the items shown in Table B.13.2, the syngas compressor ($W_s = 16,000$ kW) and recycle compressor ($W_s = 5000$ kW) make up about 20–25% of the total purchased equipment cost. This is about the same as the cost of the reactor.

QUESTIONS:
> *Compare the cost and required power of the syngas and recycle gas compressors. What can you conclude about economy of scale? (Also use Figure 13.11.)*
> *Estimate the syngas compressor cost of the ammonia plant in Figure 6.7 in 1980 and 1995. Although certainly not true, you may assume ideal gases and ideal compression. (The stream to the syngas compressor in the methanol plant had a flow rate of 10700 kmol/h, a temperature of 318 K and was at a pressure of 17 bar.)*

Table B.13.2 Cost estimate for methanol plant (1980); based on data from [44]

Item (number of items)	Estimated cost ($ thousand)
Reactor (1)	2600
Vessel (7)	230
Column (2–3)	2000
Heat exchanger (10)	2800
Pump (7)	140
Syngas compressor (1)	1880
Recycle gas compressor (1)	690
Total	10340

variable costs, on the other hand, depend on the quantity produced. Table 13.14 shows the main components of the operating costs and how they can be estimated.

The variable costs are calculated from material and energy balances and material prices. Fixed costs are partly labor-related and partly capital-related. The latter are estimated as a percentage of fixed capital investment. The indirect operating costs are the company overheads, which include administration, general research and development, and sales expenses.

The most reliable sources of cost data are company records, especially with regard to labor, utilities, overhead, and other general costs. When these are not available, costs are estimated according to generally accepted principles. Refer to [29,31–33,36, 45] for estimation methods.

Raw material costs usually constitute a major part (40–60%) of the operating costs, especially for base chemicals. The cost of raw materials follows from raw material consumption (e.g. t/t product) and the unit cost of the particular raw material (e.g. $/t). For commodity chemicals the unit costs are published in specialized journals, such as the *Chemical Marketing Reporter* and *European Chemical News* (see Appendix), and for refinery hydrocarbons in *The Oil and Gas Journal* for use in preliminary estimates. Other sources of price data, although not on a systematic basis, include *Chemical Engineering Progress*, *Chemical Engineering News*, and *Chemical Week*. For pharmaceuticals, etc. prices can be obtained from their manufacturers.

Utilities include steam, electricity, process water, cooling water, etc. Waste disposal is also often included in the utilities. Accurate consumption figures can only be obtained from mass and energy balances over major equipment items. Utility prices depend on location and proximity of their source, as well as on whether they are purchased or generated on site. Unit costs are best obtained from company records.

Operating labor costs (OLC) are based on an estimation of the number of operators for the process and the cost to the company of one operator. The number of operators depends on the type and arrangement of the equipment, the number of equipment items, the amount of instrumentation and control for the process, and company policy in establishing labor requirements. Supervision and other related costs are then estimated as a fraction of OLC. See also [29,32,45].

Depreciation is an income tax credit to replace depleted facilities and applies to fixed assets only. Depreciation is calculated as a fraction of the total permanent investment (C_{TPI}). For calculation of operating costs a straight line depreciation is common, usually over a 10 year period with zero scrap-value at the end, i.e. 10% of C_{TPI} per year. Sometimes 15% is used, while for small investments 33% is used.

The total sales revenues can be computed for the products and by-products, based on the amount produced and the selling price.

13.6.4 Profitability Measures

One of the simplest methods for estimating process profitability is the *return on investment* (ROI), which relates the (estimated) annual earnings to the total capital investment (C_{TCI}):

$$\text{ROI} = \frac{\text{annual profit (pretax or aftertax)}}{\text{total capital investment}} \cdot 100\% \qquad (\%) \qquad (8)$$

The total capital investment includes working capital. The usual practice is to calculate the ROI aftertax based, i.e. using the net annual profit. ROIs can also be calculated on a cash-flow basis (see below) or using total permanent investment only. Generally, acceptable ROIs for larger projects are in the 20% range; less than 5% is usually a No for a new project. When the selling price is unavailable, it is common to select a desired ROI and calculate the selling price required.

Another simple profitability measure is the *payback period* (PBP), the time required to recover the C_{TPI}:

Example: savings in ammonia plant

In an ammonia plant with a capacity of 357 t/d, the axial ammonia converters have been replaced by radial converters. This led to a reduction of the required loop pressure from 300 to 250 bar, while the pressure drop was reduced from 5 to 3 bar. As a result, energy savings of 0.85 GJ/t ammonia were achieved. The *operating factor* of the plant is only 0.7675 (i.e. the plant does not operate at full capacity and/or not during 365 days per year). The investment amounted to 1.1 $ million, while the energy costs were 5.35 $/GJ. Calculate the ROI and PBP for this investment (neglect tax).

Plant ammonia production: $357 \times 365 \times 0.7675 = 10^5$ t/a
Energy savings: $0.85 \times 10^5 \times 5.35 = 0.455$ $ million
ROI = 41%
PBP = 2.4 year

$$PBP = \frac{\text{total permanent investment}}{\text{annual cash flow}} \quad \text{(year)} \tag{9}$$

in which cash flow is:

$$CF = (1 - \text{tax rate}) \cdot (\text{annual revenues} - \text{anual operating costs})$$

$$+ \text{annual depreciation} \tag{10}$$

Typical PBPs for profitable investments are 3–4 years. Shorter PBPs are common for relatively small replacement or modification investments in an existing plant (e.g. the installation of a heat exchanger to recover waste heat).

Estimates of ROI and PBP are often based on the profit or cash flow in the third year of operation or on an average value. The ROI and PBP are useful as indicators when many different alternatives are screened, but they are unsuitable as profitability instruments, because changes in future cash flows and the 'time-value of money' are not taken into consideration.

The 'time-value of money' represents the fact that one dollar received now is worth more than the same dollar received after some time, the difference being the interest rate (or discount rate), r, that could have been recovered during that time span:

$$NFV = NPV \cdot (1 + r)^n \tag{11}$$

in which NPV is the *net present value* (what we have now) and NFV the *net future value* (what it will be worth after n years). For example, if we put $1000 in the bank now, at an interest rate of 4%, it will be worth $1000(1 + 0.04)^{10} = \$1480$ after 10 years.

When applying this principle to cash flows received from an investment project such a chemical plant, this means that money earned in the early years of operation is more valuable than money earned in later years. This is expressed as follows:

$$NPV_{\text{cash flow in year } n} = \frac{\text{estimated cash flow in year } n \text{ (NFV)}}{(1 + r)^n} \tag{12}$$

and demonstrated in the following example (see box).

Of course, this procedure can be extended to different investments at different times (e.g. investment for design, construction, working capital) and variable cash flows (e.g. starting with lower production capacity, decreasing selling price as a result of operation in a different market, expiration of patents, etc.), and other methods of compounding interest (e.g. per month, per day, or continuous). More elaborate information can be found in, for instance, [32].

13.7 CURRENT AND FUTURE TRENDS

Process research and development, and scale-up in particular, have come a long way since the early days. The original practice of extrapolating from laboratory scale through several intermediate stages to the industrial-scale plant has for a large part been replaced by the miniplant approach. This approach saves both time and money, which are the primary factors in the profitability and competitiveness of a process. Current trends in research and development are:

Example: choosing between two alternatives

Assume we could invest \$10 million in a plant at the beginning of year 1 (end of year 0). Then, during operation, at the end of year 1, year 2, …year 10 a cash flow of \$2 million is generated. Alternatively, we could invest the money in a plant generating \$3 million at the end of years 1 to 5 and \$1 at the end of years 6–10. $r = 15\%$. Which is the most profitable investment?

1.

$$\text{NPV} = \sum_0^n \frac{\text{NFW}}{(1+r)^n} = -C_{\text{TCI}} + 2 \cdot \left\{ \frac{1}{(1+r)} + \frac{1}{(1+r)^2} + \dots + \frac{1}{(1+r)^{10}} \right\}$$

$$= -C_{\text{TCI}} + 2 \cdot \frac{1}{r} \cdot \left\{ 1 - \frac{1}{(1+r)^{10}} \right\} = -10 + 10.04 \approx \$0 \text{ million}$$

2.

$$\text{NPV} = \sum_0^n \frac{\text{NFW}}{(1+r)^n}$$

$$= -C_{\text{TCI}} + \left\{ \frac{3}{(1+r)} + \dots + \frac{3}{(1+r)^5} + \dots + \frac{1}{(1+r)^6} + \dots + \frac{1}{(1+r)^{10}} \right\}$$

$$= -C_{\text{TCI}} + \frac{1}{r} \cdot \left\{ 1 - \frac{1}{(1+r)^{10}} \right\} + \frac{2}{r} \cdot \left\{ 1 - \frac{1}{(1+r)^5} \right\} = \$1.72 \text{ million}$$

Hence, option 2 is the best investment based on a net present value analysis. Apparently, at a discount rate of 15%, for situation 1 we have just earned back our investment. This is the so-called *discounted cash-flow rate of return* (DCFRR). The higher the DCFRR, the higher the profitability of the project.

QUESTIONS:

> *Derive the equation in option 1.*
> *Why is a discount rate of 15 percent used in this example, and by many companies, instead of the interest rate used by banks?*
> *Calculate the DCFRR for situation 2 (trial and error).*

1. Increased use of the miniplant technique combined with mathematical simulation, skipping the pilot stage, increased scale-up factors.
2. Miniaturization of research equipment, especially in catalyst testing. This can extend to 'lab-on-the-chip'. Application of high-throughput experimental procedures, heavily based on robotics and automation. Roboticized synthesis programs have proven to be successful in the development of drugs. They are referred to as combinatorial methods. More recently, it has been realized that also in the synthesis

of chemicals (e.g. ligands, fine chemicals) combinatorial methods are powerful. In process development they also have large potential. One can think of catalyst development, but also of optimization studies. Many companies have started this type of research, e.g. Symex, California and Avantium, The Netherlands.

3. Development and use of expert (knowledge based) systems[1]. These are already available for simple systems such as the optimization of unit operations. Other systems for which expert systems might be applied in the (far) future are catalyst development and optimization of entire plants.

4. In case of short-lived consumer products (dyes, detergents, pharmaceuticals, etc.), which are mostly produced in batch processes, the trend is towards Multi-Product or Multi-Purpose Plants (MPPs, see Chapter 10). On the other hand, there is a trend towards dedicated plants, due to the increased use of catalysis and higher quality demands.

5. Gaining increased insight in the relation between variability of process conditions and variability in product quality. For instance, polymers need to be extremely pure, e.g. 99.9999%. Small fluctuations in process conditions may well cause the purity to become below this value.

6. Finding novel routes in bulk chemicals production. In particular, steam cracking in fact is a crude technology associated with low selectivity and large consumption of exergy. It is expected that selective catalytic processes will replace steam cracking.

7. 'Process Intensification' will be the goal in new plant designs. Multifunctional reactors will be part of that.

The trends mentioned above all promote the saving of time and money. Other future developments will be driven by environmental legislation, safety and the development of the raw materials market. The latter includes C_1 chemistry, renewable raw materials, the development of technology based on photons (from the sun) and the usage of waste streams for the production of useful products. An explicit drive for R&D in the chemical industry will be the achievement of a sustainable society.

It is expected that the Chemical Process Industry will undergo major changes in the next decades. The plants of the future will be more efficient and their sizes will be reduced; 'Process Intensification' will lead to spectacular improvements [2]. For high selectivities and safety, miniaturization might be the way to go; reactors are miniaturized and scale-up is done by arranging the reactors in a parallel way. Advantages are the absence of large gradients (concentration and temperature), the easy scale-up and the intrinsic safety. The absence of gradients can increase productivity and selectivity tremendously.

Another development that might become important is the trend towards pipeless plants in fine chemistry and pharmaceuticals production. Conventionally, reactants and products flow from one piece of equipment to the following. It is conceivable that in many cases it is more efficient to lead the reactant or product mixtures from station

---

[1] Expert systems: The difference between simulation programs and expert systems is that in the former modeling is rigidly embodied by an algorithm, whereas in the latter, knowledge of the model is stored in the knowledge base. This knowledge not only consists of facts and relations between facts but also of a heuristic method for storage and acquisition of this information. When a human expert is faced with choosing from a large number of alternatives, (s)he eliminates those which seem unfruitful to find the optimal solution, using heuristic knowledge (rules of thumb), gained by experience in other similar situations. An expert system mimics the human expert.

to station. Those stations are designed for the various unit operations such as mixing, reaction, distillation, filtration, etc. Such a plant has advantages; for instance, the flexibility is very high. These trends will deeply influence process development and will stimulate the development of real breakthroughs in the chemical industry. It is the right time to be a chemical engineer or industrial chemist!

References

1 Anon. (2000) *CEN* May 1, 17.
2 Stankiewicz AI and Moulijn JA (2000) 'Process intensification: transforming chemical engineering' *Chem. Eng. Progr.* 96 Jan., 22–33.
3 Cybulski A, Moulijn JA, Sharma MM and Sheldon RA, *Technology and Fine Chemicals Manufacture*, to be published.
4 Vogel H (1992) 'Process development' in: Gerhartz W, et al. (eds.) *Ullmann's Encyclopedia of Industrial Chemistry*, vol. B4, 5th ed., VCH, Weinheim, pp. 438–475.
5 Douglas JM (1988) *Conceptual Design of Chemical Processes*, McGraw-Hill, New York.
6 Trambouze P, Van Landeghem H and Wauquier JP (1988) *Chemical Reactors, Design/Engineering/Operation*, Gulf Publishing Company, Houston, TX, Chapter 14.
7 Krekel J and Siekmann G (1985) 'Die Role des Experiments in der Verfahrensentwicklung' *Chem. Ing.-Technik* 62 March, 169–174.
8 Den Hollander MA, Makkee M and Moulijn JA (1997) 'Coke formation during fluid catalytic cracking' *AIDIC Conference Series* vol. 2 427–434.
9 Bisio A and Kabel RL (1985) *Scale-up of Chemical Processes*, Wiley, New York.
10 Van Damme PS and Froment GF (1981) 'Scaling up of naphtha cracking coils' *Ind. Eng. Process Des. Dev.* 20 366–376.
11 Weekman VW (1970) 'Kinetics of catalytic cracking selectivity in fixed, moving, and fluid bed reactors' *AIChE J.* 16(3) 397–404.
12 Jacob SM, Gross B, Voltz SE and Weekman VW (1976) 'A lumping and reaction scheme for catalytic cracking' *AIChE J.* 22(4) 701–713.
13 Sie ST and Krishna R (1998) 'Process development and scale up: I. Process development strategy and methodology' *Rev. Chem. Eng.* 14(1) 47–88.
14 Sie ST and Krishna R (1998) 'Process development and scale up: III. Scale-up and scale-down of trickle bed processes' *Rev. Chem. Eng.* 14(3) 203–252.
15 Fleming R (1958) *Scale-up in Practice*, Reinhold Publishing Corporation, New York.
16 Crowl DA and Louvar JF (1990) *Chemical Process Safety: Fundamentals with Applications*, Prentice Hall, Englewood Cliffs, NJ.
17 Bryce A (1999) Bhopal Disaster Spurs U.S. Industry, Legislative Action, U.S. Chemical Safety Board. ⟨http://www.chemsafety.gov/lib/bhopal01.htm⟩
18 Marrs GP, Lees FP, Barton J and Scilly N (1989) *Chem. Eng. Res. Des.* 67 381.
19 Prugh RW (1996) 'Plant safety in K-O, Safety' in: Kroschwitz JI and Howe-Grant M (eds.) *Kirk Othmer Encyclopedia of Chemical Technology*, vol. 19, 4th ed., Wiley, New York, pp. 190–225.
20 Sinnott RK (1996) *Coulson and Richardson's Chemical Engineering*, vol. 6, 2nd ed., Butterworth-Heinemann, Oxford, Chapter 9.
21 *Case Histories* (1999) Office of the Fire Marshal, Ontario. ⟨http://www.gov.on.ca/OFM/recycle8.htm⟩
22 Musser MT (1987) 'Cyclohexanol and cyclohexanone' in: Gerhartz W, et al. *Ullmann's Encyclopedia of Industrial Chemistry*, vol. A8, 5th ed., VCH, Weinheim, pp. 217–226.
23 Kletz TA (1985) *What Went Wrong? Case Histories of Process Plant Disasters*, Gulf Publishing Company, Houston, TX.

24 Cardillo P and Girelli A (1984) 'The Seveso case and safety problem in the production of 2,4,5-trichlorophenol' *J. Hazard. Mat.* 9, 221–234.

25 Major Industrial Accidents Council (1999) *Major Accidents.* ⟨http://www.miacc.ca/content/about/seveso.htm⟩

26 The European Commission (1999) *Chemical Accidents Prevention, Preparedness and Response.* ⟨http://europa.eu.int/comm/environment/seveso/index.htm⟩

27 Sharrat PN (1997) *Handbook of Batch Process Design*, Blackie, London.

28 Bibo BH and Lemkowitz SM (1995) *Chemical Risk Management*, Delft Technical University.

29 Sinnott RK (1996) *Coulson and Richardson's Chemical Engineering*, vol. 6, 2nd ed., Butterworth-Heinemann, Oxford, Chapter 6.

30 (1981) *Manual of Economic Analysis of Chemical Processes*, McGraw Hill, New York.

31 Ulrich GD (1984) *A Guide to Chemical Engineering Process Design and Economics*, Wiley, New York.

32 Seider WD, Seader JD and Lewin DR (1999) *Process Design Principles – Synthesis, Analysis and Evaluation*, Wiley, New York.

33 Peters MS and Timmerhaus KD (1991) *Plant Design and Economics for Chemical Engineers*, 4th ed., McGraw-Hill, New York.

34 Zevnik FC and Buchanan RL (1963) 'Generalized correlation of process investment' *Chem. Eng. Progr.* 59 Feb., 70–77.

35 Ward ThJ (1984) 'Predesign estimating of plant capital costs' *Chem. Eng.* 77 Sept. 121–124.

36 Cevidalli G and Zaidman B (1980) 'evaluate research projects rapidly' *Chem. Eng.* 73 July, 145–152.

37 Allen DH and Page RC (1975) 'Revised technique for predesign cost estimating' *Chem. Eng.* 68 March, 142–150.

38 Viola Jr L (1981) 'Estimate capital costs via a new, shortcut method' *Chem. Eng.* 74 April, 80–86.

39 Couper JR and Rader WH (1986) *Applied Finance and Economic Analysis for Scientists and Engineers*, Van Nostrand-Reinhold, New York.

40 Lang HJ (1948) 'Simplified approach to preliminary cost estimates' *Chem. Eng.* 55 June, 112–113.

41 Guthrie KM (1969) 'Data and techniques for preliminary capital cost estimating' *Chem. Eng.* 76 March, 114–142.

42 Guthrie KM (1974) *Process Plant Estimating, Evaluation, and Control*, Craftsman, Solano Beach, CA.

43 Sinnott RK (1996) *Coulson and Richardson's Chemical Engineering*, vol. 1, 2nd ed., Butterworth-Heinemann, Oxford.

44 Axtell O and Robertson JM (1986) *Economic Evaluation in the Chemical Process Industries*, Wiley, New York.

45 Wessel HE (1952) 'New graph correlates operating labor data for chemical processes' *Chem. Eng.* 59 July, 209–210.

General Literature

McKetta JJ and Cunningham WA (eds.) (1976) *Encyclopedia of Chemical Processing and Design*, Marcel Dekker, New York.

Kroschwitz JI and Howe-Grant M (eds.) (1991–1998) *Kirk-Othmer Encyclopedia of Chemical Technology*, 4th ed., Wiley, New York.

Gerhartz W, et al. (eds.) (1985–1996) *Ullmann's Encyclopedia of Industrial Chemistry*, 5th ed., VCH, Weinheim.

Magazines

Chemical Marketing Reporter
Chemical Engineering Magazine
European Chemical News
The Oil and Gas Journal
Chemical Engineering Progress
Chemical & Engineering News
Chemical Week

Appendix A

Chemical Industry – Figures

Table A.1 Chemical sales (only) of the chemical industry top 20 in the USA (10^9 US \$)[a]

Company[b]	1998	1997	1996
Dupont (D)	26.2	21.3	20.5
Dow Chemical (B)	17.7	19.1	19.0
Exxon (P)	10.5	12.2	11.4
General Electric (D)	6.6	6.7	6.5
Union Carbide (B)	5.7	6.5	6.1
Huntsman Chemical (B)	4.9	5.0	4.5
ICI Americas (S)	4.9	4.6	4.0
Praxair (B)	4.8	4.8	4.5
BASF (B)	4.8	4.9	4.7
Eastman Chemical (B)	4.5	4.7	4.8
BP Amoco (P)	4.5	5.9	5.7
Air Products (B)	4.4	4.1	3.7
Shell Oil (P)	4.2	4.7	4.3
AlliedSignal (D)	4.2	4.3	4.0
Ashland Oil (P)	4.1	4.0	3.7
Monsanto (L)	4.0	3.1	7.3
Celanese (B)	3.9	5.0	4.8
Rohm and Haas (B)	3.7	4.0	4.0
Chevron (P)	3.1	3.6	3.4
Occidental Petroleum (P)	3.0	4.3	4.5

[a] Source: Chemical & Engineering News; 1998 data CEN May 3, 1999; 1996/1997 data CEN June 29, 1998.

[b] Industry classification: B = base chemicals; D = diversified; L = life sciences; P = petroleum; S = specialty chemicals.

Table A.2   Chemical sales (only) of the chemical industry top 20 in the Europe (10^9 US \$)[a]

Company	1998	1997	1996
Bayer (Germany)	31.2	31.3	27.6
BASF (Germany)	30.7	31.7	29.7
Hoechst (Germany)	24.8	29.6	28.9
Roche Group (Switzerland)	17.0	12.9	11.0
ICI (UK)	15.4	18.3	17.4
Rhône-Poulenc (France)	14.7	15.3	14.5
Akzo Nobel (Netherlands)	13.9	12.2	11.3
Norsk Hydro (Norway)	12.9	12.7	12.7
Degussa[b] (Germany)	9.0	8.7	7.8
Solvay (Belgium)	8.3	8.6	7.8
DSM (Netherlands)	7.1	6.3	5.2
Air Liquide (France)	6.8	6.5	5.8
Clariant (Switzerland)	6.6	7.0	1.6
BOC International[b] (UK)	5.9	6.6	6.7
Ciba Specialty Chemicals[c] (Switzerland)	5.8	5.4	4.6[d]
Hüls (Germany)	4.7[e]	6.7	6.0
EniChem (Italy)	4.5	5.6	5.9
Borealis (Denmark)	3.0	2.8	2.4
SKW Trostberg (Germany)	3.0	3.1	2.2
Kemira (Finland)	2.7	2.7	2.5

[a] Source: CEN June 28, 1999.
[b] Fiscal year ends September 30.
[c] Formed on March 13, 1997.
[d] Pro forma results.
[e] Jan–Sept only (fiscal year recalculated to match merger partner Degussa).

Table A.3   Chemical sales (only) of the chemical industry top 20 in the Japan (10^9 US \$)[a]

Company	1998	1997	1996
Asahi Chemical	7.3	8.2	8.4
Mitsubishi Chemical	6.6	8.0	8.3
Mitsui Chemicals	4.7	–	–
Sekisui Chemical	4.5	5.5	6.2
Takeda Chemical Industries	4.9	4.9	4.9
Sumitomo Chemical	4.3	4.9	4.7
Toray Industries	4.1	4.6	4.3
Showa Denko	3.0	3.4	3.2
Dainippon Ink & Chemicals	3.3	3.6	3.7
Ube Industries	2.4	2.8	2.9
Teijin	2.1	2.4	2.6
Hitachi Chemical	1.8	2.0	2.0

[a] Source: CEN June 28, 1999.

Table A.4 Production of some of the largest volume chemicals in the USA $(10^9 \text{ t})^a$

Chemical[b]	1998	1997	1996	1988
Sulfuric Acid (I)	48340	48789	48537	43264
Ethene (O)	23614	23169	22270	16875
Lime (M)	22852	22068	21508	17326
Ammonia (I)	20070	18178	18211	17091
Phosphoric acid (I)	14626	13370	13422	11846
Chlorine (I)	13050	13130	12660	11438
Propene (O)	12979	12489	11390	9627
Sodium hydroxide (I)	11681	12164	11749	10702
Sodium carbonate (M)	11538	11986	11427	9034
Ethene dichloride (O)	11140[c]	11923[c]	5142[c]	5909
Nitric acid (I)	9524	9584	9353	8119
Ammonium nitrate (I)	8758	8742	8634	7625
Urea (O)	7984	7430	7755	7179
Polyethene, LD and LLD (P)	6715	6613	6416	4716
Polyvinylchloride (P)	6578	6388	5996	3787
Polypropene (P)	6271	6042	5439	3299
Polyethene, HD (P)	5864	5696	5612	3810
Ethylbenzene (O)	5743[c]	5432[c]	4699[c]	4504
Styrene (O)	5166	5156	5386	4075
Polystyrene[d] (P)	4284	4297	4153	3517
p- and *o*-Xylenes (O)	3950	4028	3200	2981
Ethene oxides (O)	3692	3738	3284	2700
Cumene (O)	2045	2775	2667	2021
Polyesters[d] (P)	2006	1932	1828	749
Butadiene (O)	1844	1863	1744	1437
Acrylonitrile (O)	1415	1493	1530	1112
Benzene (O)	1011	1062	960	729

[a] Source: Chemical & Engineering News, June 28, 1999.
[b] I = inorganics; M = minerals; O = organics; P = plastics.
[c] Reporting method changed in 1996.
[d] Includes styrene-acrylonitrile, acrylonitrile-butadiene-styrene and other styrene polymers.

Table A.5 Prices of some bulk chemicals in Europe and the USA (US $/t), January 1999[a]

Chemical	Europe, spot	Europe, contract	USA, contract
Ethene	320–330 (cif, nom)	404	335–340 (Oct)
Propene[b]	230–250 (cif, nom)	265	276 (Oct)
Butadiene	200–210 (fob, nom)	288	331 (Nov)
Benzene	207–212 (fob)	222	239 (Nov)
Toluene	170–175 (fob, nom)	184	204 (Nov)
p-Xylene	225–230 (fob)	315	325 (Q4)
o-Xylene	228–237 (fob)	243	287 (Q4)
Styrene	390–400 (T2, fob, nom)	467–492	507–520 (Oct)
Methanol	90–95 (T2, fob)	105	103–108 (Oct)
	80–85 (T1, cif, nom)		
MTBE	168–170 (fob, nom)	na	na
Ammonia	105–110 (CFR)	na	na

[a] Source: European Chemical News, 11–18 January 1999. fob = free-on-board; cif = cost, insurance and freight; CFR = cost and freight; T1 = imported material subject to EC common external tariffs; T2 = EC material, not dutiable; nom = initial but unfixed negotiating range; na = not applicable
[b] Polymer grade.

Table A.6 Prices of some fine chemical intermediates (US $/t), August 1998[a]

Chemical	Use	Price	Packaging
3,4-DCNB (3,4-dichloronitrobenzene)	Intermediate for dyes, pigments, drugs and agrochemicals	1900	300 kg drum
OCPNA (*o*-chloro-*p*-nitroaniline)	Intermediate for dyes and pigments	2750	50 kg bag
MCA (*m*-chloroaniline)	Intermediate for dyes, pigments, drugs and agrochemicals	3750	250 kg drum
DCA (3,4-dichloroaniline)	Intermediate for agrochemicals (diuron)	3000	50 kg bag, 200 kg drum
OPDA (*o*-phenylenediamine)		4000	

[a] Source: Shaper Chemicals Limited, Bombay, India ⟨http://x-philes.com/home/shaper/products/prices.html⟩

Appendix B

Main Symbols used in Flow Schemes

B.1 REACTORS AND OTHER VESSELS

Empty (reactor) vessel, vertical

Packing, e.g. catalyst, ion exchanger, structured/random packing

Vertical fixed-bed reactor with one bed and multiple beds

Multi-tubular reactor

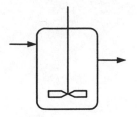

Continuous stirred-tank reactor (CSTR)

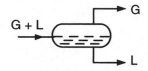

Phase separator (liquid 2, L_2 has higher density than liquid 1, L_1)

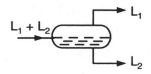

Gas(G)/liquid(L) separator

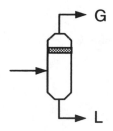

Knock-out drum

B.2 COLUMNS

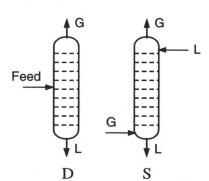

Tray column (basic symbol) above feed: rectifying section below feed: stripping section e.g. distillation column (D), stripping/scrubbing column (S, stripping of gases from liquid, part above feed emitted)

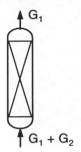

Packed-bed column e.g. adsorption (shown) of gas 2 (G_2) producing pure gas 1 (G_1) from a mixture , catalytic distillation (then only small part of column occupied by packed bed)

B.3 HEAT-TRANSFER EQUIPMENT

Heat exchanger (basic symbol) process fluids (arrow shows direction of heat flow)

Cooling (usually with air or water)

Heating (also used without circle)

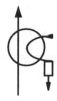

Heating by steam condensation. Usually as column reboiler

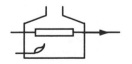

Furnace (heater)

B.4 MISCELLANEOUS

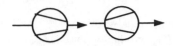 Compressor & Turbine/expander

 Pump

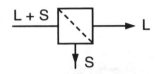 Filter (L = liquid; S = solid)

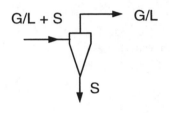 Cyclone

 Storage tank

Index